Lecture Notes in Physics

Springer
Berlin
Heidelberg
New York
Barcelona
Budapest
Hong Kong
London
Milan
Paris
Santa Clara
Singapore
Tokyo

The Editorial Policy for Proceedings

The series Lecture Notes in Physics reports new developments in physical research and teaching – quickly, informally, and at a high level. The proceedings to be considered for publication in this series should be limited to only a few areas of research, and these should be closely related to each other. The contributions should be of a high standard and should avoid lengthy redraftings of papers already published or about to be published elsewhere. As a whole, the proceedings should aim for a balanced presentation of the theme of the conference including a description of the techniques used and enough motivation for a broad readership. It should not be assumed that the published proceedings must reflect the conference in its entirety. (A listing or abstracts of papers presented at the meeting but not included in the proceedings could be added as an appendix.)
When applying for publication in the series Lecture Notes in Physics the volume's editor(s) should submit sufficient material to enable the series editors and their referees to make a fairly accurate evaluation (e.g. a complete list of speakers and titles of papers to be presented and abstracts). If, based on this information, the proceedings are (tentatively) accepted, the volume's editor(s), whose name(s) will appear on the title pages, should select the papers suitable for publication and have them refereed (as for a journal) when appropriate. As a rule discussions will not be accepted. The series editors and Springer-Verlag will normally not interfere with the detailed editing except in fairly obvious cases or on technical matters.
Final acceptance is expressed by the series editor in charge, in consultation with Springer-Verlag only after receiving the complete manuscript. It might help to send a copy of the authors' manuscripts in advance to the editor in charge to discuss possible revisions with him. As a general rule, the series editor will confirm his tentative acceptance if the final manuscript corresponds to the original concept discussed, if the quality of the contribution meets the requirements of the series, and if the final size of the manuscript does not greatly exceed the number of pages originally agreed upon. The manuscript should be forwarded to Springer-Verlag shortly after the meeting. In cases of extreme delay (more than six months after the conference) the series editors will check once more the timeliness of the papers. Therefore, the volume's editor(s) should establish strict deadlines, or collect the articles during the conference and have them revised on the spot. If a delay is unavoidable, one should encourage the authors to update their contributions if appropriate. The editors of proceedings are strongly advised to inform contributors about these points at an early stage.
The final manuscript should contain a table of contents and an informative introduction accessible also to readers not particularly familiar with the topic of the conference. The contributions should be in English. The volume's editor(s) should check the contributions for the correct use of language. At Springer-Verlag only the prefaces will be checked by a copy-editor for language and style. Grave linguistic or technical shortcomings may lead to the rejection of contributions by the series editors. A conference report should not exceed a total of 500 pages. Keeping the size within this bound should be achieved by a stricter selection of articles and not by imposing an upper limit to the length of the individual papers. Editors receive jointly 30 complimentary copies of their book. They are entitled to purchase further copies of their book at a reduced rate. As a rule no reprints of individual contributions can be supplied. No royalty is paid on Lecture Notes in Physics volumes. Commitment to publish is made by letter of interest rather than by signing a formal contract. Springer-Verlag secures the copyright for each volume.

The Production Process

The books are hardbound, and the publisher will select quality paper appropriate to the needs of the author(s). Publication time is about ten weeks. More than twenty years of experience guarantee authors the best possible service. To reach the goal of rapid publication at a low price the technique of photographic reproduction from a camera-ready manuscript was chosen. This process shifts the main responsibility for the technical quality considerably from the publisher to the authors. We therefore urge all authors and editors of proceedings to observe very carefully the essentials for the preparation of camera-ready manuscripts, which we will supply on request. This applies especially to the quality of figures and halftones submitted for publication. In addition, it might be useful to look at some of the volumes already published. As a special service, we offer free of charge LaTeX and TeX macro packages to format the text according to Springer-Verlag's quality requirements. We strongly recommend that you make use of this offer, since the result will be a book of considerably improved technical quality. To avoid mistakes and time-consuming correspondence during the production period the conference editors should request special instructions from the publisher well before the beginning of the conference. Manuscripts not meeting the technical standard of the series will have to be returned for improvement.

For further information please contact Springer-Verlag, Physics Editorial Department II, Tiergartenstrasse 17, D-69121 Heidelberg, Germany

Henrik Flyvbjerg John Hertz
Mogens H. Jensen Ole G. Mouritsen
Kim Sneppen (Eds.)

Physics of Biological Systems

From Molecules to Species

Springer

Editors

Henrik Flyvbjerg
Höchstleistungsrechenzentrum (HLRZ)
Forschungszentrum Jülich
D-52425 Jülich, Germany
and
Department of Optics and Fluid Dynamics
Risø National Laboratory
DK-4000 Roskilde, Denmark

Mogens H. Jensen
Niels Bohr Institute, Blegdamsvej 17
DK-2100 Copenhagen Ø, Denmark

Ole G. Mouritsen
Department of Chemistry, Building 206
The Technical University of Denmark
DK-2800 Lyngby, Denmark

John Hertz
Kim Sneppen
NORDITA, Blegdamsvej 17
DK-2100 Copenhagen Ø, Denmark

Cataloging-in-Publication Data applied for.

Die Deutsche Bibliothek - CIP-Einheitsaufnahme

Physics of biological systems : from molecules to species / Henrik Flyvbjerg ... (ed.). - Berlin ; Heidelberg ; New York ; Barcelona ; Budapest ; Hong Kong ; London ; Milan ; Paris ; Santa Clara ; Singapore ; Tokyo : Springer, 1997
(Lecture notes in physics ; 480)
ISBN 3-540-62475-9
NE: Flyvbjerg, Henrik [Hrsg.]; GT

ISSN 0075-8450
ISBN 3-540-62475-9 Springer-Verlag Berlin Heidelberg New York

Typesetting: Camera-ready by the authors
Cover design: *design & production* GmbH, Heidelberg
SPIN: 10550502 55/3144-543210 - Printed on acid-free paper

Preface

Last year when we planned a combined summer school and workshop,[1] we gradually realized that we would have something too good for the brief lifespan of a summer school if our ambitious invitations were accepted. We wanted a school which would introduce first-year physics graduate students and researchers alike to modern biological physics, assuming they possessed interest, curiosity, and *no prior knowledge* about biological physics. Simultaneously, this school should do the same for students and researchers in biology with no special background in physics, but some grasp of mathematics and computing. To this end we had invited speakers to cover four main themes,[2] chosen to differ enough to be independent subjects, each representing a fresh start, yet close enough for a special interest in one to carry a general interest in others with it.

The invited lecturers confirmed they would come, and we had something others might benefit from as well. We consequently suggested a publication project to the lecturers, easing their tasks by offering them 'scientific secretaries', typically a lecturer's own graduate student or junior collaborator, as well as editorial assistance from us.

We are glad the lecturers agreed to the project because they did a formidable pedagogical job during the school, conveying their material and enthusiasm to a diverse audience coming from medicine, biology, molecular biology, biochemistry, materials science, chemistry, physics, and mathematics. We were thus fully confirmed that we had an group of authors that could produce a book which is timely and unique, both with respect to the range of topics covered and with respect to the quality of the presentation. They did, and we hope this book will be a source of inspiration as well as information to students and researchers wanting an introduction to modern physics of biological systems.

---

[1] *Physics of Biological Systems: From Molecules to Species.* Held at Krogerup Højskole, Humlebæk, Denmark, August 14–27, 1995.

[2] *Protein Structure and Dynamics*, *Membrane and Cell Biophysics*, *Neurons and Sensory Processing*, and *Growth and Evolution.*

We are grateful to NorFa, the European Union, the Niels Bohr Foundation, and NORDITA for sponsoring the School and Workshop, and to Springer-Verlag, Heidelberg, for encouraging us from the start to produce this book.

Copenhagen,
November 1996

Henrik Flyvbjerg, John Hertz, Mogens Høgh Jensen
Ole G. Mouritsen, Kim Sneppen

Table of Contents

List of Contributors

Austin, Robert H.
Physics Department, Princeton University, Princeton, NJ 08544-0708, USA,

Bai, Linsen
Beckman Institute and Department of Physics, University of Illinois at Urbana-Champaign, Urbana, IL 61801, USA,

Bak, Per
Department of Physics, Brookhaven National Laboratory, Upton, NY 11973, USA,

Ben-Jacob, Eshel
School of Physics and Astronomy, Raymond & Beverly Sackler Faculty of Exact Sciences, Tel-Aviv University, Tel-Aviv 69978, Israel,

Bialek, William
NEC Research Institute, 4 Independence Way, Princeton, NJ 08540, USA,

Billings, Eric M.
Laboratory of Structural Biology, Division of Computer Research and Technology, National Institutes of Health, Bethesda, MD 20892, USA,

Bryngelson, Joseph D.
Physical Sciences Laboratory, Division of Computer Research and Technology, National Institutes of Health, Bethesda, MD 20892, USA,

Cohen, Inon
School of Physics and Astronomy, Raymond & Beverly Sackler Faculty of Exact Sciences, Tel-Aviv University, Tel-Aviv 69978, Israel,

Czirók, Andras
Department of Atomic Physics, Eötvös University, Budapest, Puskin u 5-7, 1088 Hungary,

Duke, Thomas
Cavendish Laboratory, Madingley Road, Cambridge CB3 0HE, U.K.,

Flyvbjerg, Henrik
Höchstleistungsrechenzentrum (HLRZ), Forschungszentrum Jülich, D-52425 Jülich, Germany; *and* Department of Optics and Fluid Dynamics, Risø National Laboratory, DK-4000 Roskilde, Denmark,

Frauenfelder, Hans
Center for Nonlinear Studies, Los Alamos National Laboratory, Los Alamos, New Mexico, USA,

Gittes, Frederick
Department of Physiology and Biophysics, University of Washington, Box 357290, Seattle, WA 98195-7290, USA,
Grüner, Walter
Institut für Molekulare Biotechnologie e.V. Jena, Germany,
Hill, Nick A.
Department of Applied Mathematical Studies, University of Leeds, Leeds LS2 9JT, U.K.,
Holt, Gary
California Institute of Technology, Computation and Neural Systems Program, Pasadena, CA 91125, USA,
Howard, Jonathon
Department of Physiology and Biophysics, University of Washington, Box 357290, Seattle, WA 98195-7290, USA,
Kessler, John O.
Physics Department, University of Arizona, Tucson, AZ 85721, USA,
Lu, Hui
Beckman Institute and Department of Nuclear Engineering, University of Illinois at Urbana-Champaign, Urbana, IL 61801, USA,
Luthey-Schulten, Zan
School of Chemical Sciences, University of Illinois, Urbana, IL 61801, USA,
Paczuski, Maya
Department of Physics, Brookhaven National Laboratory, Upton, NY 11973, USA,
Peliti, Luca
Dipartimento di Scienze Fisiche and Unità INFM, Università "Federico II", Mostra d'Oltremare, Pad. 19, I-80125 Napoli, Italy,,
Reidys, Christian
Institut für Molekulare Biotechnologie e.V. Jena, Germany,
Sackmann, Erich
Physics Department (Biophysics Group E22), Technische Universität München, D-85747 Garching, Germany,
Schulten, Klaus
Beckman Institute and Department of Physics, University of Illinois at Urbana-Champaign, Urbana, IL 61801, USA,
Schuster, Peter
Institut für Molekulare Biotechnologie e.V. Jena, Germany,
Softky, William
National Institutes of Health, Mathematics Research Branch, 9190 Wisconsin Ave. #350, Bethesda, MD 20814, USA,
Weber, Jacqueline
Institut für Molekulare Biotechnologie e.V. Jena, Germany,
Wolynes, Peter G.
School of Chemical Sciences, University of Illinois, Urbana, IL 61801, USA

Introduction: Physics — and the Physics of Biological Systems

Henrik Flyvbjerg[1,2] and Ole G. Mouritsen[3]

[1] Höchstleistungsrechenzentrum (HLRZ), Forschungszentrum Jülich, Germany
[2] Department of Optics and Fluid Dynamics, Risø National Laboratory, Denmark
[3] Department of Chemistry, Technical University of Denmark, DK-2800 Lyngby, Denmark

1 The nature of physics

In a historical perspective, physics is rather a generic approach to the material world than a particular set of disciplines and methods. Since Galilei and Newton established its principles of empirical analysis and mathematical model building, physics has inspired and invaded other fields of science. It has enriched chemistry with chemical physics, materials science and metallurgy with materials physics, astronomy with astrophysics, and biology with biological physics. We point this out not to put physics above other fields of science. Its connection with other fields is a two-way street, and while other fields have been enriched, they have not been made obsolete. As for biology, physics has little or nothing to say about most of its many great questions—but it has much to say about some of them. What is clear, logically and historically, is that because physics provides the description of the elementary constituents of the material world and the interactions among these constituents, together with well-proven methods for mathematical model building on empirical grounds, there is no field in science which may not be invaded by physics at some stage and to some degree, and repeatedly so.

Such invasions have their own demographics: physics is done by physicists—the invasions are carried out by people. The saturation in academic employment in the western world in the eighties and nineties has put a brake on this process, however (Gruner et al., 1995). The subdisciplines pursued by physics faculty today are to a high degree the disciplines that were vigorous and exciting in the early seventies. This static state of affairs is slowly changing now, and the change will become rapid as the great retirement wave sets in. As part of this change we see an increasing number of physics students and scientists with a strong and genuine interest in the physics of biological systems. This interest reflects the fact that physics is very much alive and ready to take up the challenge of the complex world of living systems. The present volume substantiates the last statement, and is itself a result of this growing interest.

A frequently asked question is what *is* biological physics, or physics of biological systems? Many different answers have been offered. We believe the present volume provides some kind of answer, though necessarily one that is far from complete, since it is given by example. A number of traditional physics journals, as well as a couple of new ones, provide similar answers by their increasing number of articles which are classified as biophysics and biological physics. Simultaneously, an increasing number of authors of articles published in traditional biological and biochemical journals are affiliated with physics institutions. A vivid illustration of the situation is provided by the American Institute of Physics, a major physics publisher, which recently published a collection of seminal papers in biological physics that had appeared mostly in physics journals (Mielczarek et al., 1993). Also a new series of handbooks, Handbook of Biological Physics, has just been launched by Elsevier (Hoff, 1995).

2 Physics of biological systems

In the present volume, the physics of biological systems is illustrated with cases taken from most length and time scales: DNA, proteins, motor proteins, biological fibers and membranes, entire cells, their organization into organs, such as the eye and the brain, as well as information processing in these organs. It is discussed how biological molecules on the micro-scale are developed by evolution, and how cells are organized in cultures that display specific growth patterns as part of their life process. Finally, the very longest time scales of life—the life times of species and genera—are discussed within a model for the evolutionary dynamics of extensive ecological systems.

2.1 DNA

The entire genomic information of a living organism is contained in a small number of chromosomes, each of which is a single molecule of DNA. Modern genetics is concerned with understanding how this information is organized and with deciphering the message encoded in the long sequence of nucleotides that form the links of the DNA chain. In order to examine a long chromosome it has to be cut into pieces that each can be studied in detail. In many molecular biology experiments, e.g. those implied by the Human Genome Project, DNA molecules have to be sorted according to size. *Thomas Duke* and *Robert H. Austin* show how concepts from physics can be used to device microtechnologies and miniature laboratories that can accomplish this formidable molecular sorting problem. Using microfabrication techniques, miniature obstacle courses can be engraved on a silicon chip. DNA molecules, propelled through the device by an electric field, move with a complex dynamics. By carefully designing the pattern of obstacles, the migration speed of the DNA molecules can be taylored to provide excellent separation within a specified range of molecular sizes.

2.2 Proteins

Proteins are biological polymers made up of long strings of amino acids that in the functional state are organized in a hierarchy of structures, ranging from the primary amino-acid sequence, over secondary structure, to tertiary and higher orders of structure. The structure and dynamics of a protein are determinants of its function. It has been repeatedly stated that one of the most outstanding problems in modern biology is understanding the folding of a protein into its functional state. Biopolymers are challenging objects of study for physicists because they are simple, yet display complex phenomena, and they are structured, yet display a substantial degree of apparent disorder, heterogeneity, and complexity.

The protein section of the book consists of four very different introductions to proteins. Each introduction reads well independently, and there should be something to satisfy every taste for how to approach a complex subject. Read together, these introductions bring out how richly facetted their subject of study is. The complexity of protein dynamics is described by *Hans Frauenfelder* who introduces the concept of substates and considers the kinetics and the relaxation phenomena that are associated with the protein's dynamic pathway through the complex landscape of substates. The case of a particular protein, myoglobin, which is responsible for the oxygen transport in muscles, is discussed in detail. A series of results obtained for the dynamical processes of this protein, using a number of different physical experimental techniques, are described as well as their interpretation by means of reaction theories.

Peter Wolynes and *Zan Luthey-Schulten* discuss a statistical physics approach to the protein-folding problem. Rather than building on a detailed interaction potential, they advocate a phenomenological theory which tries to capture the essentials of the relevant free energy in terms of a statistical energy landscape which the protein explores during the folding process. This energy landscape approach, which permits use of powerful concepts from thermodynamics such as order parameters and entropy, offers experimentalists a framework within which they can analyze folding experiments.

The grand question of how to determine protein structure from a knowledge af the interatomic interactions is addressed by *Joseph Bryngelson* and *Erich Billings*. After introducing the basic structural elements of proteins and the different fundamental forces and interactions that operate over different ranges and which underlie the ultimate determination of the protein structure, they show how concepts from polymer physics have general implications for the prediction of protein structures.

Klaus Schulten, *Hui Lu*, and *Linsen Bai* scrutinize the theory and molecular models of protein motion by means of computational techniques, including molecular dynamics simulations combined with mathematical analysis of the simulation data by means of principal component and normal mode analysis. The potential of this computational approach is illustrated by applications to a specific membrane protein, bacteriorhodopsin, which acts as a light-sensitive proton pump in a salt-loving bacterium.

2.3 Motors, membranes, microtubules

This section of the book gives three examples of physics contributing to molecular and cell biology, three examples of organelles are discussed and understood in physics terms.

In cell biology there exists a widespread mechanism for generating movement, which is employed for phenomena ranging from the active organization and rearrangement of organelles within cells to the large scale motion of muscles and flagella. As part of this mechanism, *Jonathon Howard* and *Frederick Gittes* describe certain motor proteins that cyclically hydrolyze the fuel molecule ATP as they move along either actin filaments or microtubules, which are two components of the cytoskeleton in eukaryotic cells. At least three separate families of motor proteins function in this way. The form of the cytoskeleton can become highly specialized through evolution, so that this basic motor/filament arrangement can result in a dramatic variety of mechanical effects. A discussion is presented of the specific characteristics of the motor protein kinesin, which is a remarkable molecular motor responsible for organelle transport along microtubules. Various lines of experimentation are determining the mechanical properties of this molecular machine.

Biological membranes are important stratified and composite structures in all cells. They not only function as means of compartmentalization but most processes of biological importance take place in association with membranes. A central element of a membrane is the lipid bilayer which is a bimolecular layer formed of lipids that are amphiphilic molecules. *Luca Peliti* gives an elementary introduction to the theory of amphiphilic membranes. The Helfrich Hamiltonian, describing the elastic free energy of a membrane, is introduced and its implications for the shapes of isolated closed lipid bilayers (vesicles) are described. The role of fluctuations in the renormalization of the elastic constants and their relevance for the phase behavior of amphiphile solutions in water are discussed.

Erich Sackmann describes the physical and material properties of membranes with particular focus on the lipid-bilayer elasticity and how it governs the self-organization of lipid/protein-bilayers, how it stabilizes domain structures and shapes of cells, and how it controls shape transitions (e.g. bud- and pit-formation) and shape instabilities (vesicle fission). It is demonstrated that many complex shape transitions of cell membranes can be mimicked by single lipid bilayer vesicles, suggesting that these processes are governed by the universal minimum bending energy concept of closed shells composed of stratified membranes. The essential role of the coupling between curvature and phase separation in mixed membranes for the formation and stabilization of local pits and buds or the fission of budded vesicles is demonstrated. The softness of membranes leading to surface undulations is argued to have fundamental consequences for the material exchange at membrane surfaces and the control of the adhesion of vesicles (or cells) on solid substrates, a phenomenon of potential relevance for understanding cell motility.

Microtubules are extremely rigid protein polymers which provide rigidity in eucaryotic cells in many different contexts. But microtubules have a dynamic capacity as well. During mitosis, they rapidly polymerize, depolymerize, and repolymerize in all directions, thereby performing a random search of the cells interior, "looking" for its chromosomes, to which they attach like "boat hooks" when they find them. Only when all chromosomes have been found does the cell divide, with exactly one sister chromatid ending up in each daughter cell, pulled there by the "boat hook" holding on to it. Thus Nature uses a random process to assure a deterministic outcome of the one process which tolerates no errors, the transmittal of the genetic code to daughter cells. But how does it work, the built-in "random number generator" that microtubules are equipped with? *Henrik Flyvbjerg* describes how two random processes in their polymerization process are poised against each other to produce the relatively rare events of change from the polymerizing to the depolymerizing state. Another aspect of microtubule dynamics, their self-assembly from solution, is shown to be so well studied experimentally that an "inverse problem" can be formulated and solved: the kinetic pathway of assembly is deduced mathematically from experimental time series.

2.4 Neurons, brains, and sensory signal processing

The different sensory systems, and ultimatively the whole brain, have long held a special fascination for physicists. The performance of these devices is impressive, often approaching the fundamental physical limits of the thermal and quantal noise.

William Softky and *Gary Holt* pose some questions related to the functioning of the whole brain as well as its neural components. These questions, which are formed on the basis of anatomical and physiological facts and which involve neural coding, information transfer, and the bandwidth of the transmission, are very different from the problems physics has successfully solved in the past, it is pointed out, both with respect to complexity and experimental accessibility. Answers to these questions are attempted in terms of the spatio-temporal properties of simple models of the neuronal circuitry.

William Bialek discusses how the brain reaches decisions and makes estimates which are near the limits imposed by noise sources. Making accurate estimates and reliable decisions is not merely a matter of avoiding extraneous noise sources. Maximally accurate estimates are possible only if the input data is processed in a particular way allowing for an optimal estimate that is a unique function of the inputs. If it is assumed that the brain acts as an optimal processor of sensory information, then the theory of these optimal devices furnishes a predictive theory of what the brain should be computing. It is shown that this problem of optimal estimation can be recast as a statistical mechanics problem.

2.5 Evolution, Micro- and macro-scale

Peter Schuster, *Jacqueline Weber*, *Walter Grüner*, and *Christian Reidys* present a view on molecular evolutionary biology along the lines of Darwinian dynamics in which molecules such as DNA and RNA play the role of species that evolve in a certain fitness landscape. The evolutionary dynamics is described within the general framework of complex dynamical systems that develop spatio-temporal patterns and possibly chaotic states. A comprehensive model of this type of micro evolution is presented which operates in an abstract space combining the sequence space of genotypes being DNA and RNA sequences, the shape space of phenotypes, and the concentration space of biochemical reaction kinetics.

Many natural phenomena in both living and non-living systems display spontaneous emergence of patterns, e.g. growth of snow flakes, aggregation of soot particles, formation of corals, growth of bacterial colonies as well as cell differentiation during embryonic development. Motivated by this kind of pattern-formation phenomena, *Eshel Ben-Jacob*, *Inon Cohen*, and *Andras Czirók* describe cooperative microbial behavior under stress and how the bacteria's response to such stress may lead to genetic evolution. Bacterial colonies exhibit a far richer behavior than patterning of non-living systems, reflecting the additional levels of complexity involved. The building blocks of the colonies are themselves living systems, each with its own autonomous self-interest and internal degrees of freedom. At the same time, efficient adaptation of the colony to the imposed growth conditions requires adaptive self-organization – which can only be achieved via cooperative behavior of the individual bacteria. The patterns exhibited by the colonies suggest that they should be viewed as adaptive cybernetic systems or multi-cellular organisms possessing impressive capabilities for coping with hostile environmental conditions. To this end, bacteria have developed a number of communication channels, from direct and indirect physical and chemical interaction to genetic communication via exchange of genetic material. It is proposed that the bacteria are 'smart' because they develop new patterns, which are better adapted to given growth conditions, by designed genetic changes.

John O. Kessler and *Nick A. Hill* consider the complementary and competing effects of forces and symmetry-breaking, biological behavior and consumption, as well as geometric constraints on the hydrodynamics of pattern formation in suspensions of swimming micro-organisms, such as bacteria and algae. It is shown how one can measure and incorporate the stochastic nature of the swimming velocity into mathematical models of these complex dynamical systems, and it is demonstrated that such systems can benefit the organisms' local environment by, for example, enhancing the transport of oxygen throughout a suspension of bacteria.

Finally, *Per Bak* and *Maya Paczuski* present some new theoretical ideas about biological macro evolution, contrasting the uniformitarian view of evolution as a continuous process of change by mutations to that of punctuated equilibrium involving discontinuous events leading, e.g., to mass extinctions. It is usually believed that Darwin's theory leads to a smooth gradual evolution, so that mass extinctions must be caused by external shocks. Bak and Paczuski however argue that mass extinctions can arise from the intrinsic dynamics of Darwinian evolution. Species become extinct when swept by intermittent avalanches propagating through the global ecology. These ideas are made concrete through studies of simple mathematical models of co-evolving species. The models exhibit self-organized criticality and describe some general features of the extinction pattern found in the fossil record.

References

Gruner, S. M., Langer, J. S., Nelson, P., Vogel, V. (1995): What future will we choose for physics? Physics Today **48**, 25–30

Hoff, A. J., ed. (1995): *Handbook of Biological Physics*, Elsevier, Amsterdam

Mielczarek, E., Greenbaum, E., Knox, R. S., eds. (1993): *Biological Physics*, AIP Press, New York

Part I

DNA

Microchips for Sorting DNA

Thomas Duke[1] and Robert H. Austin[2]

[1] Cavendish Laboratory, Madingley Road, Cambridge CB3 0HE, U.K.
[2] Physics Department, Princeton University, Princeton, NJ 08544-0708, U.S.A.

1 Importance of DNA Sorting in Biology

The entire genomic information of a living organism is contained in a handful of chromosomes, each of which is a single molecule of DNA. Modern genetics is concerned with understanding how this information is organized and with deciphering the message encoded in the sequence of nucleotides that form the links of the DNA chain. Inevitably, this involves cutting the enormously long chromosomes into pieces that are small enough to examine in detail. It is no surprise, then, that an efficient method of sorting DNA fragments is of paramount importance to biological research. This need has been emphasized at the current time by the launch of the Human Genome Project. This vast enterprise aims at two goals. The first is to create a library containing hundreds of thousands of DNA fragments, covering the entire genome. When the chromosome of origin of each fragment has been identified, together with the location at which the fragment belongs, this collection will constitute a detailed map of the genome. Once accomplished, researchers wishing to study a particular gene may simply access the library to obtain a clone, rather then laboriously isolating the gene themselves. The second goal is to establish the precise nucleotide sequence of each entry in the library. The value of this information may be appreciated when one knows that genetic diseases can be caused by the modification or absence of just two or three nucleotides in a gene. As indicated in Fig. 1a, these two parts of the project require the ability to analyse DNA molecules with sizes that differ by many orders of magnitude.

The task of separating DNA molecules according to size is not an easy one because a DNA mixture is a tangled mass of long, flexible polymers. Imagine that, given a pot of boiling spaghetti, you were asked to separate all the broken pieces from the whole ones. While you might manage to pick them out one by one, a reliable automatic procedure is less evident. All the more so, considering that big DNA fragments correspond to spaghetti over a kilometer long! Centrifugation is one possible approach, but even in an ultra-centrifuge the force acting on DNA molecules proves too small to provide efficient fractionation. So electrophoresis, in which the DNA is driven by an electric field, is the preferred method (Figure 1b). Straightforward electrophoresis in solution does not work, however, because molecules of different length all migrate at the same speed (Olivera et al. 1964). A variety of tricks is required to achieve good separations. Passing the DNA through a gelified medium, for example, works well in many instances (Southern 1979). Indeed, gel electrophoresis has been phenomenally successful and is now so ubiquitous that it is impossible to leaf through a molecular biology journal without seeing images of gels. It plays a major role in the Human Genome Project, where it is used both to sequence DNA and to make genetic maps. However, in many respects gel electrophoresis is unsatisfactory and one wonders whether there may not be a more effective alternative.

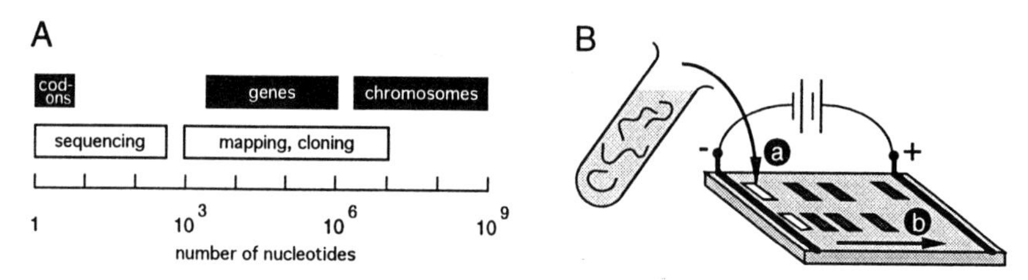

Fig. 1. A: Range of sizes of important elements in the organization of the genome and the typical sizes of DNA fragments analysed in sequencing and in mapping. In sequencing, the locations at which a particular nucleotide appears in a DNA fragment are deduced by a precise measurement of the spectrum of lengths of a set of single-stranded chains which are synthesized using the DNA fragment as a template, starting at one end of the molecule and finishing up at one of the various positions where the nucleotide occurs. Mapping makes use of enzymes that recognize and cut the chromosome at specific sequence motifs which occur at random intervals along the DNA, to create a set of *restriction fragments* which typically contain many thousands of nucleotides. B: A typical electrophoresis procedure. a: The mixture of DNA fragments is introduced into the electrophoresis matrix, which is traditionally a gel but may alternatively be an etched silicon chamber. b: An electric field is applied and the DNA molecules migrate towards the anode. Molecules move at a speed that depends on their size so that, after an interval of time, fragments of different length resolve into spatially segregated bands. The sizes of the molecules in the mixture may be inferred from the spatial distribution of the bands, or alternatively by examining the temporal distribution of molecules arriving at the anode.

The advent of new technologies in the physical sciences permits novel approaches to old problems. In recent years, the technique of microlithography has been developed to engineer microscopic features on the surface of silicon chips. This technology is being adapted to create specially-designed electrophoretic chambers—miniature obstacle courses engraved on a silicon wafer (Volkmuth and Austin 1992). In part, this work is motivated by the immediate need to accomplish better, more efficient, fractionation of DNA molecules. In part, by the more grandiose conception that microfabrication techniques promise a new level of convenience to biological researchers. One day, all the individual steps of an experiment—isolation, manipulation and analysis—could be accomplished without test-tubes. A miniature laboratory on a single chip.

Electrophoresis is, in many respects, an ideal case where physics can help the biological sciences. On the practical side, technologies provide the capacity to engineer precisely controlled electrophoretic environments and observe the dynamics of individual molecules. On the theoretical side, DNA isolated from the cell behaves as a purely physical system and reliably obeys the laws of classical dynamics and statistical mechanics. So theory and experiment can advance hand in hand to create new solutions to the challenges posed by DNA sorting.

2 Physical Description of DNA

The enormous length of DNA molecules confers several advantages when it comes to a physical description. First, their detailed chemical structure (even their celebrated double-helical nature) can be neglected. They can simply be treated as long chains and a minimum number of physical parameters characterizing length, flexibility and charge density suffice to describe them. Second, the comportment of *individual* molecules can be adequately described by *statistical* methods. Brownian motion continually modifies the molecular conformation so that, throughout the course of a typical experiment, a DNA molecule explores a representative set of all possible configurations. This 'self-averaging' means that physical properties such as the migration speed can be deduced if the probability distribution of configurations is known.

Measuring along the polymer backbone, a DNA molecule has an overall length L proportional to the number of nucleotides that it contains. This can be very long: even the smallest chromosome, containing a few million nucleotides, has $L \sim 1\,\mathrm{mm}$—a macroscopic size! However, in solution the polymer has a much smaller overall dimension. This is because the DNA is bent this way and that, as it gets buffeted by the water molecules and consequently becomes randomly coiled. Were you to travel along the backbone, after a short distance you would lose track of the original direction in which you were headed. This length scale evidently depends on the inherent flexibility of the DNA and on the magnitude of the Brownian forces. It is called the 'Kuhn length' b. For double-helical DNA $b \approx 100\,\mathrm{nm}$, equivalent to 300 nucleotides, while the more flexible denaturized single strand has $b \approx 10\,\mathrm{nm}$. Since the molecule looks rigid on scales shorter than the Kuhn length, but is randomly coiled on larger scales, it may be conveniently pictured as a random walk made up of $N = L/b$ straight segments of length b. The overall size of the coil R may be estimated from the statistics of the random walk:

$$\langle R^2 \rangle = Nb^2 = Lb \; . \tag{1}$$

For a fragment containing a million nucleotides, $R \sim 5\,\mu\mathrm{m}$, big enough to resolve with a light microscope. One might wonder whether electrostatic interactions affect these arguments, for DNA is an acid and is consequently negatively charged in aqueous solution. However, the charge is screened at very short range by the ions in solution, so electrostatic interaction between different parts of the molecule can be neglected. In the presence of an external electric field, the electro-hydrodynamic coupling of the moving counter-ions and the polymer is such that the DNA moves as if each Kuhn segment has an effective charge q somewhat smaller than its actual ionic charge. For double-helical DNA, $q \sim -100e$.

The arguments above apply to the case where no external forces are acting on the polymer. During electrophoresis, however, the electro-hydrodynamic forces driving the motion might also deform the chain. What typical magnitude of force would need be applied to stretch the DNA? Suppose that one could grasp a molecule by the ends and pull them apart. As the end-to-end separation R increases, the number w of corresponding random configurations of the DNA decreases, resulting in an entropy loss. Equilibrium is reached when $kT\frac{\mathrm{d}\ln w}{\mathrm{d}R}$ is equal to the applied force F. The statistics of the random walk (the probability distribution of the end-to-end vector is approximately Gaussian with variance given by (1)) then leads to

$$F \sim \frac{kTR}{Lb} \quad \text{for} \quad R \ll L \ . \tag{2}$$

At small extensions, the DNA acts as a Hookean spring, with spring constant kT/Lb. The longer the DNA, the easier it is to deform. Equation (2) also indicates that the typical force required to stretch the DNA to almost its full length is

$$F_{\text{stretch}} \sim kT/b \ . \tag{3}$$

For double-stranded DNA, $F_{\text{stretch}} \sim 0.1\,\text{pN}$. As we shall see, this value is not so high, and it is relatively easy to stretch out DNA molecules.

3 Reptation of DNA in Sequencing Electrophoresis

A DNA molecule in solution incessantly changes its form, at random, by thermal agitation. Its dynamics is most readily described by the Langevin equation, which relates the instantaneous velocity of each part of the chain to the forces acting on it. The velocity is proportional to the sum of the forces (which in electrophoresis includes the electrostatic interaction with the external field, as well as random Brownian forces) and inversely proportional to the friction coefficient. This local description is appropriate because the flow of counter-ions screens the hydrodynamic interaction between different parts of the molecule. The fluid friction may be characterized by a uniform friction coefficient, ζ, per Kuhn segment. In free solution, both the total force acting on a DNA molecule and the total friction opposing its motion are proportional to its length. As a result, the electrophoretic mobility $\mu \equiv v/E$ is independent of size, $\mu = \mu_0 = q/\zeta$, and molecules cannot be fractionated (Olivera et al. 1964). The simple expedient of using a gel matrix to support the electrophoretic solution, however, can result in excellent separations (Southern 1979). How does the gel network help? One might think that it is just acting as a sieve, blocking the big molecules and letting the little ones through. But it is not so simple, because DNA is flexible and even very long molecules can slither through small holes.

A long polymer, which when randomly coiled is too big to fit in a single open space in a gel, must thread its way from one pore to another. The description of its dynamics is based on Edwards' hypothesis (Doi and Edwards (1986)) that the polymer is effectively confined to the contorted tube formed by this sequence of pores (Figure 2). The tube length S is proportional to the polymer length, but shorter by a factor that depends on the pore size. De Gennes (1971) realized that the surrounding network hems in the molecule so effectively that it can only wriggle longitudinally along the tube. Thus it diffuses by sliding backwards and forwards like a two-headed snake. The ends search new paths through the network, but the rest of the molecule must follow wherever they lead. Because of the reptilian imagery, this was christened 'reptation' dynamics (de Gennes 1971). The same model can be applied to electrophoresis where, in addition to the Brownian motion, the polymer is driven by the electric field. This biases the motion towards one end of the tube and the polymer moves more like a one-headed snake. The biased reptation model (Lerman and Frisch 1982, Lumpkin and Zimm 1982) provides an excellent description of how fractionation of single-stranded DNA is achieved in sequencing electrophoresis.

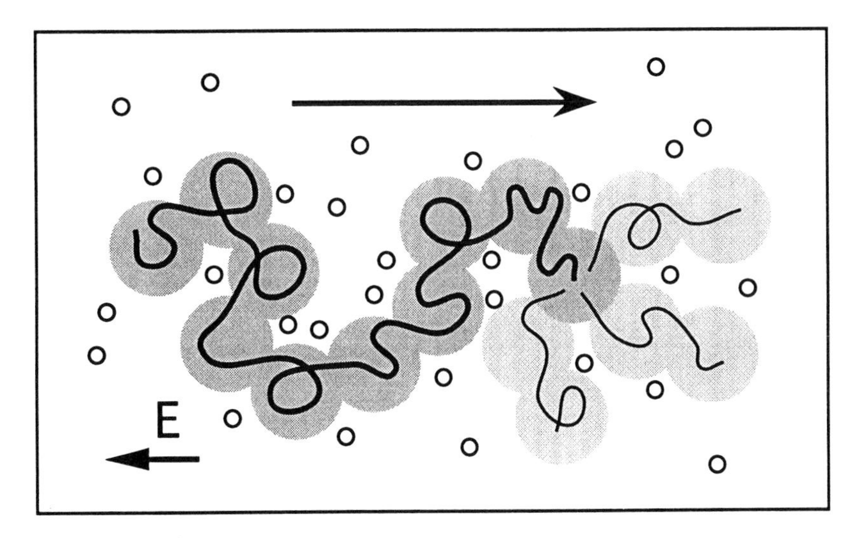

Fig. 2. DNA in a gel is effectively confined to a tube (shaded) by the surrounding gel fibers (drawn in cross-section as circles). Under the influence of an electric field, the DNA's diffusion is biased towards one end of the tube (in the direction of the arrow). The leading end of the molecule is free to explore alternative paths through the gel network (light shading).

The most important observation is that only a component of the field acts along the local contour of the tube, so that the total force pushing the DNA along is proportional to the projection (in the field direction) of the end-to-end separation, $R_{\|}$. The opposing friction, on the other hand, is linear in the tube length, S, so that the curvilinear velocity, v_s, along the tube contour is $v_s = \mu_0 R_{\|} E/S$. Since the end-to-end distance is S when measured along the tube, but $R_{\|}$ when measured in real space, the velocity v at which the DNA migrates through the network is smaller by a factor $R_{\|}/S$ than the curvilinear velocity. Thus the mobility $\mu = v/E$ is

$$\mu = \mu_0 \frac{\langle R_{\|}^2 \rangle}{S^2} . \tag{4}$$

This equation indicates that knowledge of the statistics of the polymer conformation permits a calculation of the mobility (Lumpkin and Zimm 1982). If the polymer remains randomly coiled, then $R_{\|} \propto N^{\frac{1}{2}}$ and $S \propto N$. This difference in scaling leads to a length-dependent mobility:

$$\mu/\mu_0 \propto N^{-1} . \tag{5}$$

Separation occurs because the DNA, constrained to follow the random route taken by the head, is obliged to follow a tortuous path through the gel. If, on the other hand, the polymer becomes oriented in the field direction so that both $R_{\|}$ and S are linear in N then no fractionation occurs:

$$\mu/\mu_0 \propto N^0 = \text{constant} . \tag{6}$$

In sequencing electrophoresis, one finds that (5) holds for short DNA molecules and (6) for long fragments. How does the orientation arise? It is a consequence of a delicate interplay between fluctuations of the molecule, which release its end section from constraints, and the field-induced drift that seeks to re-imprison the liberated section in the tube (Duke et al. 1992). One can imagine the free end rapidly exploring alternative routes through the gel, influenced by both the field and the randomizing effect of thermal agitation. The longer the terminal section, the greater the total electric force acting on it and the more it tends to be oriented. But the length of the section that has time to escape depends on the drift speed, and this in turn depends on the orientation of the entire molecule. This interdependence leads to an orientation transition, similar in many respects to phase transitions in equilibrium statistical mechanics, except that in this case it is a purely dynamic phenomenon. Long polymers are oriented, but short ones remain randomly coiled. The critical length at which the transition occurs varies inversely with the field strength (Duke et al. 1992).

Interesting though the physics may be, for practical purposes the orientation is a nuisance, for it limits the length of fragment that can be sequenced. Presently, 500 nucleotides is the maximal extent of sequence that can be determined at one go. This means that the sequence of a restriction fragment has to be painstakingly reconstructed from about one hundred separate pieces. A number of strategies have been proposed to augment the size of fragments that can be sequenced. For example, it has been suggested that a length of uncharged chain could be attached to the ends of all of the DNA molecules in the mixture so that they will not become oriented. Another possibility is to replace the gel with a more temporary network, like the one formed by concentrated polymer solutions; since the obstacles to motion are continually changing, the DNA is not held back long enough for orientation to build up. Yet another approach is to modify the geometry of the obstacles; this can be accomplished by substituting specially-designed solid-state devices for the random network of the gel.

4 Obstacle Courses on Microchips

The advent of optical microlithography, permitting precise engineering on the microscopic scale, has led to the development of a whole range of 'nanotechnologies'. This technique may be adapted to make miniature solid-state electrophoresis chambers (Volkmuth and Austin 1992). Two-dimensional obstacle courses of almost any pattern can be created on a silicon wafer in the following way. The surface of the chip is etched away to a depth of about 100 nm, through a mask which protects selected regions so that they remain raised. The structure is then sealed with a cover slip and the gap between silicon and glass filled with saline solution, into which a mixture of DNA molecules can be injected. The unetched regions provide obstacles to the free movement of DNA molecules migrating in the fluid. Figure 3 shows the most straightforward example of such a device—a regular lattice of cylindrical columns.

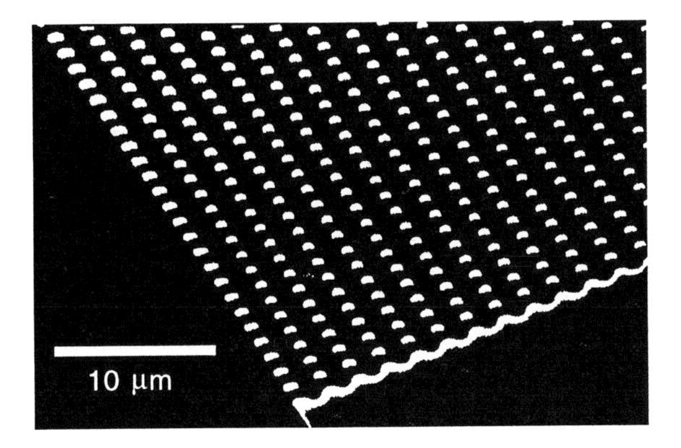

Fig. 3. An electron micrograph of an array of obstacles, engraved on a silicon chip. The posts have diameter 1 μm, height 0.15 μm and centre-to-centre spacing 2 μm.

The advantages of this solid-state device over a conventional gel can be readily appreciated. The structure is completely regular and well-controlled and the geometry can be chosen at will. A gel, by contrast, is random and ill-characterized. Also, the scale may be varied to cover a range that is not feasible using gels. A gel with pores 1 μm in size, for example, would be too fragile to handle. Furthermore, the motion of DNA molecules in the device can easily be examined. The DNA can be stained with a fluorescent dye and observed under a light microscope. Since the arrays are 2-dimensional, the DNA can be kept in focus over a wide field of view. Videomicroscopy allows ready comparison between experiment and theory, greatly aiding the design of devices to perform specific tasks.

5 Complex Dynamics of Long DNA Molecules

Sorting restriction fragments requires the ability to fractionate long, double-stranded DNA molecules. A natural question arises in comparison with the task of sequencing single-stranded DNA. Is it possible to scale up all dimensions—the contour length, the Kuhn length and the pore size—and maintain the same dynamical behaviour? Does biased reptation describe the motion of a 50,000-nucleotide fragment in the microlithographic array of Fig. 3, for example? Alas, no! At large scales the dynamics becomes much more complicated. The tube hypothesis is no longer valid, and consequently the reptation model is inappropriate. This is because when the obstacles are widely spaced, the entropy loss associated with a loop of DNA squeezing between a pair of obstacles can be outweighed by the change in electrostatic energy, even at very low field strengths (Figure 4). Consequently, there is no free energy barrier to prevent the DNA leaking out of the sides of the tube.

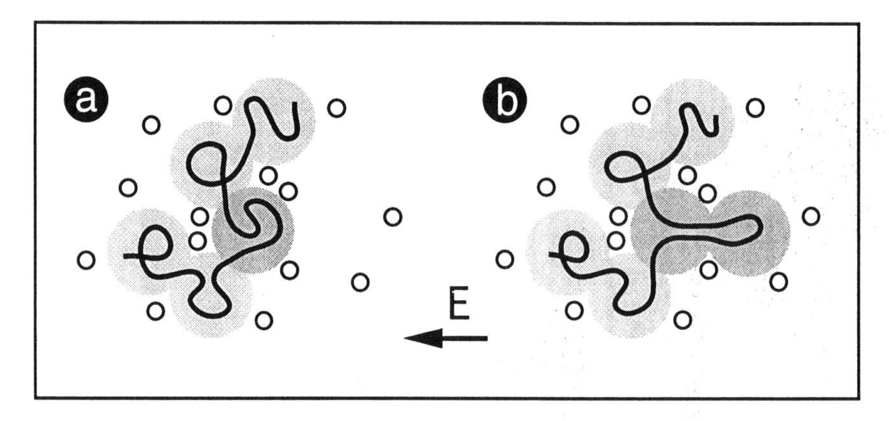

Fig. 4. A DNA molecule confined in a tube (a), and a loop of the molecule leaking out of the tube (b). In an applied electric field, the latter configuration may have a lower free energy, particularly when the average separation a of the obstacles is large. The shaded section of the molecule has coil size $R = a$. From (1), it has length $l = a^2/b$ and charge qa^2/b^2 so that the force acting on it is $F = qEa^2/b^2$. From (2), the coiled section will be deformed into a loop that leaks out of the tube when $F > kTa/lb$. Thus when the field exceeds the critical value $E^* = kTb^2/a^3q$, the DNA is not confined to a tube and does not reptate. For sequencing electrophoresis in small-pored gels, $E^* \sim 10^4$ V/cm and reptation occurs in normal experimental conditions. In the array of Fig. 3, $E^* \sim 10^{-2}$ V/cm and the DNA does not reptate.

The electrophoretic dynamics of long DNA molecules in a regular array of obstacles was first described by Deutsch (1988), and experiments confirm that the theoretical predictions apply in gels with large pores too (Schwartz and Koval 1989, Smith et al. 1989). The DNA executes an episodic type of motion that, when viewed under a microscope, appears almost animate (Figure 5). Typically, a molecule is oriented along the field and migrates in that direction. But frequently a loop of the molecule, instead of obediently following the route taken by the head, passes a different way around an obstacle and sets off in pursuit. Loop and head advance simultaneously, until the former unravels as the tail of the molecule passes along it. This leaves the DNA hooked over an obstacle in a long **U** shape. In this configuration, the DNA can be very highly extended. As mentioned previously, the force required to stretch double-stranded DNA, given by (3), is about 0.1 pN. In a typical field, $E = 1\,\mathrm{V/cm}$, this corresponds to the electrostatic force acting on a 50,000-nucleotide piece of DNA. Molecules longer than this, when hooked on an obstacle, are almost fully extended by the electric field pulling on both arms. At this point, the migration of the DNA can be almost completely arrested. This is particularly the case if the arms of the **U** are of nearly equal length, for then the net force acting to move the molecule, which is proportional to the end-to-end vector R, is very small. However, the slightly greater force tugging on the longer arm slowly hauls the molecule around the obstacle and, just like a heavy rope slipping round a pulley, the DNA gathers speed and eventually pulls free. Whereupon it immediately starts to contract elastically, and rapidly returns to a situation where the whole cycle can repeat itself.

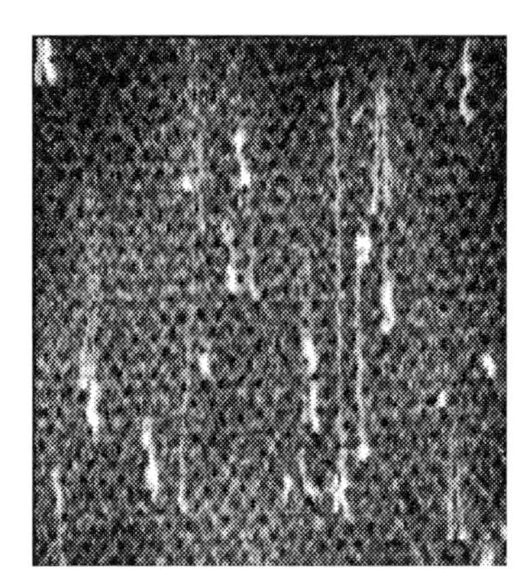

Fig. 5. Fluorescently stained DNA molecules moving through the array of Fig. 3, observed under an optical microscope. Molecules in the upper right hand corner are draped over a post in **U** configurations. Other molecules are elongated, having recently disengaged from a post. Yet others have contracted to compact conformations and will soon hook again. The longest DNA molecules in the picture contain approximately 200,000 nucleotides. The field is $E \approx 2\,\mathrm{V/cm}$.

Does this episodic motion result in a useful separation of fragments according to size? Experimentally there is evidence for only weak fractionation (Volkmuth et al. 1994). The reasons are two-fold and can be appreciated by considering the dynamics of hooking and unhooking. First, hooking is a random process, and a very wide distribution of hooking times is possible, depending on the asymmetry of the **U** shape formed. It is easy to see (Figure 6) that if DNA of length L gets hooked so that its ends are initially a distance $R_{\|}$ apart, the time it takes to slide off the obstacle is

$$t_{\text{hook}} \sim L \exp(L/R_{\|})/\mu_0 E \ . \tag{7}$$

This exponential distribution of times leads to a big variance in the migration velocity of individual molecules, causing poor electrophoretic resolution. Second, experimental observations suggest that the probability distribution of the parameter $R_{\|}/L$ characterizing hook asymmetry is independent of molecular length (Volkmuth et al. 1994). In this case, averaging (7) over the ensemble of different hooks, the mean time that DNA requires to disengage from an obstacle varies linearly with its length. But since the molecules usually hook again as soon as they pull free, the mean distance travelled between hooking events also varies linearly with molecular size. As a result, the mean velocity is length-independent. Hooking slows down all molecules by the same amount.

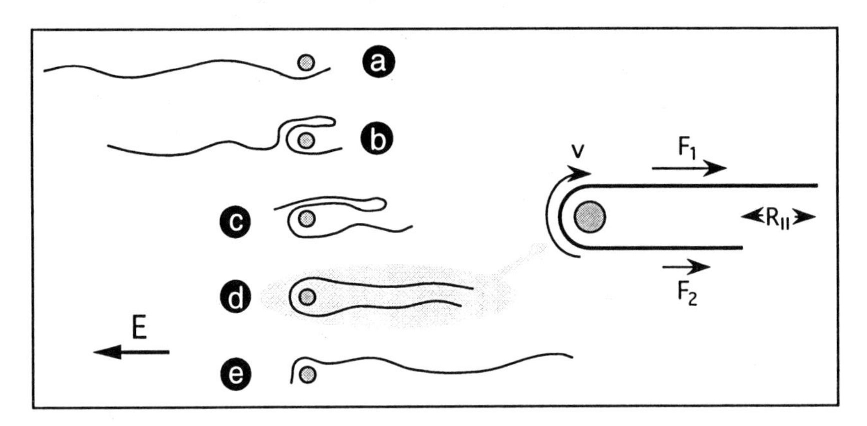

Fig. 6. Long DNA molecules undergo the episodic motion sketched in a–e. The duration of configuration d, in which the DNA is hooked over an obstacle and stretched into a long **U** shape, can be long. A simplified model for the disengagement of the molecule is illustrated. The total force influencing the DNA to slide around the obstacle is equal to the difference in the forces acting on the two arms $F = F_1 - F_2 = qER_{\|}/b$. The friction opposing the motion is $N\zeta$, proportional to the DNA length. The instantaneous sliding velocity is thus $v = \mu_0 E R_{\|}/L$. Integrating, the total disengagement time is $t_{\text{hook}} \sim L \exp(L/R_{\|})/\mu_0 E$.

These arguments suggest an obvious solution: Impose a particular frequency of hooking events. This may be done by changing the geometry of the array. For example, suppose that the majority of the obstacles are removed from the regular lattice, leaving only single rows of posts separated by long open spaces. In such a device, the DNA molecules hook only when they encounter a line of obstacles, so the distance travelled between hooking events is constant, independent of the molecular size. The average hooking time, however, remains proportional to the

DNA length, so each encounter with a row delays a longer molecule more than a shorter one. The desired fractionation is achieved.

This simple example demonstrates the versatility of microfabricated arrays. Variability of the geometry permits straightforward solutions that would be impossible to implement using gels. In fact, this particular design is expected to work much better as a sequencing tool than as a method of separating long restriction fragments. The reason is that the timescale for relaxation of big DNA molecules can be extremely long. A fragment containing 100,000 nucleotides, for example, typically takes a few seconds to relax to a random coil after it has been stretched. If the DNA does not have time to recoil as it traverses from one row of posts to the next, it might slip through the obstacles without hooking. The tiny fragments used in DNA sequencing relax very quickly, however, so the timescale does not impose a limitation. By miniaturizing the silicon arrays, using electron beam lithography to create obstacles on the scale of nanometers, it may be possible to manufacture a sequencing device that rivals the current gel-based methods.

The long timescales associated with the diffusive motion of big DNA molecules have other consequences. One important one is that, as the dimensions of a regular array of obstacles are scaled up, the episodic dynamics of the DNA does not continue indefinitely. When the distance between obstacles exceeds a few microns, the DNA simply lacks the time to diffuse around an obstacle and create a loop before its leading head has moved on. In this case, hooking is suppressed and the DNA migrates directly down the channels of the array. Although this does not give rise to a length-dependent mobility, the uniformity of the motion is a significant improvement on the episodic behaviour of DNA in fine arrays and in gels. In Sect. 7, we shall consider how to take advantage of this.

6 Pulsed-Field Gel Electrophoresis: Separation of Restriction Fragments

While standard gel electrophoresis fails to fractionate double-stranded DNA molecules containing more than about ten thousand nucleotides, it has been known for some time how to separate fragments containing up to ten million nucleotides. Schwartz and Cantor (1984) demonstrated that this remarkable improvement can be achieved by pulsing the electric field. The technique works by playing on the transient behaviour of the molecules following a sudden field switch. Unlike the steady-state mobility, this is length-dependent: small molecules can readjust rapidly to changes in the field, but long ones are too sluggish to do so.

The episodic motion of the DNA discussed in the previous section gives a clear insight into how pulsed field electrophoresis works at the molecular level. Consider, for example, a repetitive pulsing pattern in which the field is applied initially for time T in the forward direction, then time $T/2$ in the reverse direction, and so on (Carle et al. 1986). Small molecules quickly readjust at each switch, then migrate steadily in the direction of the field and, since the forward pulse lasts longer than the reverse, they advance through the gel. But what about longer molecules whose average hooking time is close to T? At the end of the forward pulse, they are typically hung over an obstacle in an extended **U**. When the field reverses, the arms of the **U** retract, driven by both the field and elastic recoil, so that the molecule becomes a compact coil. Subsequently, when the field reverts to its former direction, the arms extend again, but there is not sufficient time for the DNA to disengage from the obstacle before the field switches once more. Consequently, the molecule remains hung on the same obstacle for many successive field cycles and makes no progress.

Many varieties of pulsed-field techniques exist, but all of them work by a similar mechanism, which may be regarded as a sort of resonance. Molecules of a characteristic length have a typical hooking time that is comparable with the pulse time T. The pulsed field drives the internal modes of these chains, making them stretch and recoil, rather than forcing their movement through the gel (Duke and Viovy 1992). Consequently, they may be separated from shorter molecules which do migrate. A useful feature of the pulsed-field technique is that the range of fractionation can be extended simply by increasing the pulse time.

Although pulsed-field gel electrophoresis has simplified the mapping of the genome, it suffers from a number of drawbacks that prevent it from becoming a standard laboratory procedure. First, there is no general prescription for choosing the pulsing parameters that provide fractionation in a particular range. Consequently, automation of the technique is problematic. Second, the resolution is poor, owing to the inherent randomness of the dynamics that gives rise to the separation. Third, DNA longer than ten million nucleotides cannot be analysed by this method. These huge molecules get permanently trapped in the gel (Viovy et al. 1992). Precisely what causes their arrest remains in debate. One possibility is that they get tied in knots around the gel fibers. The accumulated electric force acting on a mega-nucleotide molecule may induce enough tension in the chain to pull a knot tight. Subsequently, no alteration of the field conditions will set the molecule free.

7 Efficient Pulsed-Field Fractionation in Silicon Arrays

In the previous section we saw that pulsing the field permits fractionation of restriction fragments, but the resolution is poor owing to the irregular motion of the molecules. In Sect. 5, it was mentioned that the dynamics of long DNA molecules in a regular array of obstacles becomes very uniform when the lattice spacing is sufficiently large. Combining these two pieces of information, we suspect that applying pulsed fields in coarse arrays may yield something interesting.

Figure 7 shows DNA moving in a regular array under transverse pulsed-field conditions. The field alternates with switch time T between two directions separated by an obtuse angle. Since the field pulls first one way, then the other, the repeated pulsing maintains the molecules in a highly extended state. This tautness prevents loops from growing in the middle of the molecule so that the motion is always led by an end. Since the molecules remain linear, the dynamics is remarkably simple. The total force F pushing a DNA molecule along its contour is proportional to the projection of the end-to-end vector in the field direction. So immediately following a field switch through an *obtuse* angle, F *changes sign.* The molecule sets off in the new direction led by what was previously its back end! This head-tail switch causes the DNA to retrace part of the route that it took during the previous pulse. If one looks along the bisector of the fields, which is the direction of net migration, the molecules first move backwards before they advance; longer molecules backtrack further and this gives rise to fractionation.

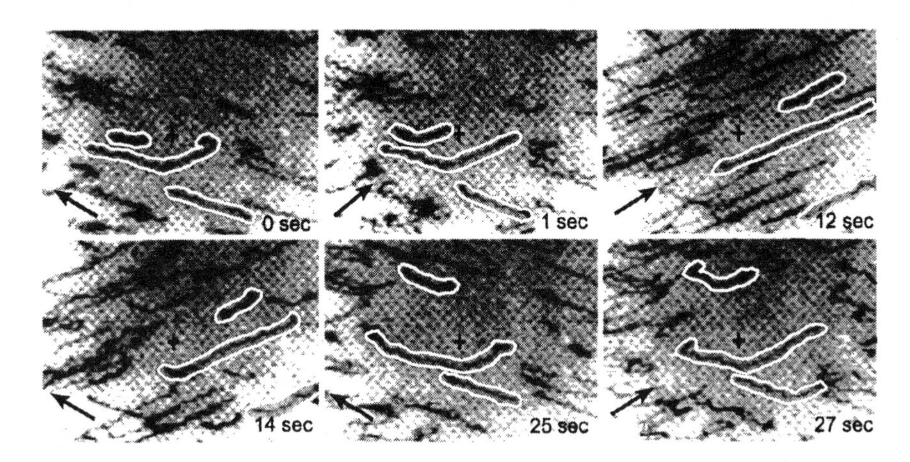

Fig. 7. DNA undergoing obtuse-angle pulsed-field electrophoresis in a microfabricated array. The field alternates every $T = 12.5\,\mathrm{s}$ between two directions separated by 135° (the arrows indicate the instantaneous direction of motion of the DNA). Fractionation can be observed in this sequence of frames taken from a video. The approximate sizes of the highlighted molecules are 135,000, 90,000 and 60,000 nucleotides. The shortest molecule advances furthest in the vertical direction during one full period of pulsing. The longest molecule backtracks along its path and makes almost no progress.

DNA fragments which are too long to realign completely before the end of a pulse keep retracing their tracks and fail to advance through the array. So only those molecules whose reorientation time is shorter than the pulse time can be fractionated. The total time that it takes a molecule to reorient is proportional to its length and inversely proportional to the field strength. This implies that the upper limit of the separation range increases linearly with both the pulse time and the field strength. The recipe for setting the field parameters to obtain the desired range of fractionation could hardly be simpler. Molecules smaller than the limiting size first backtrack and then move forward as they reorient, so that they make no progress during this period and advance only during the remaining fraction of the cycle. Consequently, their average migration speed falls linearly with increasing length. The backtracking mechanism leads to a clean, linear fractionation of the DNA.

This way of switching the field through an obtuse angle to fractionate long DNA seems almost too simple. But often the simplest methods are the best. In fact, the backtracking mechanism was previously proposed by Southern et al. (1987) to explain how pulsed-field gel electrophoresis works. As we saw, though, the dynamics of DNA in gels is much more complicated owing to the way that loops of the molecule can easily squeeze between the fibers. Microfabricated arrays, however, offer the possibility to realize the desired dynamical behaviour—by sufficiently spacing the obstacles, hooking can be suppressed so that the molecules remain linear. The resulting uniformity of the motion implies less dispersion in the migration velocity leading to a reduction in band width and improved resolution. Also, the problem of trapping of mega-nucleotide DNA in gels should be alleviated by using an array, since its 2-dimensional geometry and open structure precludes the possibility of DNA getting tied in knots.

8 Summary

Physics has played a useful role in tackling a problem of great technical importance in biology. The swift separation of DNA molecules is essential in many of the manipulations performed in molecular genetics. More efficient fractionation methods are urgently required to ensure the successful completion of the Human Genome Project. The new technology of microfabrication offers a novel approach to DNA fractionation. Tiny electrophoretic chambers, of versatile design, may be etched from silicon chips. By investigating the dynamics of DNA molecules migrating in these devices it has been possible to identify particular designs that are ideally-suited to specific separation tasks. These devices represent the first step on a road that leads to the creation of miniature laboratories on a chip which, by automating many laborious experimental procedures, will free researchers to pursue more fruitful investigations.

Acknowledgment

We thank our colleagues E. C. Cox, W. D. Volkmuth and S. S. Chan. This work was supported by National Science Foundation Grant MCB-920217, Office of Naval Research Grant N00014-91-J-4084 and National Institutes of Health Grant HG00482. T. Duke is a Royal Society University Research Fellow.

References

Carle, G. F., Frank, M., Olson, M. V. (1986): Electrophoretic separations of large DNA molecules by periodic inversion of the electric field. Science **232**, 65–68

de Gennes, P. G. (1971): Reptation of a polymer chain in the presence of fixed obstacles. J. Chem. Phys. **55**, 572–579

Deutsch, J. M. (1988): Theoretical studies of DNA during gel electrophoresis. Science **240**, 922–924

Doi, M., Edwards, S. F. (1986): *The Theory of Polymer Dynamics* (Oxford Univ. Press, Oxford), 188–191

Duke, T. A. J., Viovy, J. L. (1992): Simulation of megabase DNA undergoing gel electrophoresis. Phys. Rev. Lett. **68**, 542–545

Duke, T. A. J., Semenov, A. N., Viovy, J. L. (1992): Mobility of a reptating polymer. Phys. Rev. Lett. **69**, 3260–3263

Lerman, L. S., Frisch, H. L. (1982): Why does the electrophoretic mobility of DNA in gels vary with the length of the molecule? Biopolymers **21**, 995–997

Lumpkin, O. J., Zimm, B. H. (1982): Mobility of DNA in gel electrophoresis. Biopolymers **21**, 2315–2316

Olivera, B. M., Baine, P., Davidson, N. (1964): Electrophoresis of the nucleic acids. Biopolymers **2**, 245–257

Schwartz, D. C., Cantor, C. R. (1984): Separation of yeast chromosome-sized DNA molecules by pulsed field gradient gel electrophoresis. Cell **37**, 67–75

Schwartz, D. C., Koval, M. (1989): Conformational dynamics of individual DNA molecules during gel electrophoresis. Nature **338**, 520–522

Southern, E. M. (1979): Measurement of DNA length by gel electrophoresis. Anal. Biochem. **100**, 319–323

Southern, E. M., Anand, R., Brown, W. R. A., Fletcher, D. S. (1987): A model for the separation of large DNA molecules by crossed field gel electrophoresis. Nucleic Acids Res. **15**, 5925–5943

Smith, S. B., Aldridge, P. K., Callis, J. B. (1989): Observation of individual DNA molecules undergoing gel electrophoresis. Science **243**, 203–206

Viovy, J. L., Miomandre, F., Miquel, M. C., Caron, F., Sor F. (1992): Irreversible trapping of DNA during crossed-field gel electrophoresis. Electrophoresis **13**, 1–6

Volkmuth, W. D., Austin, R. H. (1992): DNA electrophoresis in microlithographic arrays. Nature **358**, 600–602

Volkmuth, W. D., Duke, T., Wu, M. C., Austin, R. H., Szabo, A. (1994): DNA electrodiffusion in a 2-D array of posts. Phys. Rev. Lett. **72**, 2117–2120

Part II

Proteins

The Complexity of Proteins

Hans Frauenfelder

Center for Nonlinear Studies, Los Alamos National Laboratory, Los Alamos, New Mexico

1 Introduction

How will the physics of the future look? What will a student in the year 2096 learn in his physics courses? It is a good bet that he or she will still learn classical and quantum mechanics, thermodynamics, electromagnetism, and so on. But which fields will be active in research? My guess is that the physics of complex systems will occupy center stage and that biological physics will dominate.

Two different approaches can be distinguished (Frauenfelder (1988)): In *biophysics,* physics is the servant and physical tools and techniques are used to study biological systems. In *biological physics,* the biosystem is considered just like any other physical system, and the research aims at finding new phenomena, concepts, and laws. In practice, of course, the two fields are not cleanly separated; it is impossible to look for new laws without also learning more about function. In this chapter I will favor biological physics and try to show that new concepts can be found.

Biological systems that can be studied by physical techniques range from small molecules, such as neurotransmitters, to biopolymers, membranes, organs, to the brain and the entire living being. The vast realm of systems forces us to ask: What are the best systems for physicists to explore biological physics? I believe that biopolymers, in particular proteins, qualify; they are complex enough to show new and unexpected properties, yet they are simple enough so that we can hope to make progress.

Once we have chosen a system, we ask the next questions: What are the problems to be solved? How do we approach the problems? What tools should we use? Some vague answer to the first question can be gleaned by looking at the history of physics. In nearly every field, atomic, condensed matter, nuclear, and elementary particle physics, three fundamental properties of the systems were crucial, namely *structure, energy levels,* and *dynamics.* These three aspects are also important in biological physics, but a fourth aspect enters, namely function. Moreover, the concept of energy levels found in "simple" systems such atoms and nuclei has to be replaced by *"energy landscapes,"* as will be clear later on. We sketch the connection in Fig. 1.

In a textbook or a course, the walk in the plane of Fig. 1 is systematic, with the start at "structure" and the logical finale at "function." Reality is different, however. We start experiments or theory in one area, and are forced by the results (or their absence) to a random walk in this knowledge space. If we are lucky,

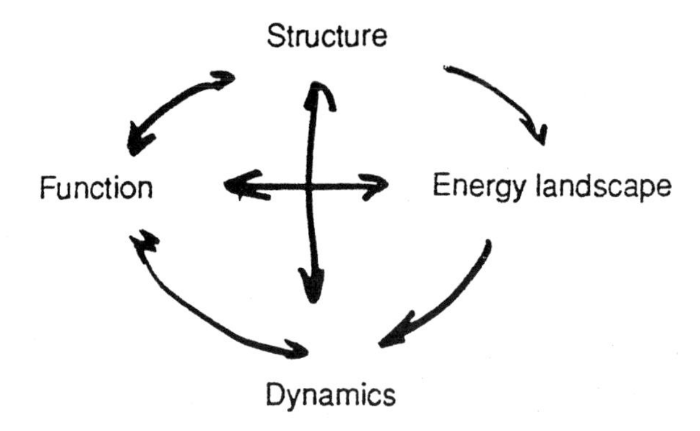

Fig. 1. Areas to be studied in biological physics.

we will find some logical connections at the end. Here I will follow a reasonably logical path, but stress that the actual steps in the discovery process were much more erratic.

The four concepts (structure, energy landscape, dynamics, function) provide a framework, but the new concepts and laws only emerge when experiments are performed and theories are beeing built to explain the facts.

The second and third questions, approach and tools, are not different from any other field of physics. We must combine experimental, theoretical, and computational expertise, and use essentially every tool available in atomic, chemical, and condensed matter physics. Two aspects, however, are different from conventional physics. First, without close collaboration with biochemists and biologists, it is easy to encounter unneccessary difficulties, such as spoiled samples. Such a collaboration requires that the physicist learns a new language, but the effort is fun and the result can be exciting. Second, in pure physics it is usually useless to ask for instance "Why are atoms the way they are?". In biological physics, however, the question why for instance an enzyme works the way it does, is legitimate and useful.

2 Structure

The two most important classes of biopolymers are nucleic acids and proteins. Nucleic acids carry the information, proteins do the work. The general features of the structure of biopolymers are well known and there exist excellent texts and monographs (Stryer (1995), Branden and Tooze (1991)). I will consequently only sketch the features of proteins that are important for the later discussions.

Proteins are built from twenty different building blocks, called amino acids. They have the same backbone, -C-C_α-N-. Bonded to the α-carbon is a side chain, and this "residue" is different for the different amino acids. Directed by the blueprint encoded in the DNA, amino acids are covalently linked to form a linear polypeptide chain, containing of the order of a few hundred amino acids. In the proper environment, the chain folds into a secondary structure, and finally into the working tertiary structure that is often globular and has a diameter of a few nanometers.

Four properties of the folded protein structure are crucial:

1. Because of the aperiodic arrangement of the amino acids in the primary sequence, the protein itself is aperiodic.
2. The protein is nearly close-packed.
3. The protein is frustrated (Toulouse (1977)). Because the protein is nearly close-packed, side chains get into conflict; different side-chains try to occupy the same space, but only one succeeds. Because of this competition, the protein cannot reach a single well-defined structure, as will be discussed later.
4. Along the main chain, the binding forces are covalent, and hence cannot be broken by thermal fluctuations. Cross connections are usually formed by weaker hydrogen bonds that can break and reform. This arrangement permits the protein to make large motions.

The number of proteins is very large. A typical biosystem will have about 10^5 different proteins; there are at least 10^6 different systems. Each protein can exist in a large number of mutations. Each investigator can therefore select his or her own protein and study it, free from any competitive pressure. Such an approach has, however, drawbacks. Comparisons will be difficult or impossible. Studies will be incomplete because an individual investigator usually has only a limited range of tools. It is fortunate that many groups have selected one particular protein, *myoglobin,* as a standard; it can therefore been called the hydrogen atom of biology.

Myoglobin is the protein that confers the red color to a good steak; it is abundant in muscles. Textbooks state that its role is the storage and transport of dioxygen (O_2) in muscles, for instance the heart. It is built from about 150 amino acids that form something like a box containing a small organic molecule, the heme group (Dickerson and Geis (1983)). At the center of the heme group is an iron atom to which the O_2 binds reversibly, $Mb + O_2 \leftrightarrow MbO_2$. A schematic cross section is given in Fig. 2.

The binding process of dioxygen to myoglobin appears at first to be very simple(Antonini and Brunori (1970)), and one would expect that all important aspects would have been solved after more than 40 years of work by many groups. Surprisingly, however, these studies have yielded many unexpected results, as we will discuss. A number of unsolved problems exist even now.

A few more remarks can be added about structure. The main features of the structure of myoglobin and its larger cousin, hemoglobin, were solved by X-ray

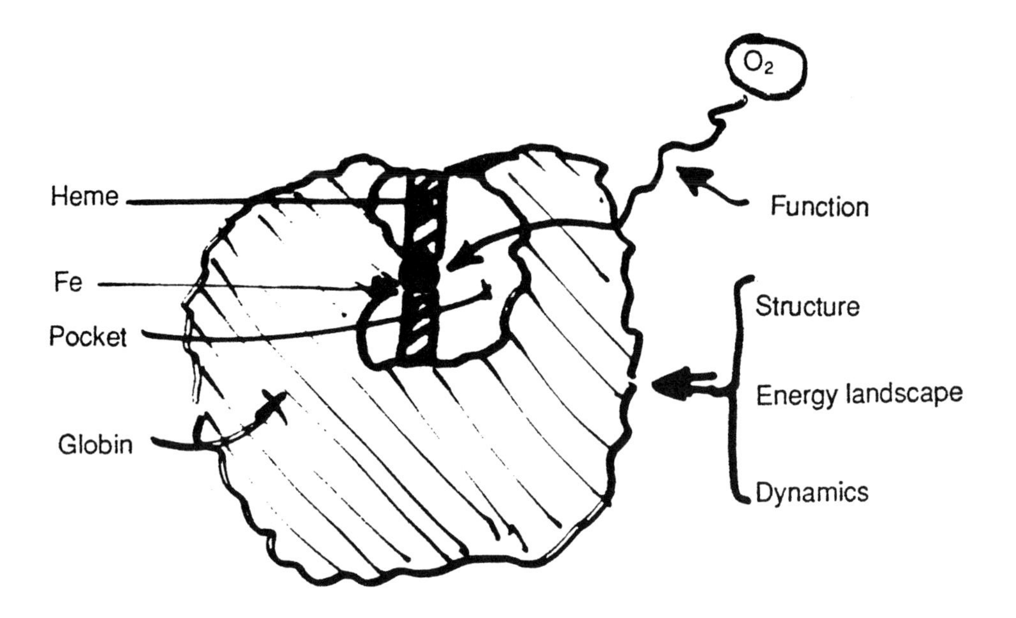

Fig. 2. A schematic cross section through myoglobin.

diffraction studies by Kendrew and Perutz. These studies give the average position of each non-hydrogen atom in the protein. The X-ray work is complemented by neutron diffraction studies and by nuclear magnetic resonance investigations. Finer details of the structure have been explored by spectroscopic methods.

Figure 2 is actually not complete. A "naked" protein does not work. A working protein is surrounded by a large number of water molecules, and this hydration shell is crucial for the motions and the function.

As a summary, we can say that we know the main features of the average structure of proteins well. This knowledge alone is not sufficient, and a feedback from the energy landscape to the structure, indicated in Fig. 1, leads to a deeper understanding of protein structure.

3 The Energy Landscape

Proteins are fundamentally different from simple systems, such as atoms or nuclei. In simple systems, the ground state has a unique energy and a unique structure. In complex systems, however, the situation is more complicated. Assume that we have a number of glass-forming liquid droplets, as sketched on top of Fig. 3. On cooling below the glass temperature, each droplet will "freeze" into a glass. The arrangement of atoms in each glass droplet will be somewhat different. Many properties, for instance energy and entropy, may also be different. To change one glass droplet into another one, the sample has to be heated and recooled. In other words, an energy barrier has to be overcome. The situation is sketched in the lower part of Fig. 3, where the energy is plotted as function of a conformation coordinate. To describe the actual structure of each glass would require a hyperspace of 3N dimensions, where N is the number of atoms. Fig. 3 actually gives a one-dimensional cross section through the hyperspace.

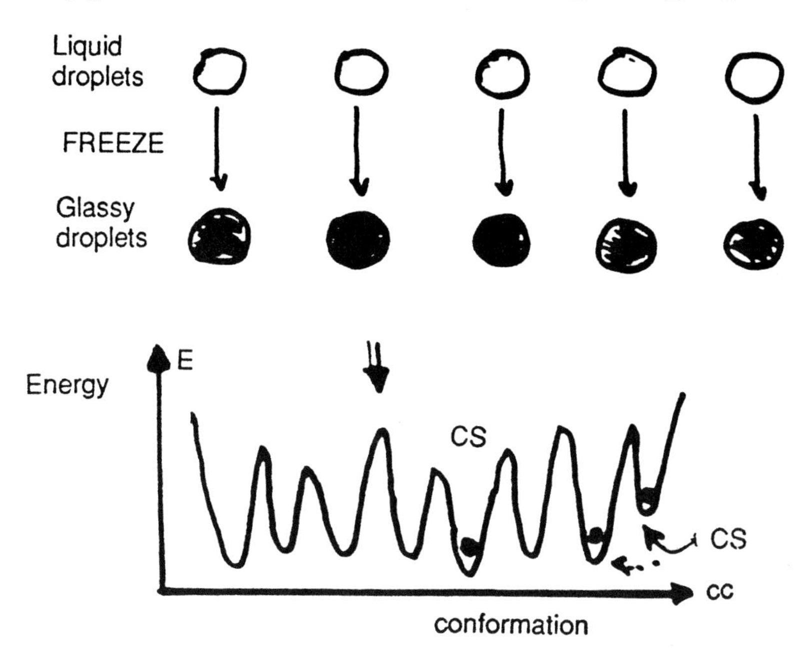

Fig. 3. If glass-forming liquid droplets are cooled, each droplet will "freeze" into a glass with somewhat different structure, energy, entropy.... The situation can be represented by a rugged energy landscape in a hyperspace. The figure shows a one-dimensional cross section through the energy landscape.

Figure 3 makes it clear why we talk about a *rugged energy landscape.* Each valley corresponds to a particular configuration, with corresponding properties. The barriers between valleys have unequal heights so that the landscape is indeed rugged. We call each valley a *conformation substate (CS). Sub*state instead of state because of its application to proteins. Proteins are in some sense machines and have usually more than one state. Each state, as we will see later, can assume a large number of substates.

Figure 3 implies that the distinction between the ground state and excited states in a rugged energy landscape is ambiguous. We could, of course, assert that only the substate with the lowest energy is the ground state, and call all others excited states. On cooling slowly enough, only this substate would be populated. There is, however, a problem. Assume that the globally lowest substate is separated by a barrier of height $H = 50\,\mathrm{kJ/mol}$ from other substates. Assuming an Arrhenius relation with a preexponential factor $A = 10^{13}\,\mathrm{s}^{-1}$, transitions over this barrier would be slower than the age of the universe at about 90 K. At 90 K, however, states within a few kJ/mol would still be appreciably populated and many would be indistinguishable from the lowest substate.

Figure 3 suggests that a similar energy landscape could also exist in proteins. As a protein folds, aperiodicity and frustration both play a role: In the folded protein, sidechains can conflict and there is no simple periodic arrangement. We therefore expect that the folded protein may exist in a number of different conformations, pictured as valleys or conformation substates in a conformation space of about 3N dimensions, where N is the number of atoms. The situation is sketched in Fig. 4.

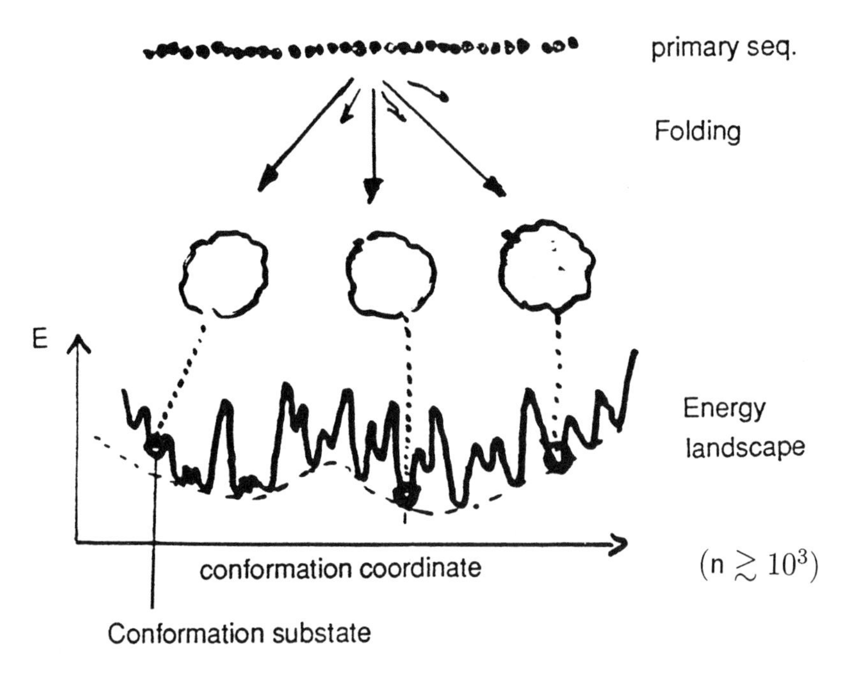

Fig. 4. A protein is expected to fold into a large number of conformation substates; its ground state is characterized by a rugged energy landscape.

If this concept is correct, we would expect proteins to move from CS to CS at high temperatures. Below a *glass* temperature, however, each protein molecule should remain frozen into a particular CS.

4 Conformation Substates

Evidence for CS comes from many experiments (Frauenfelder, Parak, and Young (1988), Frauenfelder, Nienhaus, and Young (1994)) and from molecular dynamics computations (Elber and Karplus (1985), Noguti and Go (1989), Garcia (1992)). Here I will discuss some of the clearest experiments.

4.1 Flash Photolysis

The idea underlying *flash-photolysis* experiments is straightforward (Gibson (1956)). Myoglobin with a small molecule (ligand) bound at the heme iron, for instance carboxy myoglobin (MbCO), has a different optical absorption spectrum than deoxy myoglobin (Mb). Start with MbCO. The bond between the heme iron and the CO can be broken by a light flash. The subsequent rebinding can then be followed in time by monitoring the absorption spectrum.

The first unambiguous evidence for CS came from flash photolysis experiments (Austin et al. (1974), Austin et al. (1975)). While biochemists usually perform experiments in the temperature range from 0 to 40°C, physicists by instinct extend the range down as low as possible. Indeed, it was found quite early that biological processes proceed even at cryogenic temperatures (Chance and Spencer (1959)). In the case of the binding of ligands (CO, O_2,...) to Mb, cryogenic studies have yielded exciting results. The first detailed experimental result (Austin et al. (1974)) is described in Fig. 5. As shown in Fig. 2, and also in Fig. 5, the heme in Mb is embedded in the globin and the CO in MbCO is bound to the iron. After the laser flash has broken the Fe-CO bond, the CO moves into the heme pocket. Below about 200 K, the CO does not escape into the solvent. Rebinding from the heme pocket can be described by a two-well potential. The deeper well A represents the bound state MbCO, the well B the state with the CO in the heme pocket. To bind, B $\rightarrow$ A, the system has to overcome a barrier of height H, denoted by E in Fig. 5.

At low temperature, the rate coefficient[1] can be approximated by an Arrhenius equation,

$$k_{\mathrm{BA}} = A_{\mathrm{BA}}(T/T_0)\exp\left[-H_{\mathrm{BA}}/RT\right]. \tag{1}$$

Here A is the preexponential factor, T_0 a reference temperature, usually taken to be 100 K, and R the gas constant. If H and A have unique values, rebinding is given by

$$N(t) = N(0)\exp\left[-k_{\mathrm{BA}}t\right], \tag{2}$$

where $N(t)$, with $N(0)$ normalized to 1, is the survival probability. Figure 5 shows early experimental data for $N(t)$ which make it very clear that $N(t)$ is not exponential in time. Similar results had been obtained by Goldanskii and coworkers for nonbiological systems. How do we describe the nonexponential rebinding, and how do we explain it? Figure 6 answers the first question.

We first note that the standard plot of $\log N(t)$ versus t is inadequate for nonexponential processes; the range of time covered is much too small. A plot of $\log N(t)$ versus $\log t$ provides a much better picture. In such a plot, an exponential is nearly a step function, as shown in the top-left panel. The shape of an exponential is universal and does not change if shifted in the $\log N$–$\log t$ plane. If we denote with $g(H)dH$ the probability of finding a barrier with height

[1] In much of the literature, the rate coefficient is called "rate constant." Physicists should reserve the word "constant" for true constants like c, e, h, and R, and properly call a quantity that can change by ten orders of magnitude as function of temperature a coefficient.

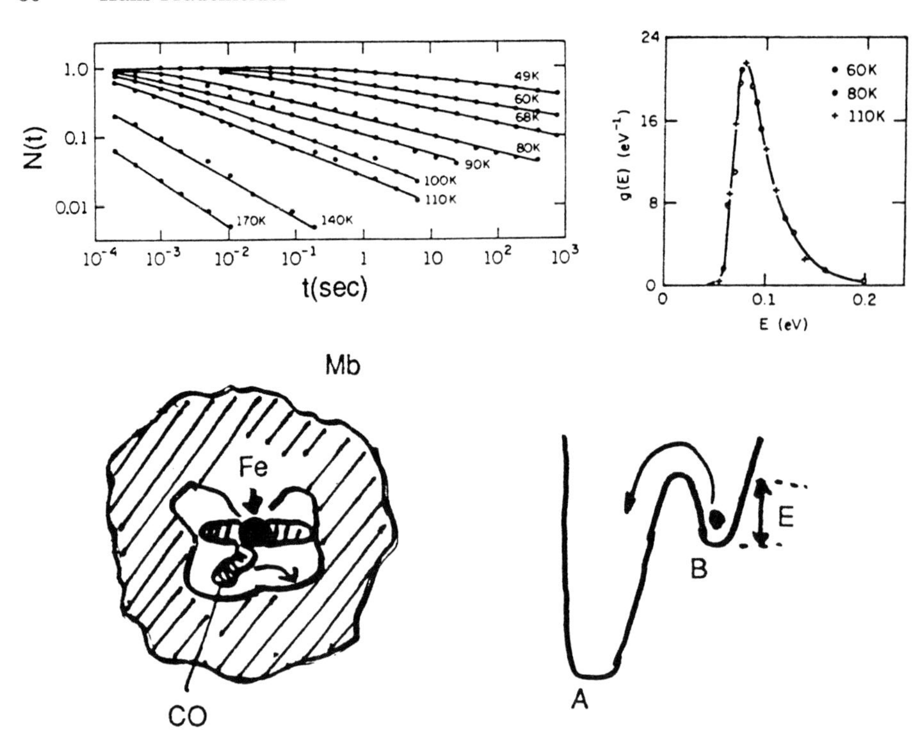

Fig. 5. The binding of CO to Mb after photodissociation - early results. Lower left: After flash-off, the CO remains in the heme pocket at low temperatures. Lower right: The rebinding can be described by a double-well potential. Upper left: The survival probability, N(t). Upper right: The distribution function g(H).

between H and $H + dH$, an exponential is given by a delta function, as sketched in the lower-left panel. The simplest way to describe a nonexponential process is to assume that the barrier in different proteins is different so that $g(H)$ is given by a distribution, as sketched in the lower-right panel. $N(t)$ then becomes

$$N(t) = \int dH g(H) \exp\left[-k_{\mathrm{BA}}(H)t\right] \quad . \tag{3}$$

Given $N(t)$, $g(H)$ can be calculated easily. Experimentally, however, it is $N(t)$ that is measured. The inversion of Eq. (3) is an inverse Laplace transform which is notoriously unstable and difficult to perform. Initially, we used crude approximations (Austin et al. (1974), Austin et al. (1975)) that gave some of the essential information, as is shown in Fig. 5. At present, the maximum entropy method provides good $g(H)$ (Steinbach et al. (1992)). Often, $g(H)$ can be adequately characterized by a Gaussian.

The explanation given so far for the nonexponential rebinding assumes that each protein molecule rebinds exponentially in time, but that different molecules

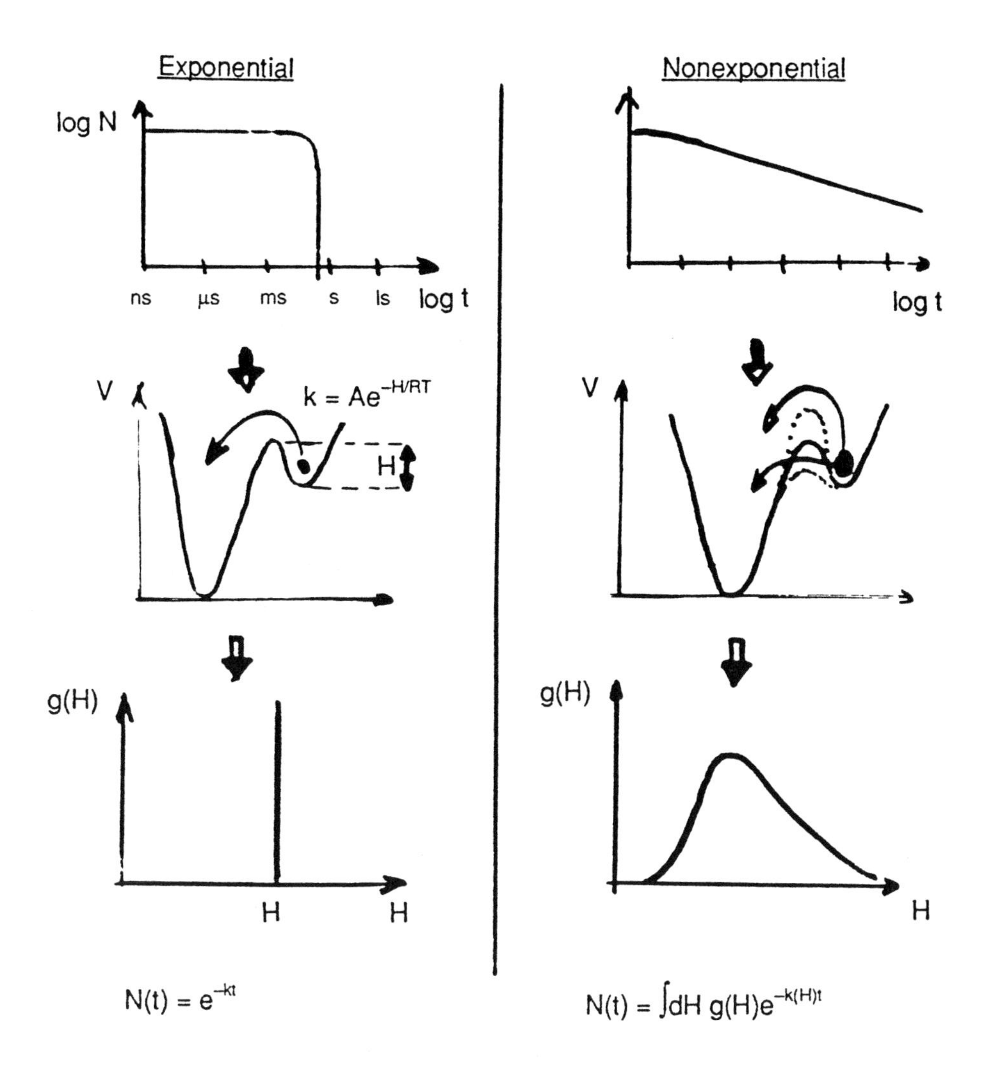

Fig. 6. Exponential and nonexponential processes.

have different barrier heights. A different explanation assumes that all protein molecules are identical, but that rebinding in each is nonexponential in time. By repeating the photolysis flash before all molecules have returned to the bound state, it is possible to distinguish between the two explanations (Austin et al. (1975), Frauenfelder (1978)). Experiments show unambiguously that the first explanation is correct; different Mb molecules have different barrier heights.

The second question—why do different protein molecules have different barriers—leads back to Fig. 4. If aperiodicity and frustration together always produce a rugged energy landscape, then the appearance of different barrier heights is nearly trivial. Different structures most likely will have different properties. Figure 7 shows the energy landscape that was proposed in 1975 (Austin et al. (1975)). The barriers between different substates are shown as thin lines and the crosses suggest the populations in the various substates. The caption is the original caption.

The figures 6 and 7 show one characteristics that I believe to be common to all complex systems, namely the occurrence of *distributions.* The figures show that the barrier heights are not sharp, but distributed. Similar distributions can be expected for many other physical properties.

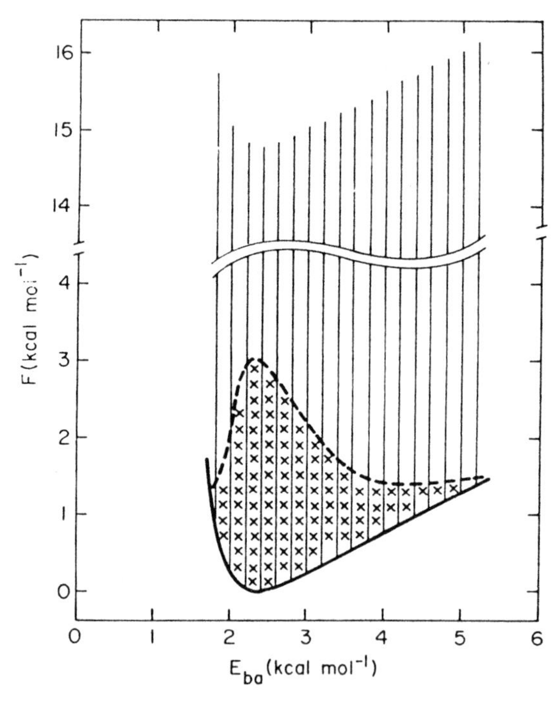

Fig. 7. Conformational energy, F, as a function of the activation energy E_{ba}. The spikes represent the energy barriers in relaxation. The crosses correspond to a Boltzmann equilibrium distribution at 200 K. (Austin et al. (1975))

Other questions also arise: Mb is a protein that is composed of α-helices. Do other α-helix proteins also show distributions? Do proteins that contain predominantly β-sheets behave similarly? The answer to both questions is yes. All α-helix proteins studied so far show nonexponential time dependent ligand binding (e.g. (Doster et al. (1987), Post et al. (1993)) as does azurin, a β-sheet protein (Ehrenstein and Nienhaus (1992)).

4.2 The Debye-Waller Factor (Frauenfelder (1989))

If substates exist, why do the beautiful illustrations of proteins in textbooks show unique structures? To answer this question, we need to return to the structure determination by X-rays (Drenth (1994), Hendrickson (1995)). The name "X-ray protein crystallography" already leads to some confusion and the role of the protein crystal is often misunderstood. Assume that we have an oriented protein, and that we can determine the differential scattering cross section of X-rays of a given wave length for many different orientations of the protein. Assume further that all atoms of the protein are at fixed positions. It is then possible to determine the position of all (non-hydrogen) atoms, and hence find the structure of the protein. Why do we need a protein crystal? There are three reasons. First, we cannot easily orient a single protein, but a protein crystal can be oriented. Second, scattering from a single protein is far too weak to be useful. A crystal acts as a low-noise high-gain amplifier that concentrates the scattered intensity into directions determined by the crystal parameters. Finally, a single protein would be destroyed quickly by the X-rays, while the destruction of a small fraction of proteins in a crystal still permits data taking.

What happens if conformation substates indeed exist? In this case, the assumption of atoms fixed to unique positions is no longer true; a particular atom can deviate considerably from its mean position, and now is characterized by a mean-square deviation $\langle x^2 \rangle$. The contribution from the atom to the scattering intensity is then reduced by the factor (Willis and Pryor (1975))

$$f_{\mathrm{DW}} = \exp\left[-B\sin^2\theta/\lambda^2\right], \tag{4}$$

where

$$B = 8\pi^2\left\langle x^2\right\rangle. \tag{5}$$

Here, θ is the scattering angle, λ is the X-ray wavelength, and B (or sometimes f_{DW}) is called the Debye-Waller or temperature factor. If CS exist, we would expect that some regions in the protein are more solid-like and some more liquid-like. The Debye-Waller factors for the atoms in a protein should consequently behave very differently from corresponding factors for solids. Indeed, the determination of the B values of proteins, first for Mb (Frauenfelder, Petsko, and Tsernoglou (1979)), supports the existence of CS (Petsko and Ringe (1984)). At present, B factors of individual atoms have been determined for many proteins.

In interpreting structures derived from X-ray scattering, two limitations must be remembered. These are outlined in Fig. 8. The B factors, and hence the $\langle x^2 \rangle$, are extracted from the experimental data by assuming a Gaussian distribution. Functionally important fluctuations, for instance the transient opening of a protein, usually occur only during a short time. They are therefore present only with a probability of 10^{-5} or less, as sketched in the upper part of Fig. 8. The conventionally determined Debye-Waller factor gives no information on these rare fluctuations.

The second problem arises if the protein can assume two considerably different structures (taxonomic substates), as indicated in the lower part of Fig. 8. Assume that the dominant structure is ten times more abundant than the other one.

Between the X-ray data and the crystallographer sits a computer ("Stupido"). The computer is usually told to come up with one structure. Obediently, it usually will, by neglecting the minority structure.

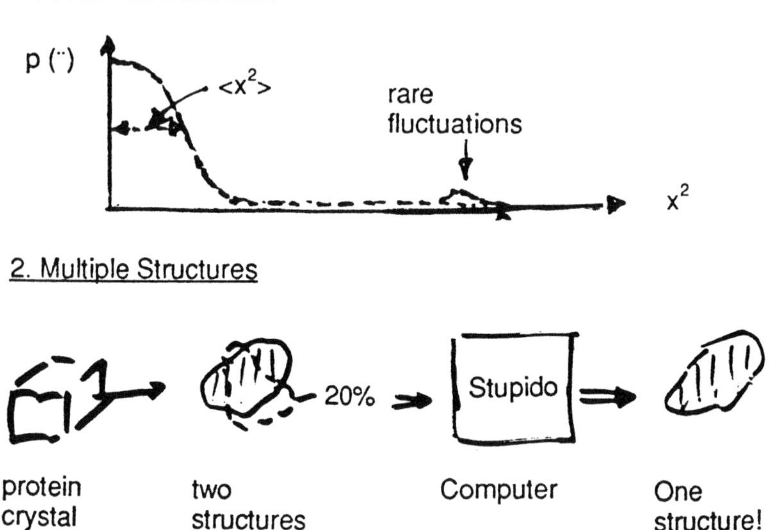

Fig. 8. Limitations of X-ray structure determinations.

The structure of proteins can also be determined by using nuclear magnetic resonance (NMR) (Wüthrich (1986), Clore and Gronenborn (1991), Wagner, Hyberts, and Havel (1992)). The NMR structures show the substates clearly.

4.3 Laser Hole Burning

The most dramatic evidence for the existence of CS comes from laser *hole burning* experiments (Jankowiak, Hayes, and Small (1993), Friedrich (1995)). The concept is shown in Fig. 9. It is to be expected that the different substates have spectral lines which are *Lorentzians* with somewhat different center frequencies (Agmon and Hopfield (1983), Agmon (1988)). (A *Lorentzians* describes the natural line shape.) The line actually observed is consequently inhomogeneously broadened, and is no longer a Lorentzian, but a *Voigtian,* i.e. a Gaussian superposition of Lorentzians. If the homogeneous line width is much smaller than the inhomogeneous one, the existence of CS can be shown directly: Irradiating with a narrow laser line at a particular position may move the proteins with lines at that wavenumber to different substates. The result is a "hole" in the spectrum.

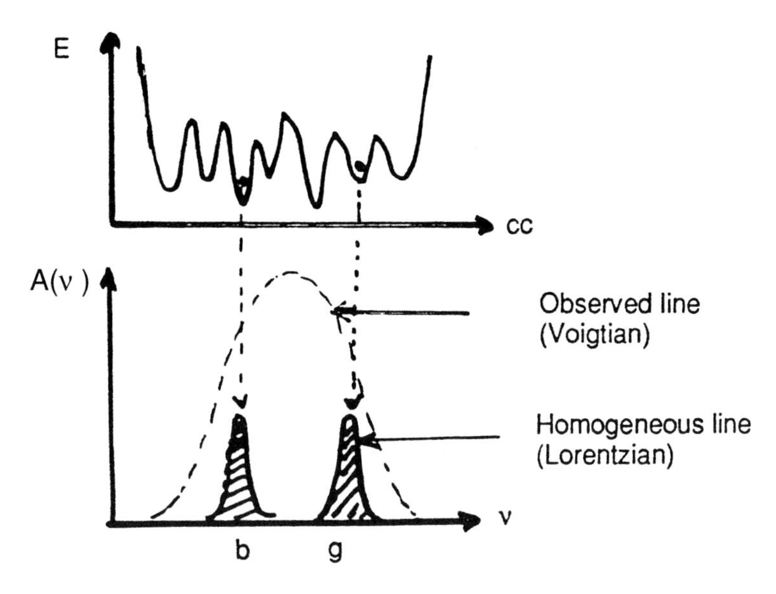

Fig. 9. Spectral lines in different substates can have different center frequencies. While the lines in individual substates are Lorentzians, the line in a protein ensemble is a Voigtian, a Gaussian superposition of Lorentzians.

Experiments clearly show such holes (Friedrich et al. (1981), Boxer et al. (1987), Zollfrank, Friedrich, and Parak (1992), Reddy, Lyle, and Small (1992), Friedrich et al. (1994)). The holes can be as narrow as 10^{-4} of the inhomogeneous width, implying that there are at least 10^4 CS present.

4.4 Connections

Figure 10 shows the connections established by the various experiments: Proteins in different conformation substates possess somewhat different structures. These CS correspond to different valleys in the energy landscape. The different CS have different barrier heights for rebinding ligands after photodissociation and hence rebind with different rates. The spectral lines in the different substates have different center frequencies. The various physical and chemical properties of a protein cannot be described by sharp values, but must be characterized by distributions.

5 The Organization of the Energy Landscape

Figure 4 shows a schematic illustration of the rough energy landscape in a protein, but all details are lacking. How are the substates arranged? How many substates are present in a given protein? What are the barrier heights between substates? A complete answer to these questions will possibly never be found,

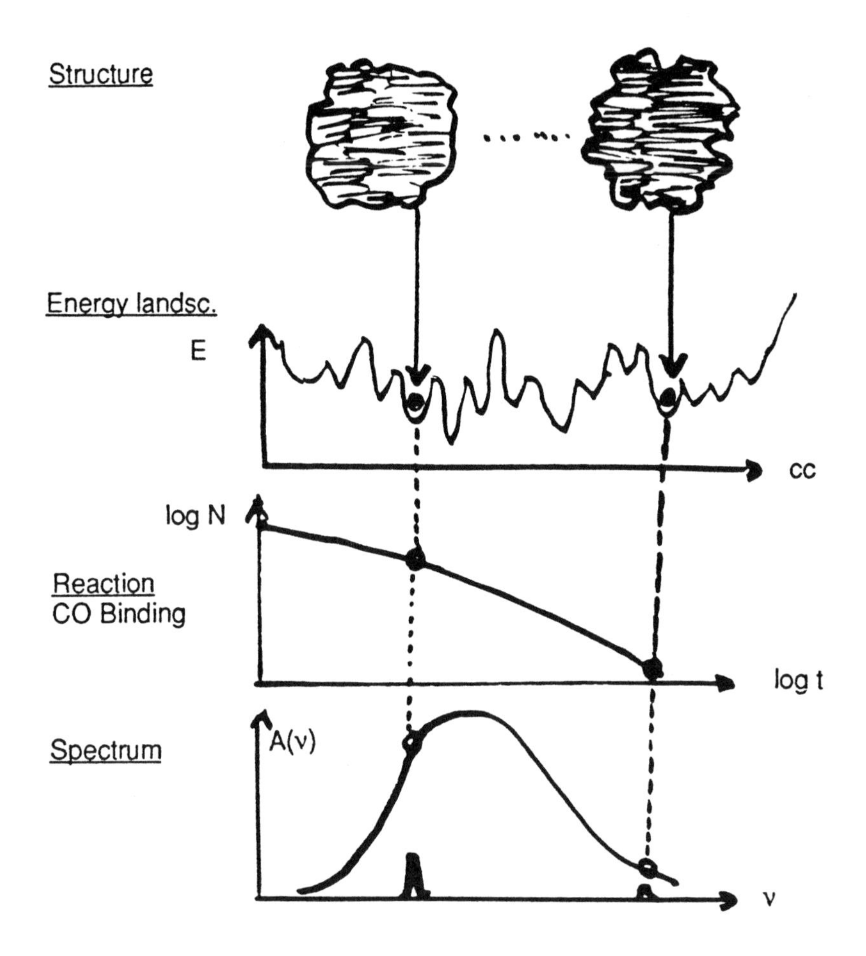

Fig. 10. Different substates (i) have different structures, (ii) correspond to different valleys in the energy landscape, (iii) have corresponding, different rebinding rates, and (iv) different center frequencies of spectral lines.

but a combination of techniques is beginning to give us some insight into the organization of the CS in at least one protein, myoglobin.

A comparison of data from different experiments suggests that Fig. 4 is far too simple (Ansari et al. (1985)). The energy landscape is actually organized hierarchically in a series of tiers. Different tiers are characterized by the average barrier height between substates of that tier. Tier 0 has the largest barriers; the ones in tier 1 are smaller, and so on. About ten years ago, the situation appeared as in Fig. 11. MbCO, myoglobin with CO bound, occurs in tier 0 in three (or possible four) different structures which we call A_0, A_1, (A_2), and A_3. These three different substates of tier 0 (CS0) can be studied individually, and we therefore call them taxonomic substates. They are distinguished by the different stretch frequencies of the bound CO (Moore, Hansen, and Hochstrasser (1988), Ormos et al. (1988)) (But see also reference (Lim, Jackson, and Anfinrud (1995))). The relative energies, entropies, and volumes of the three CS0 have been determined (Frauenfelder et al. (1990), Hong et al. (1990)) and structural information has been obtained (Yang and Phillips (1996)).

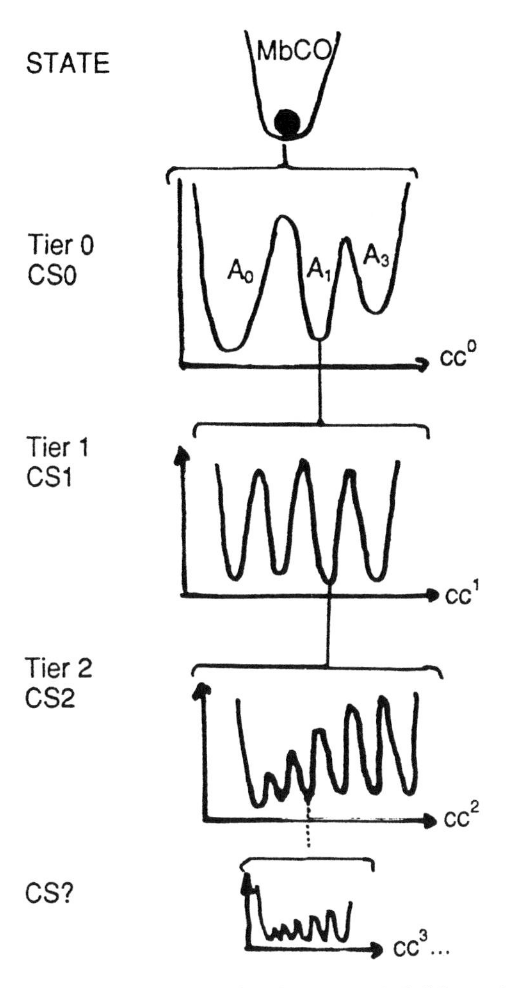

Fig. 11. The organization of the energy landscape in MbCO, as it appeared about 1985.

The existence of tier 1, CS1, in each of the preceding CS0 is shown by the inhomogeneous broadening of the individual stretch bands (Agmon (1988), Campbell, Chance, and Friedman (1987), Ormos et al. (1990)) and by the nonexponential time dependence of the CO rebinding in each CS0 (Ansari et al. (1987)).

As demonstrated by the smooth nonexponential time dependence of rebinding and by the narrow holes burned by lasers, the number of CS in tier 2 is very large. Physical properties can no longer be described by sharp values, but must be characterized statistically by distributions.

Properties of tier 2, CS2, are less well characterized. Evidence for this tier comes for instance from Mössbauer experiments (Keller and Debrunner (1980), Parak et al. (1981), Bauminger et al. (1983)), but the essential features remain to be explored.

Measurements of the specific heat of proteins show that its temperature dependence is similar to that of glasses; it deviates markedly from that of crystals and suggests the existence of tunnel (TLS) states (Goldanskii, Krupyanskii, and Fleurov (1983), Singh et al. (1984)). This observation implies that there exists a tier with very small barriers and this tier is shown as tier? in Fig. 11.

Many experiments imply that there are more tiers than shown in Fig. 11. Particularly interesting are recent results from laser hole burning (Jankowiak, Hayes, and Small (1993), Friedrich (1995)). One method for obtaining information on the energy landscape, temperature cycling, is shown in Fig. 12 (Zollfrank et al. (1991a), Zollfrank et al. (1991b), Kurita, Shibata, and Kushida (1995)). To discuss the approach, we note that the concept of glass temperature, introduced earlier, must be generalized for a protein. Transitions can still occur in tiers with low barriers when tiers will high barriers are already frozen. We consequently define a glass temperature $T_{\rm g}^{i}$ for each tier i. Below that glass temperature, motions in tier i are frozen; well above, they occur. Well below $T_{\rm g}^{i}$, a laser with frequency corresponding to the substate x in Fig. 12a depletes this substate, resulting in a persistent hole as sketched in Fig. 12b. If the temperature is cycled above the glass temperature, the substate x will be repopulated (Figure 12c) and the hole in the spectrum will disappear. The temperature at which the hole disappears thus gives an indication about the barrier height.

In a second method, the so-called spectral diffusion kernel (Black and Halperin (1977)) is measured (Gafert (1995)). First a spectral hole is burned, as in Fig. 12b. Note that the transitions shown in Fig. 12a are light-induced. At low enough temperatures, the hole will remain unchanged ("persistent"). At some temperature, however, thermal transitions occur and these can go from the substate x to other substates and broaden the hole, or they can fill in substate x, as shown in Figure 12c. The difference in width between the thermally broadened hole and the initial hole is called the spectral diffusion kernel. Its time dependence provides information about the energy landscape. The kernel can be observed at temperatures as low as 0.1 K (Kurita, Shibata, and Kushida (1995)), implying substates with very low barriers.

A third approach uses photon echos (Littau, Bai, and Fayer (1989), Meijers and Wiersma (1994), Thorn-Leeson and Wiersma (1995b)). In this technique the broadening of a homogeneous line is measured as a function of time and temperature. Application to myoglobin yields evidence for CS between 3 and 8 K (Thorn-Leeson and Wiersma (1995a)).

In a somewhat speculative and tentative way we can combine the informa-

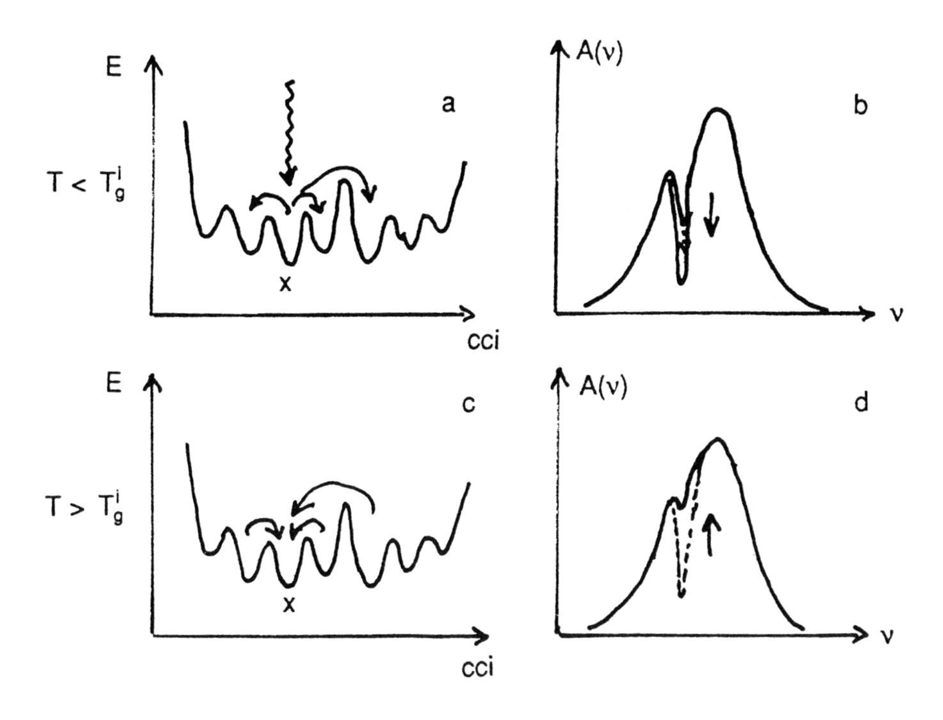

Fig. 12. Hole burning and temperature cycling (Zollfrank et al. (1991a), Zollfrank et al. (1991b), Kurita, Shibata, and Kushida (1995)).

tion from the various experiments (Frauenfelder (1995)) with theoretical and computer studies of protein folding (Frauenfelder and Wolynes (1994), Wolynes, Onucic, and Thirumalai (1995), Onuchic et al. (1995)). The result is shown in Fig. 13. The arrows at the top indicate that, in folding, the system is guided into the folding funnel. It then passes through the molten globule stage, composed of regions that are folded and regions that are still disordered (Ptitsyn (1992), Jennings and Wright (1993), Freire (1995), Fink (1995)).

Finally, the protein folds into the "native structure", the working protein. As discussed in detail above, this structure is not unique, but must still be described by a rugged energy landscape. One way to characterize the energy landscape of the folded protein is in a tree diagram, as given in Fig. 14 (Frauenfelder (1995)). As already stated, the figure is speculative, and much remains to be done with

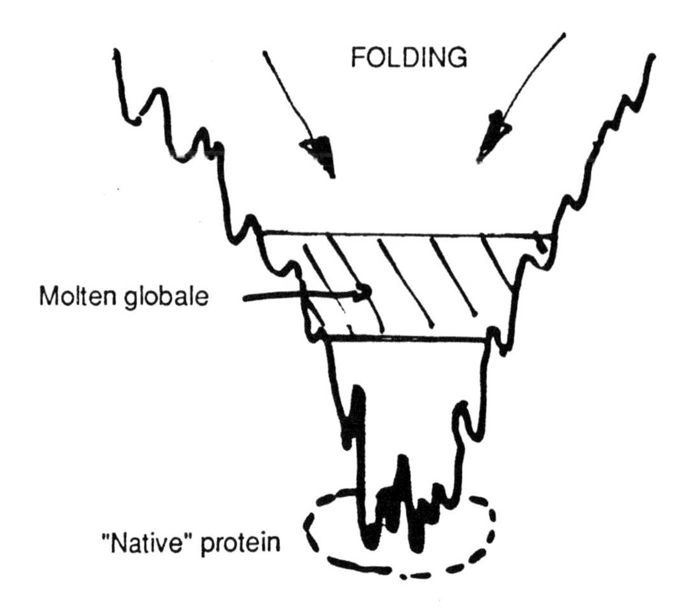

Fig. 13. A one-dimensional cross section through the energy landscape of a protein.

all available tools—experimentally, theoretically, and by computer simulation—before the organization of the energy landscape of even *one* protein is known.

In summary, the complexity of proteins is apparent from Figs. 13 and 14. The crucial concepts that are displayed in these figures, are the existence of a rugged energy landscape, conformation substates, and the hierarchy of the substates.

6 Protein Dynamics

The existence of a rugged energy landscape with a hierarchical organization implies that the motions of the protein, not unexpectedly, can be extremely complicated. Moreover, proteins can undergo chemical reactions, from the simple covalent binding of small molecules to enzymatic processes. The combination of chemical reactions in systems undergoing conformational motions raises questions that are not usually treated in texts. Here we outline some of the major issues.

6.1 Protein Motions

Motions occur in every tier shown in Fig. 14. To classify the possible motions, we look at what can happen in any one tier. The possible motions are drawn in Figure 15. First we distinguish equilibrium and nonequilibrium motions. In

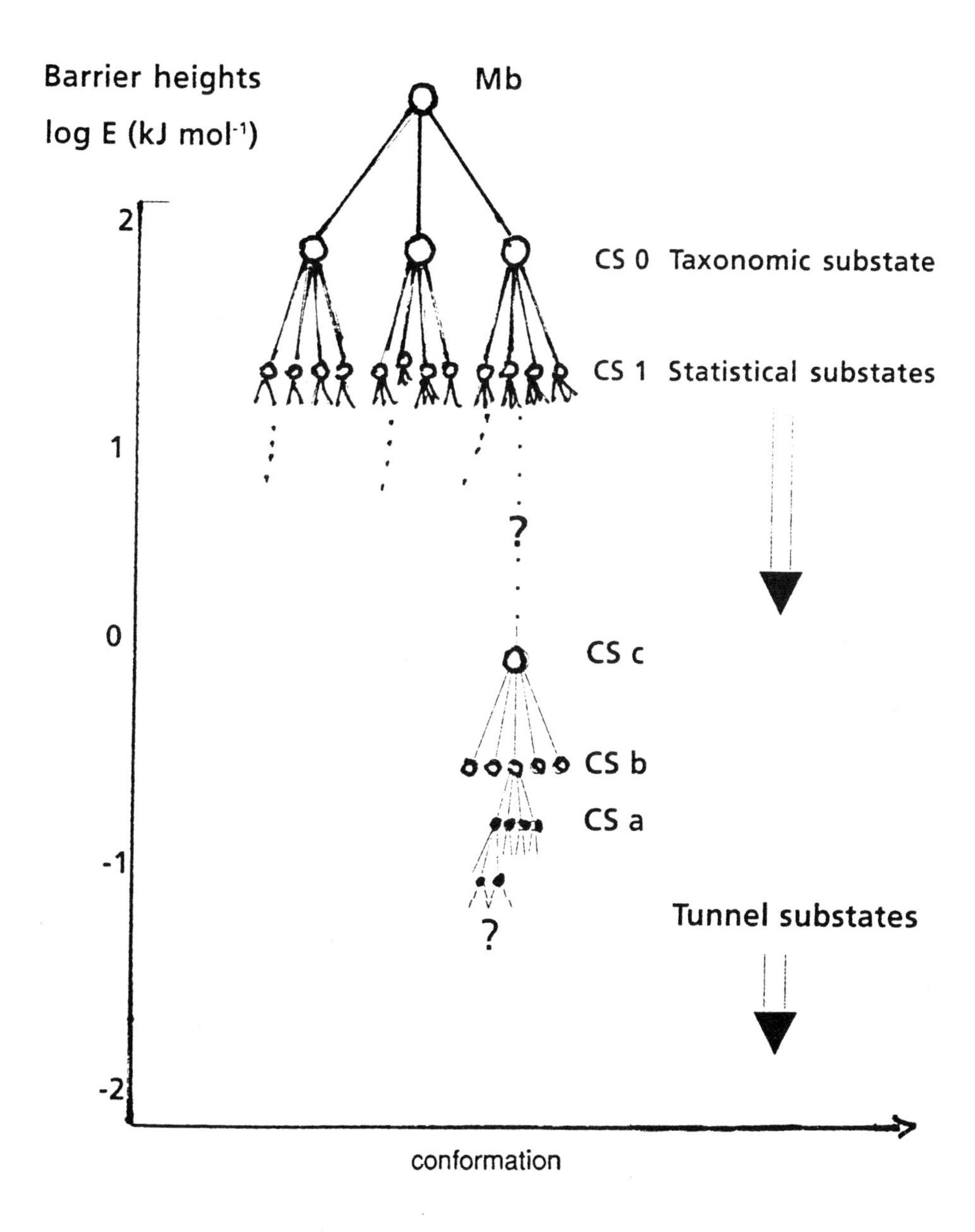

Fig. 14. Tree diagram of the rugged energy landscape in myoglobin. The vertical axis gives the logarithm of the average barrier height between CS (Frauenfelder (1995)).

equilibrium, vibrations (1) and conformational fluctuations (2) can occur. Vibrations happen in a single substate. Fluctuations change the system from one CS to another; the topology of the protein changes. If the system is lifted to a nonequilibrium substate by an external perturbation or a reaction, it will relax towards equilibrium (3). The existence of the rugged energy landscape and the hierarchy vastly complicates all motions. At very low temperatures, conformational motions will only occur in the lowest tier; the system will tunnel from one CS to another. The protein will remain frozen in a particular subset of all higher CS. As the temperature is increased, fluctuations in higher CS will begin. At room temperature, fluctuations will involve all tiers and lead to an extremely complicated behavior.

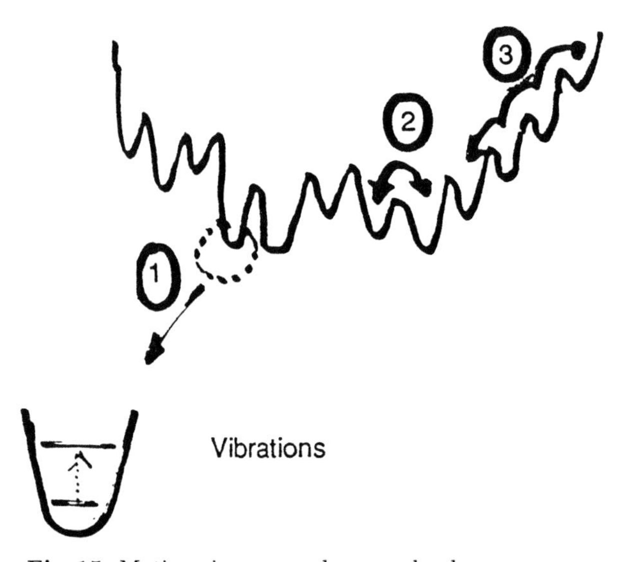

Fig. 15. Motions in a rugged energy landscape.

Fluctuations and relaxations are related by the fluctuation-dissipation theorem if the initial substate of the nonequilibrium motion is not too far away from equilibrium.

6.2 Experimental Techniques

There is no single technique that permits the exploration of protein motions in all tiers, but a number of approaches are beginning to yield some insight into the barrier heights and distributions in the various tiers. In essentially all techniques, a particular spectroscopic marker provides the information. The marker can, for instance, be the stretch frequency of the bound CO in MbCO, or the charge transfer band at 760 nm that occurs in deoxy Mb. In laser hole burning experiments, the marker can be the position and width of the hole. A nonequilibrium state is reached, for instance, by a pressure (Frauenfelder et al. (1990), Iben et al. (1989), Young and Scholl (1991)) or temperature jump and changes in the marker are recorded as function of time and temperature. The principle of a pressure jump experiment is shown in Fig. 16.

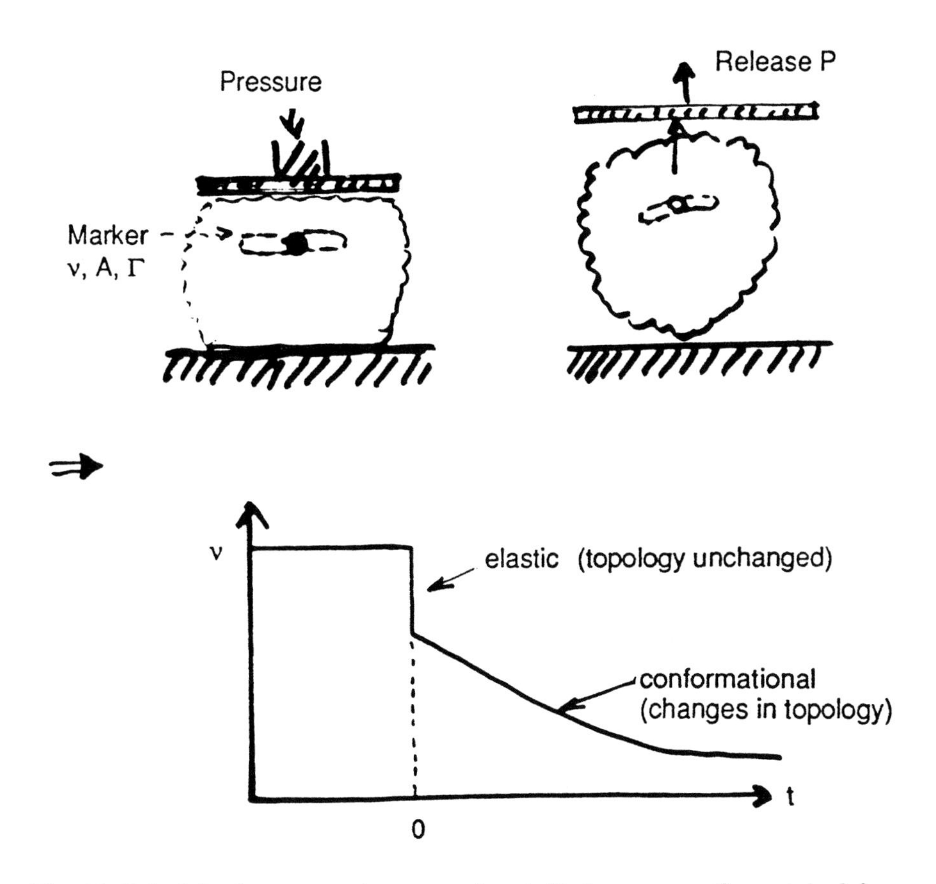

Fig. 16. Principle of a pressure jump experiment. Under pressure the protein deforms; the distribution of substates changes. On release, the protein relaxes back. The relaxation shows an elastic and a conformational component.

Under pressure, the protein changes substates and also decreases in volume. After the pressure is released, the protein relaxes back to the original substate distributions and also changes its volume. The relaxation shows a very fast elastic component and a much slower conformational change. During the elastic component, the topology of the protein does not change; the conformational relaxation represents a change in topology. In the data evaluation, both components must be taken into account (Frauenfelder et al. (1990)). Similar relaxation phenomena can be induced by a temperature jump, or by a reaction induced, for instance, by a photo flash ("protein quake" (Ansari et al. (1985)). Laser hole burning also yields information on relaxation rates, as is implied in Fig. 12c.

6.3 Some Selected Results

The experiments performed so far indicate that fluctuation and relaxation phenomena in proteins are similar to processes in glasses and spin glasses, but are much more complex (Frauenfelder, Parak, and Young (1988), Frauenfelder, Nienhaus, and Young (1994), Frauenfelder et al. (1990), Iben et al. (1989), Young and Scholl (1991), Young et al. (1991), Steinbach et al. (1991), Nienhaus, Mourant, and Frauenfelder (1992), Ansari et al. (1994), Jackson, Lim, and Anfinrud (1994)). First, as already implied in the discussion of the energy landscape, fluctuation and relaxation phenomena occur in the entire temperature range from 0.1 to 300 K. Second, the relaxation phenomena are nonexponential in time. Third, if observed over a wide enough temperature range, they do not follow an Arrhenius law.

One particularly striking observation concerns the effect of light on the relaxation. In myoglobin, light slows the binding of CO after photodissociation (Chance et al. (1986)). The slowing is caused by a light-induced conformational relaxation that increases the barrier height for the binding of CO (Nienhaus et al. (1994), Chu et al. (1995)). This effect is not yet understood; breaking of hydrogen bonds is a possible mechanism.

7 Reaction Theories

Theories are needed to describe and understand the motions and reactions in proteins. Unfortunately, most texts are about 50 years out of date. We therefore describe some of the main issues here briefly and cite references where detailed treatments are given.

We again take myoglobin as example. Figure 2 then indicates that the goal is to understand at least three different issues: (i) the passage of the ligand through the protein to the binding site, (ii) the motions of the protein that permit the passage, and (iii) the final binding step at the heme. We begin with the last step, the binding in the heme pocket.

7.1 The Reaction Surface

Figure 17 displays the main features involved in the binding step. In state B, the CO is in the heme pocket parallel to the heme plane (Schlichting et al. (1994)), the heme is domed, and the iron has spin 2 and is about 0.5 Å out of the mean heme plane (Dickerson and Geis (1983)). In the bound state, A, the heme is planar, the iron has spin 0, and the CO is bound covalently to the heme iron. The binding process can be described by the two-well potential in figure 17. Binding can occur in two different ways. At very low temperatures, the CO can tunnel through the barrier; at higher temperatures, it will move "classically" over the barrier. Of course, this separation is incorrect; quantum-mechanically, the system moves from B to A, and the two paths cannot be distinguished.

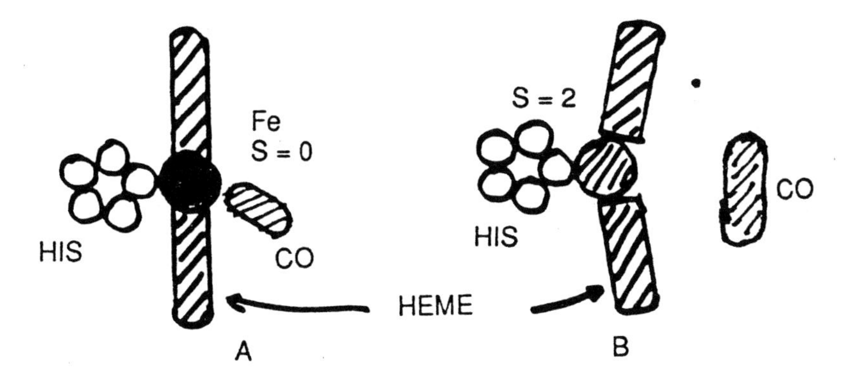

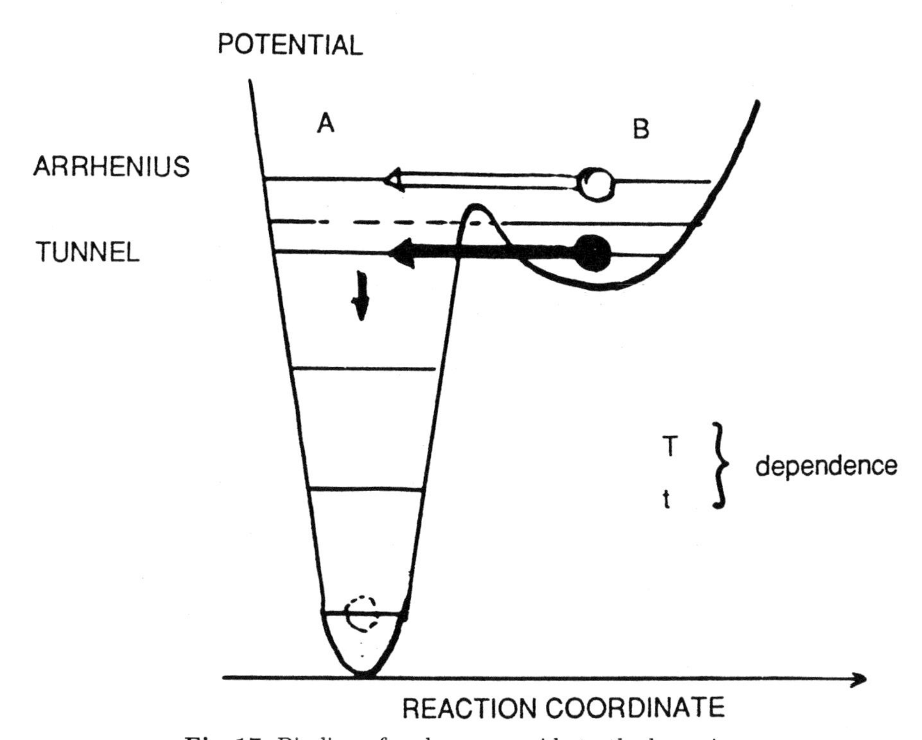

Fig. 17. Binding of carbon monoxide to the heme iron.

We have already encountered the binding process in Section 3.1. Figure 5 indicates that the binding process is nonexponential in time, because the barrier between B and A is distributed. The first crude form of the distribution function is also given in Fig. 5.

7.2 Arrhenius and Kramers Relation

In Section 4.1 we used the Arrhenius relation, Eq. (1), to characterize the rate coefficient for the transition B $\rightarrow$ A. Indeed, this relation in a slightly more sophisticated form is generally employed in chemistry to describe reactions (Berry, Rice, and Ross (1980)). The improvement over Eq. (1) consists essentially in replacing the enthalpy H by the Gibbs function (free enthalpy) $G = H - TS = E + PV - TS$, where S is the activation entropy and V the activation volume. Moreover, a fudge factor (transmission coefficient) is introduced to account for the fact that not all attempts to move over the barrier succeed, not even when sufficient energy is available. The inadequacy of this *activated-complex theory,* or *transition-state theory,* is easily realized. Consider Fig. 18. Passage over the barrier, $B \rightarrow A$, corresponds to escape from the well B. The particle in well B is coupled to the surroundings, which effectively provide a "heat bath." If this coupling is very small or zero, no transitions will occur, because particles cannot gain sufficient energy to surmount the barrier. If the coupling is very large, particles will not be able to overcome the barrier because of the very large friction. Consequently, the coupling between particle and surroundings should occur in the expression for the reaction rate, but it does not in the conventional formulation. Its effect is accounted for by the fudge factor only. A better theory is needed.

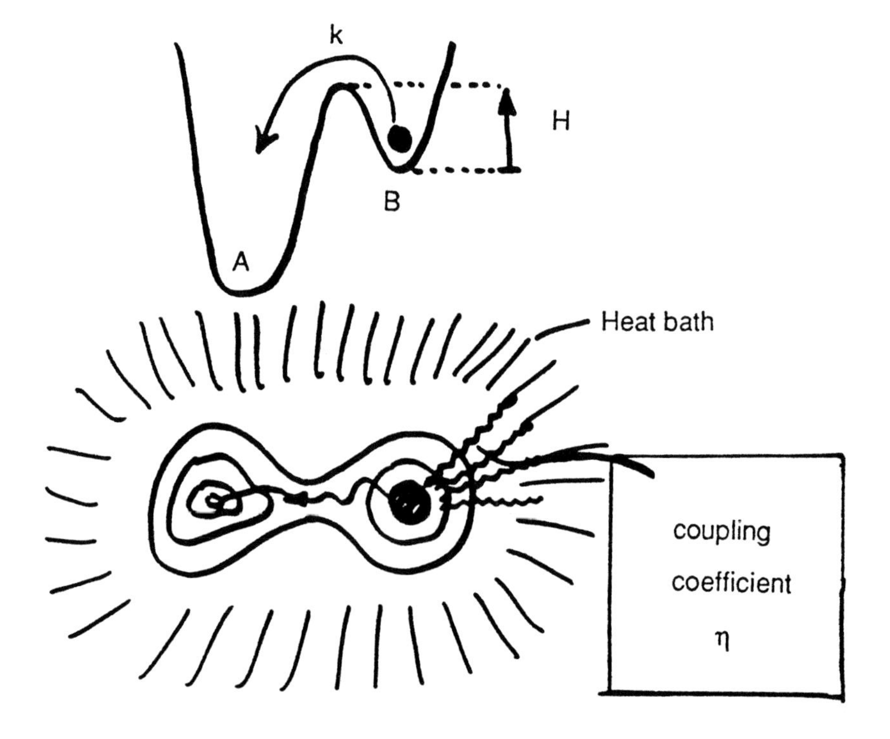

Fig. 18. Escape from a potential well.

Remarkably, such a theory has existed for over 50 years. In 1940, Kramers treated the problem (Kramers (1940)) and his result is schematically shown in Fig. 19. This figure gives the logarithm of the reaction rate coefficient as a function of the logarithm of the coupling coefficient with the heat bath. In actual applications the coupling coefficient is approximated by the viscosity. The result of Kramers has been explained (Frauenfelder and Wolynes (1985)) and improved in many papers (Hänggi, Talkner, and Borkovec (1990), Bunsengers (1991), Fleming and Hänggi (1993)).

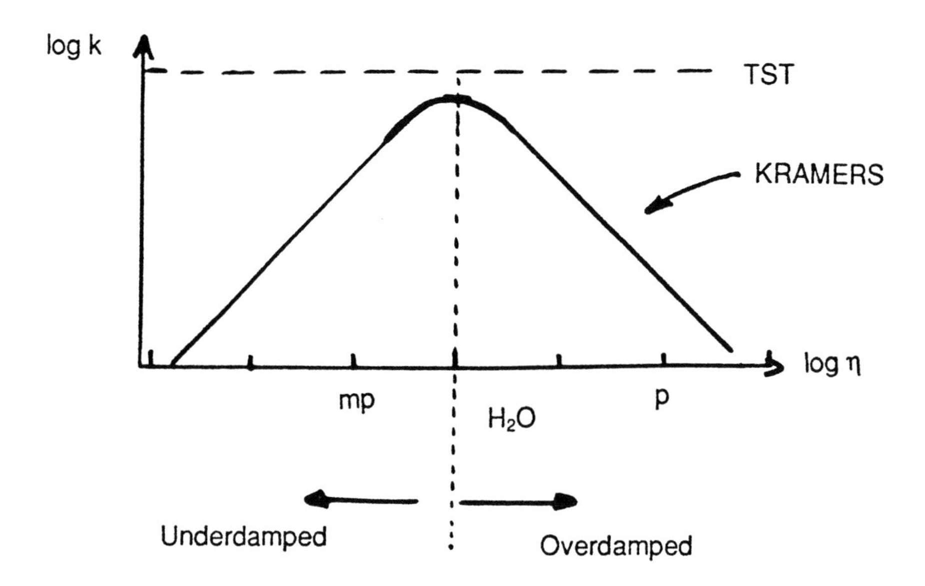

Fig. 19. The reaction rate coefficient as a function of the coupling of the system with the surrounding heat bath.

Figure 19 shows that the rate coefficient, k, increases linearly with friction (viscosity) in the underdamped regime, and decreases linearly with friction in the overdamped regime. Crudely speaking, one can say that the system undergoes Brownian motion in energy in the underdamped regime, and Brownian motion in the momentum space in the overdamped regime.

The importance of the Kramers behavior for a complex system, such as a protein, is the control that the environment can exert on reactions (and motions) by changes in viscosity. This effect has been seen in proteins (Gavish (1978), Beece et al. (1980), Settles et al. (1992), Ansari et al. (1992)).

7.3 The Molecular Tunnel Effect (Goldanskii, Trakhtenberg, and Fleurov (1989))

The rate coefficient for tunneling of a particle of mass M through a barrier of height H and width d is, in the limit $T \to 0$, approximately given by

$$k \approx A \exp\{-2\pi c [2MH]^{1/2} d/h\} , \tag{6}$$

where A is a preexponential, c is a constant of order unity, and h is Planck's constant. We can observe the tunneling of electrons over distances of nanometers because electrons are light. We are used to the tunneling of alpha particles in nuclei because the distances, fermi, are small. But can the tunnel effect shown in Fig. 17 be observed? CO is heavy, and the distances are of the order of 0.1 nm. The answer to this question is yes, and the reason is that the barrier height, H, can be selected to be very small. Figure 5 indicates that the average barrier for CO binding to Mb is about 0.1 eV, or 10 kJ/mol. The structure of Mb, as sketched, suggests that the barrier width must be of the order of 0.1 nm. Eqation (6), with $A \approx 10^9\,\mathrm{s}^{-1}$, then implies that tunneling should be barely observable. Fortunately, there exist proteins with smaller barriers for CO binding, for instance the separated beta chain of hemoglobin where the average barrier is only about 4 kJ/mol (Alberding et al. (1978)). Moreover, since the barriers are distributed, there exist much lower barriers. Indeed, rebinding measured from 4 to about 100 K, drawn in Fig. 20, clearly shows two regions. Between about 20 and 80 K, rebinding follows an Arrhenius law. At about 20 K the data deviate and exhibit the nearly temperature-independent behavior characteristic of tunneling (Alberding (1976)).

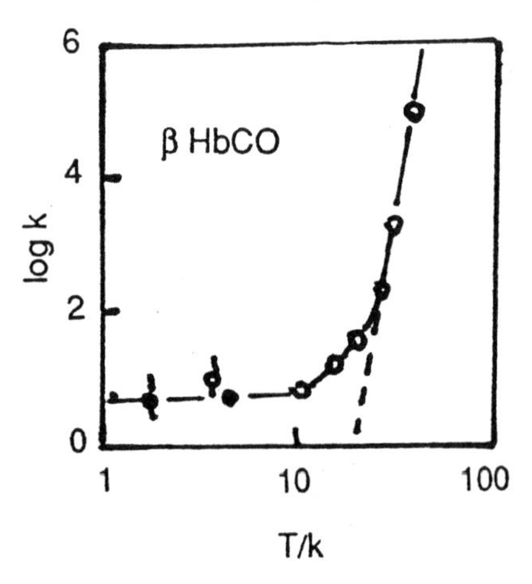

Fig. 20. Tunneling in Beta hemoglobin (Alberding (1976)). Note that $\log k$ is plotted versus $\log T$, not $1/T$.

The turnover shown in Fig. 20 did not convince everyone that quantum mechanical tunneling of such massive molecules as CO had indeed been observed.

Tunneling and Arrhenius behavior differ not only in their temperature dependence, but also in their mass dependence. The mass, M, enters the Arrhenius and Kramers relations only through the preexponential. In tunneling, however, the mass occurs in the exponent, see Eq. (6). Comparing the low-temperature binding of $^{12}C^{16}O$, $^{13}C^{16}O$, and $^{12}C^{18}O$ gives information concerning the isotope effect. The comparison is possible because the three molecules can be distinguished by their CO stretch frequencies. The isotopes to be compared can therefore be measured at the same time, avoiding difficult calibration problems. The result (Alben et al. (1980)) is convincing and surprising. From the change in rebinding rates, a value of $\delta M/M$ can be extracted. The change from ^{12}C to ^{13}C decreases the ratio of rebinding rates by about a factor of 1.5 and yields $\delta M/M \approx 0.04$, approximately as expected. The change from ^{16}O to ^{18}O, however, decreases the ratio only by about a factor 1.2, much less than anticipated. The simplest explanation is rotational binding, in agreement with the observation that the photodissociated CO lies in a plane parallel to the heme (Figure 17) (Schlichting et al. (1994)).

7.4 Large Motions

Motions that involve topological changes can be described by the Arrhenius or the Kramers relation only over small ranges in temperature. The fact that this description is inappropriate, even though it fits the data, can be seen from the fact that preexponentials can become very large, far exceeding the expected value of $10^{13}\,s^{-1}$. The deviation is often ascribed to entropy effects. However, it is well known from studies of glasses (Angell (1995)) that the temperature dependence is no longer of the Arrhenius type, but is better approximated by a Vogel-Tammann-Fulcher equation,

$$k = A \exp\{-H/R[T - T^*]\} \ , \tag{7}$$

or by a Ferry relation (Richert and Bässler (1990), Bryngelson and Wolynes (1989)),

$$k = A \exp\{-[H/RT]^2\}. \tag{8}$$

Both of these relations fit the data for large-scale motions over a wide range of temperatures (Iben et al. (1989), Young et al. (1991)). With these relations, the values for the preexponentials become more acceptable.

7.5 Reaction and Conformation Coordinate

We have so far encountered two coordinates, the reaction coordinate in Fig. 17 and the conformation coordinate in Figs. 4 and 14. In an actual protein reaction, the system will move along both coordinates. As an example, consider the motion of the dioxygen in Fig. 2. As the O_2 moves through the protein into the heme pocket and then finally binds at the heme iron, the protein concurrently changes its conformation. Studies of the process suggest the description given in Fig. 21

that combines the reaction and the conformation coordinate (Nienhaus et al. (1994)).

In the binding of CO to Mb, the CO comes from the solvent, S, and enters the protein when a fluctuation opens a path (Beece et al. (1980)). Once in the pocket, in state B^3, it encounters a large barrier, as shown in Fig. 21b. The CO therefore returns to the solvent. The shuttling between solvent and state B^3 proceeds till a major fluctuation in the protein lowers the barrier from B^3 to B^0, so that the CO can finally bind, $B^0 \rightarrow A$. In the final binding step, both transitions occur both in the reaction and the conformation coordinate; while the CO binds, the protein adjusts its conformation.The entire process is displayed in Fig. 21a in a conformation-reaction plane; in Fig. 21b, the cross section along the reaction path is shown.

In photodissociation, the system starts in state A, MbCO. After the photodissociation, the CO moves into the heme pocket, while the protein relaxes to the state B^0 (Schlichting et al. (1994)). Depending on temperature, the CO then either predominantly rebinds from B^0, or stays in the pocket till the protein relaxes from the non-equilibrium state B^0 to B^3. From there, the CO can either return to B^0 or use a fluctuation to escape into the solvent.

In this picture, we glimpse the incredible complexity of protein function, even for such a simple phenomenon as the binding of a small molecule.

Acknowledgment

This chapter was written under the auspices of the U. S. Department of Energy.

References

Agmon, N., Biochemistry **27**, 3507–3511 (1988).

Agmon, N., and J.J. Hopfield, J. Chem. Phys. **79**, 2042–2053 (1983).

Alben, J. O., D. Beece, S. F. Bowne, L. Eisenstein, H. Frauenfelder, D. Good, M. C. Marden, P. P. Moh, L. Reinisch, A. H. Reynolds, and K. T. Yue, Phys. Rev. Lett. **44**, 1157–1160 (1980).

Alberding, N., R. H. Austin, K. W. Beeson, S. S. Chan, L. Eisenstein, H. Frauenfelder, and T. M. Nordlund, Science **192**, 1002–1004 (1976).

Alberding, N., S. C. Chan, L. Eisenstein, H. Frauenfelder, D. Good, I. C. Gunsalus, T. M. Nordlund, M. F. Perutz, A. H. Reynolds, and L. B. Sorensen, Biochemistry **17**, 43–51 (1978).

Angell, C. A., Science **267**, 1924–1935 (1995).

Ansari, A., J. Berendzen, S. F. Bowne, H. Frauenfelder, I. E. T. Iben, T. B. Sauke, E. Shyamsunder, and R. D. Young, Proc. Natl. Acad. Sci. USA **82**, 5000–5004 (1985).

Ansari, A., J. Berendzen, D. Braunstein, B. R. Cowen, H. Frauenfelder, M. K. Hong, I. E. T. Iben, J. B. Johnson, P. Ormos, T. B. Sauke, R. Scholl, A. Schulte, P. J. Steinbach, J. Vittitow, and R. D. Young, Biophys. Chem. **26**, 337–355 (1987).

Ansari, A., C. M. Jones, E. R. Henry, J. Hofrichter, and W. A. Eaton, Science **256**, 1796–1798 (1992).

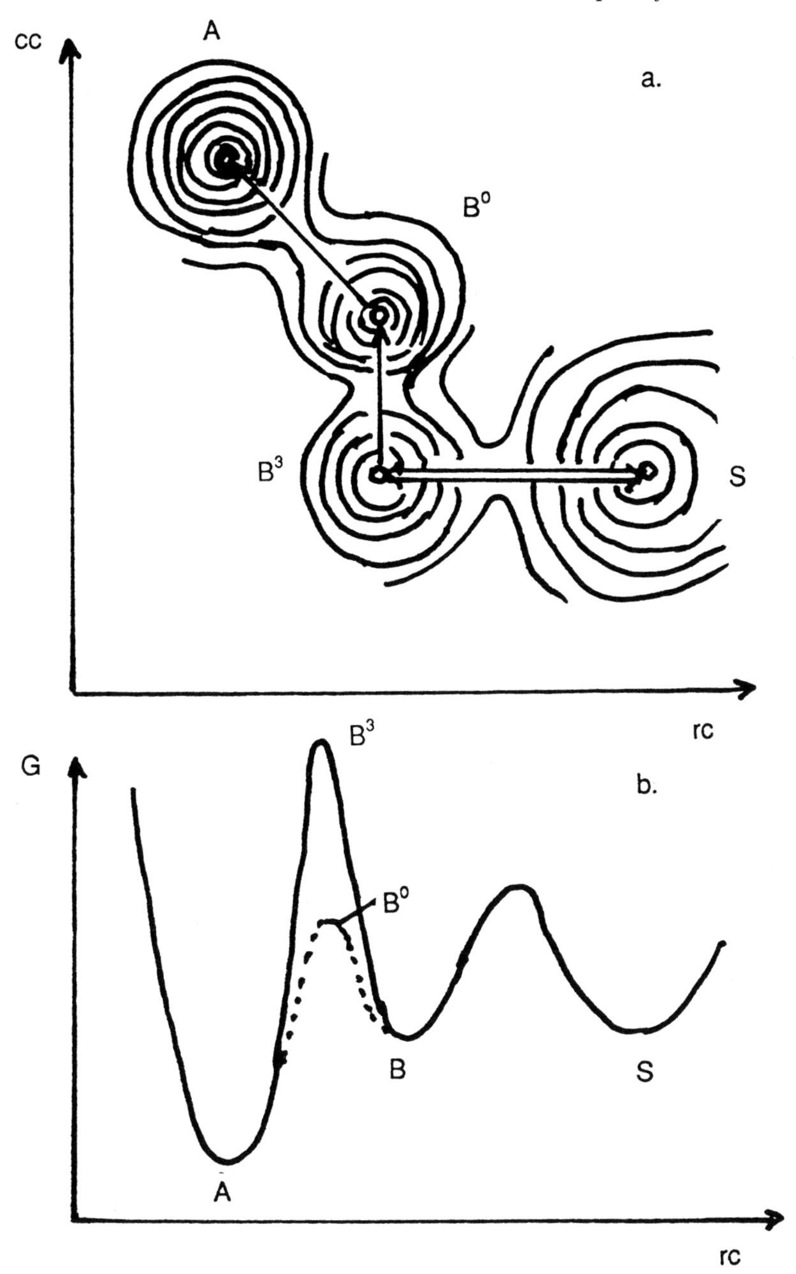

Fig. 21. (a) The binding of CO to myoglobin, represented in a conformation-reaction surface; cc denotes the conformation coordinate, rc the reaction coordinate. (b) A cross section along the reaction coordinate. Relaxation changes the barrier between B and A from the high barrier in B^3 to the much smaller barrier B^0.

Ansari, A., C. M. Jones, E. R. Henry, J. Hofrichter, and W. A. Eaton, Biochemistry **33**, 5128–5145 (1994).

Antonini, E., and M. Brunori, *Hemoglobin and Myoglobin in their Reactions with Ligands.* Alsevier, Amsterdam, 1970.

Austin, R. H., K. W. Beeson, L. Eisenstein, H. Frauenfelder, and I. C. Gunsalus, Biochemistry **14**, 5355–5373 (1975).

Austin, R. H., K. W. Beeson, L. Eisenstein, H. Frauenfelder, I. C. Gunsalus, and V. P. Marshall, Phys. Rev. Lett. **32**, 403–405 (1974).

Bauminger, E. R., S. G. Cohen, I. Nowik, S. Ofer, and J. Yariv, Proc. Natl. Acad. Sci. USA **80**, 736–740 (1983).

Beece, D., L. Eisenstein, H. Frauenfelder, D. Good, M. C. Marden, L. Reinisch, A. H. Reynolds, L. B. Sorensen, and K. T. Yue, Biochemistry **19**, 5147–5157 (1980).

Berry, R. S., S. A. Rice, and J. Ross, *Physical Chemistry*, John Wiley, New York (1980).

Black, J. L., and B. Halperin, Phys. Rev. B **16**, 2879 (1977).

Boxer, S. G., D. S. Gottfried, D. J. Lockhart, and T. R. Middendorf, J. Chem. Phys. **86**, 2439–2441 (1987).

Branden, C. and J. Tooze, *Introduction to Protein Structure.* Garland Publishing, New York, 1991.

Bryngelson, J., and P. G. Wolynes, J. Phys. Chem. **93**, 6902–6915 (1989).

Bunsenges, Ber., Phys. Chem. **95** (1991).

Campbell, B. F., M. R. Chance, and J. M. Friedman, Science **238**, 373–376 (1987).

Chance, B., K. Korszun, S. Khalid, C. Alter, J. Sorge, and E. Gabbidon, in *Structural Biological Applications of X-ray Absorption, Scattering, and Diffraction.* H. D. Bartunik and B. Chance, Eds., pp. 49–71, Academic Press, New York (1986).

Chance, B., and E. L. Spencer, Faraday Soc. Disc. **27**, 200 (1959).

Chu, K., R. M. Ernst, H. Frauenfelder, J. R. Mourant, G. U. Nienhaus, and R. Philipp, Phys. Rev. Lett. **74**, 2607–2610 (1995).

Clore, G. M., and A. M. Gronenborn, Ann. Rev. Biophys. Biophys. Chem. **20**, 29–63 (1991).

Dickerson, R. E., and I. Geis, *Hemoglobin, Structure, Function, Evolution, and Pathology.* Benjamin/Cummings, Menlo Park, 1983.

Doster, W., S. F. Bowne, H. Frauenfelder, L. Reinisch, and E. Shyamsunder, J. Mol. Biol. **194**, 299–312 (1987).

Drenth, J., *Principles of Protein X-ray Crystallography*, Springer-Verlag, New York (1994).

Ehrenstein, E., and G. U. Nienhaus, Proc. Natl. Acad. Sci. USA **89**, 9681–9685 (1992).

Elber, R., and M. Karplus, Science **235**, 318–321 (1985).

Fink, A. L., Annu. Rev. Biophys. Biomol. Struct. **24**, 495–522 (1995).

Fleming, G. R., and P. Hänggi, Eds., *Activated Barrier Crossing.* World Scientific, Singapore (1993).

Frauenfelder, H., in Methods in Enzymology, LIV Part E, S. Fleischer and L. Packer, Eds., Academic Press, New York, pp. 506–532 (1978).

Frauenfelder, H., in *Physics in a Technological World.* A. P. French, Ed., American Institute of Physics, New York, 1988.

Frauenfelder, H., Int. J. Quantum Chem. **35**, 711–715 (1989).

Frauenfelder, H., Nature Ctructural Biology, **2**, 821–823 (1995).

Frauenfelder, H., et al., J. Phys. Chem. **94**, 1024–1037 (1990).

Frauenfelder, H., G. U. Nienhaus, and R. D. Young, in *Disorder Effects on Relaxational Processes.* R. Richert and A. Blumen, Eds., Springer-Verlag, Berlin, 1994.

Frauenfelder, H., F. Parak, and R.D. Young, Ann. Rev. Biophys. Biophys. Chem. **17**, 451–479 (1988).
Frauenfelder, H.,, G. A. Petsko, and D. Tsernoglou, Nature **280**, 558–563 (1979).
Frauenfelder, H., and P. G. Wolynes, Science **229**, 337–345 (1985).
Frauenfelder, H., and P. G. Wolynes, Physics Today **47**, 58–64 (November 1994).
Freire, E., Annu. Rev. Biophys. Biomol. Struct. **24**, 141–165 (1995).
Friedrich, J., Methods Enzym. **246**, 226–259 (1995).
Friedrich, J., J. Gafert, J. Zollfrank, J. Vanderkooi, and J. Fidy, Proc. Natl. Acad. Sci. USA **91**, 1029–1033 (1994).
Friedrich, J., H. Scheer, B. Zickendraht-Wendelstadt, and D. Haarer, J. Chem. Phys. **74**, 2260–2266 (1981).
Gafert, J., H. Pschierer, and J. Friedrich, Phys. Rev. Lett. **74**, 3704–3707 (1995).
Garcia, A., Phys. Rev. Lett. **68**, 2696–2699 (1992).
Gavish, B., Biophys. Struct. Mech. **4**, 37 (1978).
Gibson, Q. H., J. Physiol. **134**, 112 (1956).
Goldanskii, V. I., Yu. F. Krupyanskii, and V. N. Fleurov, Doklady Akad. Nauk SSSR **272**, 978–981 (1983).
Goldanskii, V. I., L. I. Trakhtenberg, and V. N. Fleurov, *Tunneling Phenomena in Chemical Physics*, Gordon and Breach, New York (1989).
Hänggi, P., P. Talkner, and M. Borkovec, Rev. Mod. Phys. **62**, 251 (1990).
Hendrickson, W. A., Physics Today **48**, Number 11, 42–48 (November 1995).
Hong, M. K., et al., Biophys. J. **58**, 429–436 (1990).
Iben, I. E. T., et al., Phys. Rev. Lett. **62**, 1916–1919 (1989).
Jackson, T. A., M. Lim, and P. A. Anfinrud, Chem. Phys. **180**, 131 (1994).
Jankowiak, R., J. M. Hayes, and G. J. Small, Chem. Reviews **93**, 1471–1502 (1993).
Jennings, P. A., and P. E. Wright, Science **262**, 892–896 (1993).
Keller, H., and P. G. Debrunner, Phys. Rev. Lett. **45**, 68–71 (1980).
Kramers, H. A., Physica **7**, 284–304 (1940). Reprinted in H. A. Kramers, *Collected Scientific Works*, North-Holland Publ. Co., Amsterdam (1956).
Kurita, A., Y. Shibata, and T. Kushida, Phys. Rev. Lett. **74**, 4349–4352 (1995).
Lim, M., T. A. Jackson, and P. A. Anfinrud, Science **269**, 962–966 (1995).
Littau, K. A., Y. S. Bai, and M. D. Fayer, Chem. Phys. Lett. **159**, 1–6 (1989).
Meijers, H. C., and D. A. Wiersma, J. Chem. Phys. **101**, 6927–6943 (1994).
Moore, J. N., P. A. Hansen, and R. M. Hochstrasser, Proc. Natl. Acad. Sci. USA **85**, 5062–5065 (1988).
Nienhaus, G. U., J. R. Mourant, K. Chu, and H. Frauenfelder, Biochemistry **33**, 13413–13430 (1994).
Nienhaus, G. U., J. R. Mourant, and H. Frauenfelder, Proc. Natl. Acad. Sci. USA **89**, 2902–2906 (1992).
Noguti, T., and N. Go, Proteins **5**, 97–138 (1989).
Onuchic, J. N., P. G. Wolynes, Z. Luthey-Schulten, and N. D. Socci, Proc. Natl. Acad. Sci. USA **92**, 3626–3630 (1995).
Ormos, P., A. Ansari, D. Braunstein, B. R. Cowen, H. Frauenfelder, M. K. Hong, I. E. T. Iben, T. B. Sauke, P. J. Steinbach, and R. D. Young, Biophys. J. **57**, 191–199 (1990).
Ormos, P., D. Braunstein, H. Frauenfelder, M. K. Hong, S.-L. Lin, T. B. Sauke, and R. D. Young, Proc. Natl. Acad. Sci. USA **85**, 8492–8496 (1988).
Parak, F., E. N. Frolov, R. L. Mössbauer, and V. I. Goldanskii, J. Mol. Biol. **145**, 824–833 (1981).
Petsko, G. A., and D. Ringe, Ann. Rev. Biophys. Bioeng. **13**, 331–371 (1984).

Post, F., W. Doster, G. Karvounis, and M. Settles, Biophys. J. **64**, 1833–1842 (1993).
Ptitsyn, O. B., in *Protein Folding.* Ed. T. E. Creighton, pp. 243–300 (1992).
Reddy, N. R. S., P. A. Lyle, and G. J. Small, Photosynthesis Research **31**, 167–194 (1992).
Richert, F., and H. Bässler, J. Phys. Condens. Matter **2**, 2273 (1990).
Schlichting, I., J. Berendzen, G. N. Phillips Jr., and R. M. Sweet, Nature **371**, 808–812 (1994).
Settles, M., F. Post, D. Müller, A. Schulte, and W. Doster, Biophys. Chem. **43**, 107–116 (1992).
Singh, G. P., H. J. Schink, H. v. Lohneysen, F. Parak, and S. Hunklinger, Z. Phys. B. **55**, 23–26 (1984).
Steinbach, P. J., K. Chu, H. Frauenfelder, J. B. Johnson, D. C. Lamb, G. U. Nienhaus, T. B. Sauke, and R. D. Young, Biophys. J. **61**, 235–245 (1992).
Steinbach, P. J., et al., Biochemistry **30**, 3988–4001 (1991).
Stryer, L., *Biochemistry.* W. H. Freeman, New York, 1995.
Thorn-Leeson, D., and D. A. Wiersma, Nature Structural Biology **2**, 848–851 (1995).
Thorn-Leeson, D., and D. A. Wiersma, Phys. Rev. Lett. **74**, 2138–2141 (1995).
Toulouse, G., Comments Phys. **2**, 115–119 (1977).
Wagner, G., S. G. Hyberts, and T. F. Havel, Ann. Rev. Biophys. Biomol. Struct. **21**, 167–198 (1992).
Willis, B. T. M., and A. W. Pryor, Thermal Vibrations in Crystallography. Cambridge University Press, Cambridge (1975).
Wolynes, P. G., J. N. Onucic, and D. Thirumalai, Science **267**, 1619–1620 (1995).
Wüthrich, K., *NMR of Proteins and Nucleic Acids.* John Wiley, New York (1986).
Yang, F., and G. N. Phillips, Jr., J. Mol. Biol. **256**, 762–774 (1996).
Young, R. D., H. Frauenfelder, J. B. Johnson, D. C. Lamb, G. U. Nienhaus, R. Philipp, and R. Scholl, Chemical Phys. **158**, 315–327 (1991).
Young, R. D., and R. Scholl, J. Non-Crystalline Solids **131–133**, 302–309 (1991).
Zollfrank, J., J. Friedrich, and F. Parak, Biophys. J. **61**, 716–724 (1992).
Zollfrank, J., J. Friedrich, J. M. Vanderkooi, and J. Fidy, Biophys. J. **59**, 305–312 (1991).
Zollfrank, J., J. Friedrich, J. M. Vanderkooi, and J. Fidy, J. Chem. Phys. **95**, 3134–3136 (1991).

The Energy Landscape Theory of Protein Folding

Peter G. Wolynes and Zan Luthey-Schulten

School of Chemical Sciences, University of Illinois, Urbana, Illinois 61801

1 Introduction

Prediction of a protein's structure and folding mechanism from its sequence has been described as the determination of the second half of the genetic code (Gierasch and King (1990)). Proteins and nucleic acids are the simplest information bearing components of biological systems to have individual identities. Although the ongoing effort to map the entire DNA genome of several species including man has provided biologists with more than a hundred thousand protein sequences, the structures of only a few thousand are known. More importantly the rules or energy functions to turn these pieces of one-dimensional information (sequences) into three-dimensional structures (folded proteins) are just becoming clear. Proteins are made up of amino acids that can have several conformations so that the folding process takes place on a rough multi-dimensional potential surface. Our understanding of the physics of protein folding is impeded by the complexity of the process. On the one hand, many of the features of protein folding dynamics are like those for any random heteropolymericsystem. The folding route is not unique and passes through numerous misfolded structures whose structures and energies are unrelated. While on the other hand, protein sequences have evolved to allow proteins to fold to unique *native states* with certain *local structural motifs* and to carry out selective functions. For example, the sequence of the oxygen carrying molecule, *myoglobin*, encodes the structure shown in Fig. 1. It has a high degree of symmetry built up from the repetitive local helical units. How does this symmetrical structure come about? Why is the folding and unfolding of a protein apparently a reversible process?

The scientist seeking to answer these questions, must understand the properties of proteins as complex, heterogeneous systems. One can think of them as disordered systems, but one must equally try to uncover those non-random features of proteins that are essential for their folding kinetics and native structures. Although conventional potential functions have been developed to describe the interactions in a protein near its native state, they cannot yet be practically applied over the millisecond time scale needed to simulate the folding routes and kinetics. This seemingly hopeless situation is reminiscent of a similar challenge facing physicists in the days before BCS theory was invented to explain superconductivity. Physicists were faced with the difficulty of describing a complex effect on very different dynamic scales: Super-conductivity involves an effect on the level of 10^{-7} eV, while the energies of electronic structures were known only

```
SLSAAEADLAGKSWAPVFANKNANGLEFLVALFGKFPDSANFFADFKGKSV
ADIKASPKLNDVSSRIFTRLNEFVNNAANAGKMSAMLSQFAKDHVGFGVGS
AQFDNVRSMFPGFVASVAAPPAGADAAWTKLFGLIIDALKAAGA
```

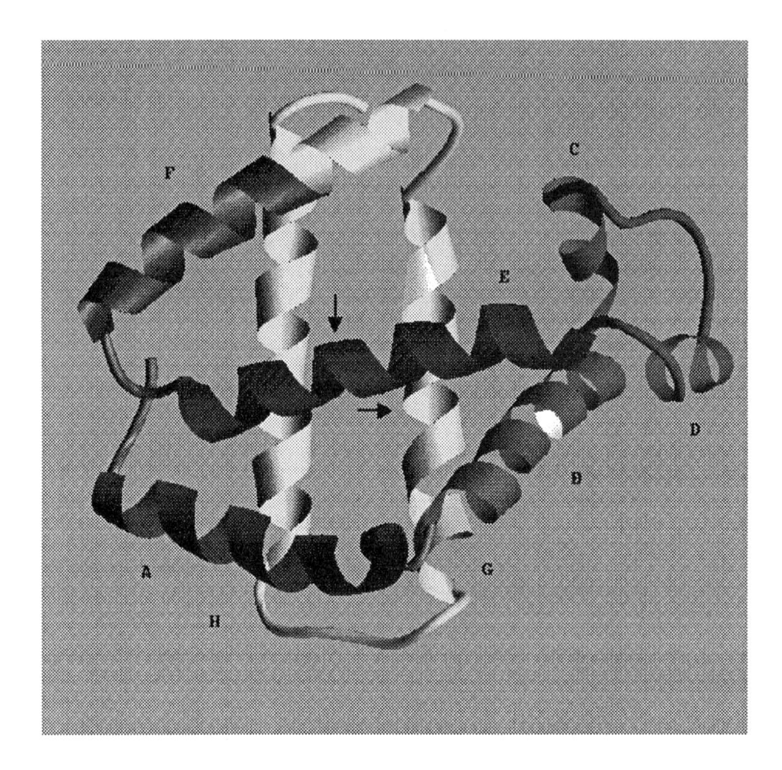

Fig. 1. Sequence and structure of myoglobin

down to 0.1 eV at that time. This gave Bardeen, Cooper and Schrieffer license to craft a phenomenological theory of super-conductivity that was then parameterized and verified by experimental data. Following their example, we have attempted to describe the diverse behavior associated with protein folding using the statistical energy landscape approach, a phenomenological theory requiring only a few energy parameters in the simplest form: δE_s, the *stability gap* between the ground state of the protein—also referred to as its *native* or *folded* state—and the mean of the excited (misfolded) states, and ΔE, the roughness of the energy landscape. This approach has had considerable success in guiding us to consider good order parameters and appropriate collective reaction coordinates to interpret thermodynamic and kinetic protein folding experiments, and suggesting detailed, but simple, energy functions for protein structure prediction (Friedrichs et al. (1991), Bryngelson et al. (1995)).

This chapter covers the basics of the *statistical energy landscape* approach, uses it to interpret what are common and specific features observed in protein folding experiments, and describes how this approach leads to optimized energy functions for protein structure prediction. Reviews that deal with the energy landscape and other approaches to these topics can be found in Bryngelson et al. (1995), Dill et al. (1995), Garel et al. (1995).

2 The Protein Folding Energy Landscape

Are proteins random objects? Figure 2a shows sequences of the protein *lysozyme* from various species. This protein has essentially the same structure and function in all species. Nevertheless, when the sequences are compared to a reference sequence, e.g. that of chicken, the degree of sequence identity is relatively low. Sections of the echidna (tacac) lysozyme sequence are only 30% identical with chicken's.

Two English texts with only 30% identity would appear totally unrelated from a strictly orthographical point of view, but could have the same meaning. Likewise, proteins with structural homologs with such low sequence identities are not uncommon. The difficulty in extracting the meaning from protein sequences, lies in discerning which features are common to all sequences, which features are specific to protein-like sequences, and which features are specific to a given structure. How the ensemble of protein-like sequences is embedded in the ensemble of random heteropolymers, is sketched in Fig. 2b. Within the space of random heteropolymers based on the twenty naturally occurring amino acids, we need to further differentiate between *thermodynamically foldable* sequences and the subset of *kinetically foldable* sequences that make up the proteins of Nature. Families like the lysozyme sequences belong in this last category. Proteins of Nature, while only marginally stable and easy to denature with either heat or pH, must fold on a time scale that is relevant for the biological processes occurring in cells. This time is relatively short, less than a minute, which seems paradoxical given the large number of conformations that a protein can theoretically be in during folding. Quantitatively this can be understood by studying the formation of local structure, such as helix formation in collapsed polymers and understanding the funneled nature of the landscape for topological rearrangements (Luthey-Schulten et al. (1994)).

2.1 Energy Landscape of a Random Heteropolymer

Many aspects of the folding process can be understood from studying the energetic properties of a *random heteropolymer* (RHP) using lattice models. The simplest of these makes use of two kinds of residues (e.g. hydrophobic and hydrophilic amino acids) randomly distributed in the polymer chain. This is a useful model for visual illustration, although it has some special properties that make it different from the more general 20 amino acid case. Both from theoretical calculations and simulations we know two basic facts about the random heteropolymer:

- Modest structural change gives rise to large change in energy.

```
                71
 lyc_chick  DYGILQINSR WWCNDGRTPG SRNLCNIPCS
 lyc_macmu  DYGIFQINSH YWCNNGKTPG AVNACHISCN
 lyc_human  DYGIFQINSR YWCNDGKTPG AVNACHLSCS
  lyc1_rat  DYGIFQINSR YWCNDGKTPR AKNACGIPCS
 lyc_bovin  DYGIFQINSK WWCNDGKTPN AVDGCHVSCR
lyc1_tacac  DYGILQINSR YWCHDGKTPG SKNACNISCS

                                         131
 lyc_chick  ALLSSDITAS VNCAKKIVSD GNGMNAWVAW
 lyc_macmu  ALLQDNIADA VTCAKRVVSD PQGIRAWVAW
 lyc_human  ALLQDNIADA VACAKRVVRD PQGIRAWVAW
  lyc1_rat  ALLQDDITQA IQCAKRVVRD PQGIRAWVAW
 lyc_bovin  ELMENDIAKA VACAKHIVSE .QGITAWVAW
lyc1_tacac  KLLDDDITDD LKCAKKIAGE AKGLTPWVAW
```

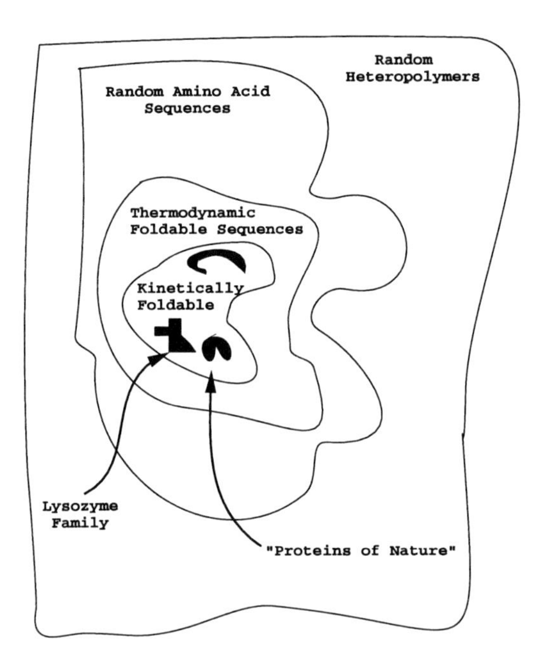

Fig. 2. (a) Partial sequences from the family of lysozyme proteins. (b) How proteins of Nature are embedded in the ensemble of random heteropolymers.

– Low energy states are very different in structure.

In the general case, the energy is a sum of random interactions that give rise to a rough energy landscape like the Alps. Since the energy contributions can either be stabilizing or destabilizing, the RHP is a frustrated system. In a leap of faith, Bryngelson and Wolynes (Bryngelson and Wolynes (1987)) already in 1987 applied Derrida's random energy model (REM) for spin glasses to proteins (biopolymers), in particular to the misfolded states of a protein. The basic validity of this approach has since been borne out by numerous analytical and numerical studies making the REM the zeroth order approximation in our understanding of biopolymers. A brief review of the basic features of the REM for the RHP is shown in Fig. 3. As a result of the random interactions, the density

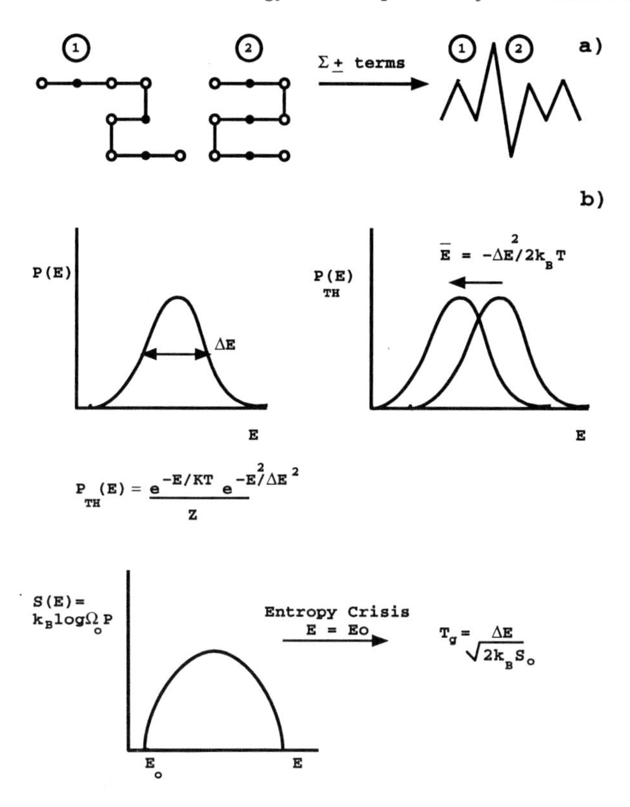

Fig. 3. (a) Flat rugged landscape of RHP. (b) Thermodynamics of REM approximation applied to a RHP.

of states is approximately a Gaussian distribution P with a variance ΔE. The thermally weighted probability is again a Gaussian distribution, centered about the mean $E = -\Delta E^2/k_B T$. The density of states cannot really be Gaussian in its tail, since there is a state of lowest energy. The entropy S is defined as

$$S(E) = k_B \log \Omega_o \, P \tag{1}$$

where Ω_o is the number of conformational states of the polymer, k_B is Boltzmann's constant. If a system is cooled, its energy decreases. The system runs out of entropy when the average energy falls below a critical value $E \le E_o$ such that $S(E_o) = 0$. This entropy crisis occurs at a temperature T_g where

$$T_g^{-1} = \sqrt{2k_B S_o/\Delta E^2} \tag{2}$$

where $S_o = k_B \log \Omega_o$. Below T_g, the kinetics of the system exhibits glassy-like behavior, showing dependence on the systems history. Above T_g, the system behaves like a viscous liquid. Transition rates between different low energy states lead to a strong, generally non-Arrhenius, temperature dependence of the rate of exploring configuration space (e.g. the logarithm of the rate varies like $1/T^2$ instead of $1/T$ so that the motions are much slower than expected.) A more detailed discussion of the kinetics on a REM landscape is given below.

The low entropy of the system at the glass transition makes it tempting to identify T_g with the folding temperature. Indeed, Shakhnovich and Gutin have estimated that a large fraction of random sequences would have unique thermodynamically stable native states below T_g. The fraction of these sequences with significant thermal occupation of the native state is independent of chain length, and is given by (Shakhnovich and Gutin (1990))

$$\mathrm{Prob(N.S.)} = \frac{\sin(\pi T/T_g)}{\pi T/T_g}\epsilon^{T/T_g} \quad \text{for} \quad T < T_g \tag{3}$$

where ϵ is related to the Boltzmann probability of the ground state E_o, $p_o = \exp(-E_o/k_B T)/Z > 1-\epsilon$. Most of these sequences would still fold too slowly in this temperature range because of glassy dynamics, i.e. because they get trapped in other collapsed configurations. Instead we must examine the exponentially small fraction which can fold above T_g, and ask what is the simplest energy landscape that these sequences would have.

2.2 Simplest Viable Protein Folding Landscape

Postulate A: The energy landscapes of proteins are rugged because of the possibility of making inappropriate contacts between residues. It is reasonable to assume that when non-native contacts are made, the energy contributions are random, and these contributions to the protein's energy can be treated just as for a random heteropolymer. In the ensemble of misfolded states with little native structure, the energetics can be described crudely by the REM shown in Fig. 3. Low energy structures will appear unrelated and conformational changes are associated with a fluctuation $\sqrt{\Delta E^2}$ in the energy. *Postulate B: The Principle of Minimal Frustration.* There is a smooth overall slope to the energy landscape because of harmonious cooperativity. Native contacts and local conformational energies are more stabilizing than expected. This more realistic model considers the protein to be a "minimally frustrated heteropolymer." This means that the rugged landscape of real protein folding is not globally flat with totally unpredictable fluctuations, as it would be for a random heteropolymer, but has a preferred direction of flow. It can be described as a rugged funnel, shown in Fig. 4, whose shape can be estimated using theory and experiment, as we shall see. At the bottom of the funnel is a unique native state. Obvious order parameters to describe the position of an ensemble of states in the funnel are Q, the percent of native-like contacts and A, the percent of correct dihedral angles in the protein backbone. Other order parameters such as total helicity may also be used to classify an ensemble of states as is done in describing the helix-coil transition in helical proteins.

Through these order parameters, the folding landscape is stratified. Within each stratum, we can define an average energy, $\bar{E}(Q)$, although there are still many states with different energies. To describe their distribution and properties, we apply a REM model. The late stages of protein folding will have few states, all of which are highly similar to the native state. These states could be given specific names, if necessary, and have their lifetimes measured. These states are analogous to the taxonomic sub-states discussed by Frauenfelder (Frauenfelder et al. (1990). As indicated in Fig. 5, some routes can dead-end in a low energy misfolded conformation, from which the protein has to partially unfold to reach the native state. In the early stages of folding, corresponding to a nearly denatured protein with $Q \approx 0$, there will be many states, and the ensemble language is clearly most appropriate. The hopping rate, R, between micro-states at each stage is roughly

$$R \approx e^{-\Delta E^2(Q)/k_{\mathrm{B}}T^2} \ . \tag{4}$$

This determines the ability of the ensemble to flow between different strata. The complete statistical mechanical treatment requires knowledge of all thermodynamic variables as a function of the order parameters. In particular, the functional dependence of the thermal average energy $\bar{E}(Q)$, the ruggedness $\sqrt{\Delta E^2(Q)}$, the density of states $\Omega(E,q)$, or, equivalently, the entropy S(E,Q), and the local glass transition temperature, $T_{\mathrm{g}}(Q)$. Following Bryngelson and Wolynes, we derive these quantities using the simplest form of the random energy approximation in which correlations within a stratum is neglected.

According to Postulate A, the energy of a given misfolded state arises from the contributions of many random terms, so the probability distribution of energies in the interval δE at any position in the funnel is a Gaussian centered about the mean energy

$$P(E)\delta E = \frac{1}{\sqrt{2\pi \Delta E^2(Q)}} \exp\left\{ -\frac{(E-\bar{E}(Q))^2}{2\Delta E^2(Q)} \right\} \delta E \ . \tag{5}$$

If there are γ configurations per residue for a protein in its unfolded state, then the total number of configurations for a protein with N residues is $\Omega_{\mathrm{o}} = \gamma^N$. In models using a reduced description of the protein that only includes the backbone coordinates, γ is less than 5, and when corrections are made for the excluded volume effect in compact configurations, $\gamma = \gamma^* \approx 1.5$ (Flory (1954), Bryngelson et al. (1995)). As the structures become more similar to the native protein, the total number of configurations will decrease, since only a single backbone conformation represents the native state. If $\Omega_{\mathrm{o}}(Q)$ is the number of structures that have a fraction Q of the contacts similar to those in the native structure, then a rough approximation to this behavior is

$$\Omega_{\mathrm{o}}(Q) = \gamma^{*N(1-Q)} \ . \tag{6}$$

The corresponding entropy will also decrease as the native structure is approached,

$$S_{\mathrm{o}}(Q) = k_{\mathrm{B}} \log \Omega_{\mathrm{o}}(Q) = k_{\mathrm{B}} N(1-Q) \log \gamma^* \ . \tag{7}$$

The density of conformational states with energy E and similarity Q is then

$$\Omega(E,Q) = \Omega_{\mathrm{o}}(Q) P(E) \ , \tag{8}$$

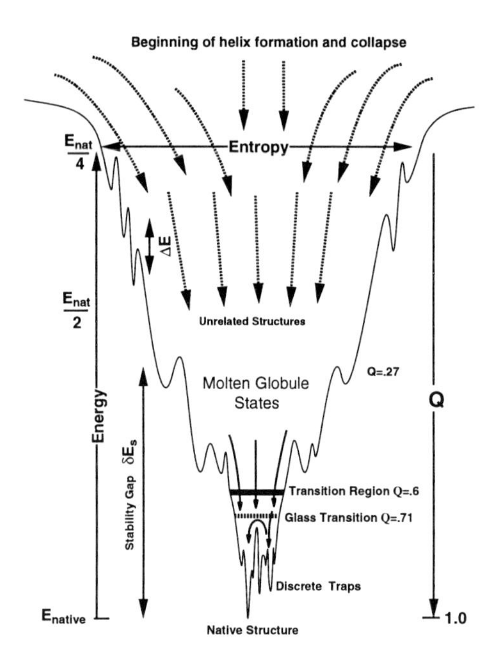

Fig. 4. Energy landscape parameters for a realistic protein folding funnel

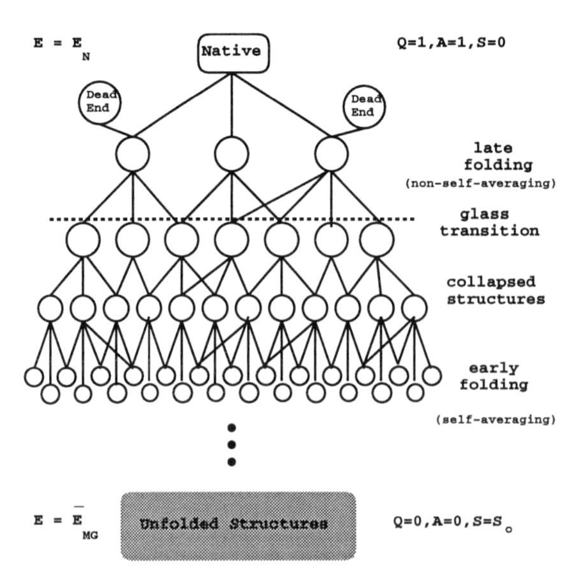

Fig. 5. Protein conformations tree structure and corresponding thermodynamic properties ordered according to various reaction coordinates describing the folding process

and the total entropy

$$S(E,Q) = S_{\mathrm{o}}(Q) - k_{\mathrm{B}}\frac{(E - \bar{E}(Q))^2}{2\Delta E^2(Q)} - k_{\mathrm{B}}\log\left(\frac{\sqrt{2\pi\Delta E^2(Q)}}{\delta E}\right) \tag{9}$$

where δE is large relative to the spacings between energy levels, but small relative to $\sqrt{\Delta E^2(Q)}$. The last term is neglected, since even in the worst case it is varying logarithmically, giving only a small correction. At thermal equilibrium, the *most probable energy*, $E_{\mathrm{m.p.}}$, can be determined using the thermodynamic definition of the temperature

$$\frac{1}{T} = \frac{\partial S}{\partial E} , \tag{10}$$

or, more directly, we can find the maximum of the thermally weighted canonical probability $p(E,Q) \propto \Omega(E,Q)e^{-E/k_{\mathrm{B}}T}$, as shown in Fig. 3,

$$E_{\mathrm{m.p.}}(Q) = \bar{E}(Q) - \frac{\Delta E^2(Q)}{k_{\mathrm{B}}T} . \tag{11}$$

The number of thermally occupied states and the entropy associated with this most probable energy are

$$\Omega(E_{\mathrm{m.p.}},Q) = \exp\left[\frac{S_{\mathrm{o}}(Q)}{k_{\mathrm{B}}} - \frac{\Delta E^2(Q)}{2(k_{\mathrm{B}}T)^2}\right] \tag{12}$$

$$S(E_{\mathrm{m.p.}},Q) = S_{\mathrm{o}}(Q) - \frac{\Delta E^2(Q)}{2k_{\mathrm{B}}T^2} . \tag{13}$$

Combining 11 and 13, the free-energy of the misfolded structures with configurational similarity Q at fixed temperature T is

$$\begin{aligned} F(Q,T) &= E_{\mathrm{m.p.}}(Q) - TS(E_{\mathrm{m.p.}},Q) \\ &= \bar{E}(Q) - \frac{\Delta E^2(Q)}{2k_{\mathrm{B}}T} - TS_{\mathrm{o}}(Q) . \end{aligned} \tag{14}$$

Folding is considered to be a two-state reaction, *unfolded* $\rightarrow$ *folded,* so that under some thermodynamic condition, the free energy has a double minimum. As seen in Fig. 6, one minimum lies near the folded state, $Q \approx 1$, and the other is at the position $Q_{\mathrm{min}} \approx 0$, where the free energy of the misfolded states has a minimum. This minimum can either be the *random coil* state or a *collapsed* state with some degree of ordering. To a first approximation, we can neglect the entropy of the folded state, so that its free energy is equal to its internal energy, E_N. At the folding temperature T_{f}, the probability of being in the folded state is equal to the probability of being in the misfolded state. Equating the free energy of the folded and misfolded states at the *folding temperature,* T_{f}, $F_{\mathrm{native}} = F(Q_{\mathrm{min}},T_{\mathrm{f}})$, one obtains expressions for the slope of the funnel,

$$\delta E_{\mathrm{s}}/T_{\mathrm{f}} = S_{\mathrm{o}} + \Delta E^2(Q_{\mathrm{min}})/2k_{\mathrm{B}}T_{\mathrm{f}}^2 , \tag{15}$$

as well as for the folding temperature

$$T_{\mathrm{f}} = \frac{\delta E_{\mathrm{s}} + \sqrt{\delta E_{\mathrm{s}}^2 - 2\,S_{\mathrm{o}}\,\Delta E^2(Q_{\mathrm{min}})/k_{\mathrm{B}}}}{2\,S_{\mathrm{o}}} \quad . \tag{16}$$

in terms of the stability gap $\delta E_{\mathrm{s}} = \bar{E}(Q_{\mathrm{min}}) - E_N$. Since Q_{min} is close to the unfolded state, we will consider the folding temperature as being referenced to a set of states with little structural similarity to the native state, $Q \approx 0$.

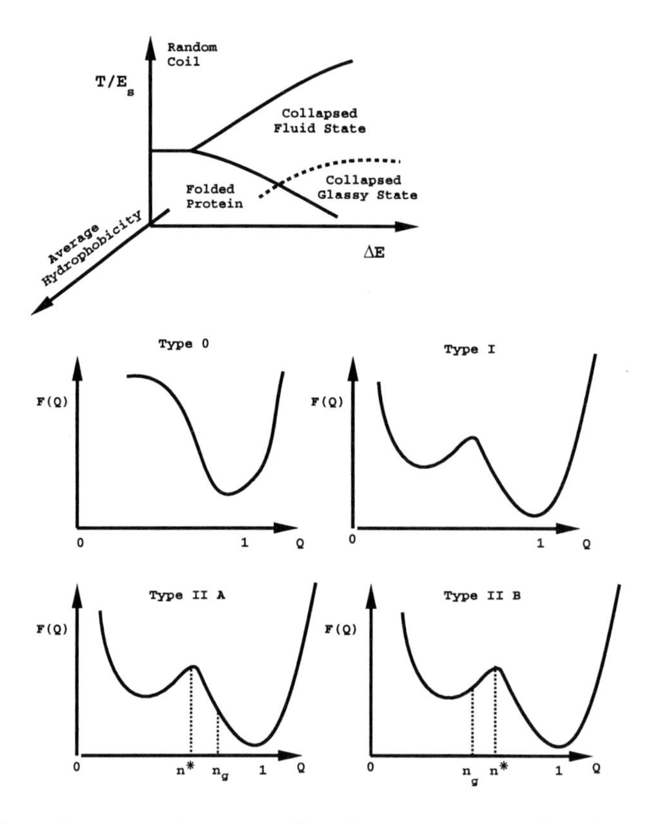

Fig. 6. Phase diagram and corresponding free energy curves for a folding protein

Recall that a transition to a glassy state occurs at the temperature where so few states are available that the system remains frozen in one of a few distinct states. Within each stratum, this transition is characterized by an entropy crisis where $S(T_{\mathrm{g}}, Q) = 0$. Using (13), the local glass transition temperature is

$$T_{\mathrm{g}}(Q) = \sqrt{\frac{\Delta E^2(Q)}{2k_{\mathrm{B}} S_{\mathrm{o}}(Q)}} \quad . \tag{17}$$

Local glass transition temperatures are manifested in the folding times and collapse times measured in lattice calculations which are summarized in Fig. 7.

Analytical and numerical studies on lattice models have shown that the ratio of T_f/T_g can be used to distinguish fast and slow folding sequences. This ratio also plays a central role in developing energy functions to predict protein structures. Calculating this ratio, using the set of states with the least structural similarity to the folded state, gives

$$\frac{T_f}{T_g} = \sqrt{\Lambda} + \sqrt{\Lambda - 1} \tag{18}$$

where

$$\Lambda = \frac{k_B}{2S_o}\left(\frac{\delta E_s^2}{\Delta E^2}\right) = \left(\frac{\delta E_s^2}{2\Delta E^2}\right)\frac{1}{N \log \gamma^*} \, . \tag{19}$$

For a protein to fold, Λ, and consequently T_f/T_g, must be greater than 1. Since S_o, ΔE^2, and E_N all depend linearly on the chain length N, T_f/T_g is independent of length and sensitive to the interaction energies.

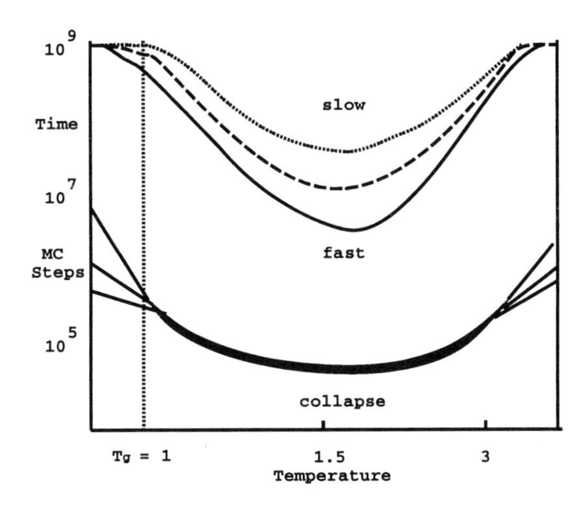

Fig. 7. Folding and collapse times of designed 27-mers

A phase diagram is a useful tool for summarizing which states of a protein are involved in the various folding scenarios. So far our analysis has only used a single parameter, Q, to characterize the changes in free energy and the differences between the native and unfolded states. Clearly there are other parameters besides the number of correct contacts that could be used to compare structures and describe the partial ordering that occurs as the protein folds. For example, in a folded protein the core consists primarily of hydrophobic residues, and the surface of hydrophilic residues. This ordering is due to the hydrophobic effect arising from folding a protein in water. Variations in the solvent properties will have profound effects on the interaction energies of the hydrophobic groups. The phase diagram in Fig. 6 shows the possible thermodynamic states of a protein as a function of temperature and the roughness of the energy landscape. The phase diagram is actually a slice through a more complicated diagram along a line of some average hydrophobicity of the sequence. Since average hydrophobicity itself depends on solvent and temperature, under some conditions the coexistence curve between the random coil and the folded state disappears, and the folded state becomes accessible only after non-specific collapse. The thermodynamic dependence of hydrophobic forces is one of the main complicating features in relating the theoretical phase diagrams (that assume temperature independent forces) to experiment. This is most manifest in the phenomena of cold and pressure induced denaturation. Approximate analytical expressions for coexistence curves between the collapsed phase and the random coil and frozen phase were derived by Sasai and Wolynes (1990) using the associative memory Hamiltonian given at the end of this chapter to describe the interaction energies between residues. The glass transition, which occurs after the collapse of the system, is a continuous transition. This portion of the phase diagram is typical of random heteropolymers (Dinner et al. (1994)). Recent lattice simulations of Socci and Onuchic (1995) on protein-like sequences probe the phase diagram as a function of temperature and average hydrophobicity and provide qualitatively the same picture.

This phase diagram can be used to understand the free energy behavior in the various folding scenarios. In general, the free energy is either unimodal or bimodal. Type 0A and Type I scenarios dominate the left-hand part of the phase diagram in which no glass transition occurs, and the system is at a temperature such that the global minimum is the native folded states with Q=1. Direct folding from the random coil state favors these scenarios since glassy states can only occur in nearly collapsed chains. Type II scenarios occur in the right-hand part near the coexistence curves for the folded, collapsed, and collapsed frozen states. In Type IIA the glass transition occurs after the thermodynamic barrier and in Type IIB the folding protein becomes glassy before the barrier is reached. Theoretical calculations of Takada and Wolynes (Takada and Wolynes (1996)) have provided a more detailed description of the crossover between the different activated folding scenarios.

3 Simple Models of Folding Kinetics of MFHP

A protein folding along the funnel shown in Fig. 4 moves through an ensemble of partially ordered structures characterized by the similarity measure Q. The gradient of the free energy determines the instantaneous drift velocity down the funnel. The roughness at any stage acts like a set of speed bumps slowing this drift. Superimposed on the drift are stochastic fluctuations in Q reflecting individual escapes from traps. A simple description of the overall dynamics within the folding funnel arising from these effects is obtained using a diffusion equation. At a given temperature, the population of the various structural strata changes with time according to

$$\frac{\partial P(Q,t)}{\partial t} = \frac{\partial}{\partial Q}\left\{D(Q,T)\left[\frac{\partial P(Q,t)}{\partial Q} + P(Q,t)\frac{\partial \beta F(Q,T)}{\partial Q}\right]\right\} \qquad (20)$$

In general, the local configurational diffusion coefficient, D, depends on the roughness of the energy surface, which determines the escape time from traps. At sufficiently high temperatures, it crudely follows a *Ferry law* typical of glasses,

$$D(Q,T) = D_o \exp\left[-\Delta E^2(Q)/(k_B T)^2\right] \ . \tag{21}$$

The functional form of $D(Q,T)$ at temperatures near the glass transition temperature is a bit more complicated, and depends on the nature of local moves. The above form is motivated by the following qualitative argument. The diffusion coefficient is inversely proportional to the lifetime, $\tau(Q)$, of a micro-state with similarity Q to the native state. If the micro-state is deep, it will be long-lived and the diffusion coefficient becomes small. To avoid being trapped in this micro-state characterized by a roughness $\Delta E(Q)^2$, motion must take place over an energy barrier $\bar{E}(Q) - E_{\mathrm{m.p.}}(Q) = \Delta E^2(Q)/k_B T$ in the time τ_o it takes for a large segment of the chain to move. This gives an escape time from the local traps that is super-Arrhenius,

$$\tau(Q) = \tau_o \exp\left[\Delta E^2(Q)/(k_B T)^2\right] \ . \tag{22}$$

In the case of fast downhill folding at a fixed temperature, the Type 0 scenario in Fig. 6, a kinetic folding bottleneck occurs at a region $Q^{\ddagger}_{\mathrm{kin}}$ with the maximum lifetime or the smallest diffusion coefficient. This maximum lifetime is also a simple estimate of the overall folding time

$$\tau_f \approx \tau_{\max}(Q^{\ddagger}_{\mathrm{kin}}) \tag{23}$$

For a bistable system, as in Type I and Type IIA scenarios in Fig. 6, the overall folding time τ_f will be determined by the difficulty to overcome the free energy barrier and a prefactor that depends on the ruggedness of the energy landscape,

$$\tau_f \approx \langle \Delta Q^2_{\mathrm{MG}} \rangle D^{-1}(Q^{\ddagger}) e^{\Delta F\ddagger/k_B T} \tag{24}$$

where $\Delta F\ddagger$ is the free energy barrier measured from the unfolded minimum to the top of the thermodynamic barrier. $\langle \Delta Q^2_{\mathrm{MG}} \rangle$ is the mean square fluctuation of the configuration coordinate in the *molten globule* state. This equation suggests that an Arrhenius plot of folding time versus inverse temperature would be curved, and such behavior is frequently observed in protein folding experiments (Creighton (1994)) and lattice simulations, as seen in Fig. 7. As the temperature is decreased, the escape time will increase until the local glass transition temperature $T_g(Q)$ is reached. For $T < T_g(Q)$, the protein has kinetic access to very few structures, and the protein is effectively frozen into a single or several low energy states. In this case the kinetics are dominated by the details of the specific landscape, and the expressions for the folding time and the diffusion coefficient have to be modified.

Using a simple three-dimensional lattice model and Monte Carlo dynamics, the collapse and folding times for designed sequences of protein-like heteropolymers have been studied, and a sketch of these results is given in Fig. 7 (Socci and Onuchic (1994), Bryngelson et al. (1995)). In all cases, the polymers are

27 monomers long, and possess a maximal compact non-degenerate native state with 28 contacts. The simulations start out with the polymer in a fully extended form, and the protein is considered folded when all 28 correct contacts are made, and collapsed when any 25 contacts are. The times given in the curves are the number of Monte Carlo steps required to first reach either the folded or a collapse conformation. The lower curve shows that the collapse time is independent of sequence or self-averaging over a wide range of temperatures until the glass transition temperature is reached. The folding curve exhibits a much greater spread in the times. The fast folder is the sequence with the least frustration (lowest energy) and the largest T_f/T_g ratio, while the slow folder has highest energy and $T_f/T_g < 1$. The folding temperature, T_f, is defined in Fig. 8 as the temperature where the probability of the occupancy of the native structure is one-half. At the glass transition temperature, the fast folder is over 90% in the native state while the slow folder has a native state population less than 20%. Clearly, at this low temperature the native state is thermodynamically more stable, but the system is now getting trapped in local minima so that it is no longer kinetically accessible. For a protein of significant length to be foldable on biological time scales, the folding temperature must be greater than the glass transition temperature.

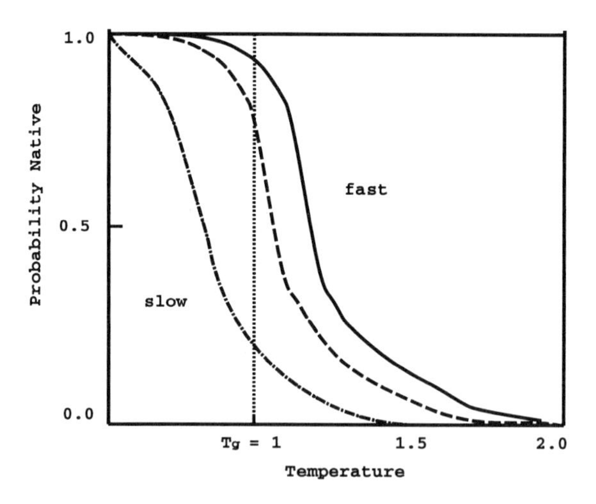

Fig. 8. Temperature dependence of the probability density of the native state for designed 27-mer

Onuchic et al. developed a law of corresponding states to relate simulations of small lattice models to real proteins. The correspondence analysis made use of a theory of helix formation in collapsed polymers that related the configurational entropy, S_o, to the amount of helical structure. For a 60-amino acid chain at 60% helicity, $S_o \approx 40k_B$, which corresponds approximately to a conformational entropy of $0.6k_B$ per monomer unit. Assuming reconfiguration times on a rough energy landscape are on the order of the folding time of milliseconds observed experimentally, expression (23) suggests that the roughness of the landscape at the folding temperature $\Delta E^2/2k_B T_f^2$ ranges from 11 to 18. These two quantities allow Onuchic et al. to estimate the slope of the funnel using 15, $\delta E_s/k_B T_f \approx 58$. They also estimate the dimensionless ratio of the energy gradient to the ruggedness, $(\delta E_s/k_B T_f)/(\sqrt{\Delta E^2/2k_B T_f^2}) \approx 14$, as well as the ratio $T_f/T_g \approx 1.6$. Comparable values for T_f/T_g are obtained for lattice simulations of 27mers with a three-letter code, implying that every lattice bead corresponds to about two amino acids. The funnel in Fig. 4 reflects the correspondence we have just advanced. At T_f folding proceeds via a Type IIB scenario with the transition state at $Q = 0.60$ and a glass transition at $Q = 0.71$. Although the funnel was originally using the correspondence with the lattice simulations, one significant aspect of its form was dramatically confirmed by NMR measurements of Huang and Oas (1995) on the sub-millisecond folding of the λ repressor. In their experiments the thermodynamic bottleneck for folding occurs when approximately half the native contacts are made, as indicated by the sensitivity of the folding rate to added denaturant.

4 Protein Structure Prediction Using Optimized Energy Functions

Nature seems to have designed its proteins to have T_f/T_g greater than 1, so how can we use this criterion to design energy functions for protein structure prediction, and what are the further problems to be faced? Molecular biology gives us the sequence of a protein, $\{q_i\}$, and in the cases where the protein has been well crystallized, or is small enough for structure determination by NMR, the mean positions, $\{r_i\}$, of all the atoms are also known. Considerable effort has been invested in developing energy functions based on standard bonding and van der Waals interactions. These have been shown to be well parameterized to describe the motion of the atoms about or near their crystallographic coordinates. It is not yet possible to test whether they are sufficient to describe the folding process since the natural process is very slow, on the time scale of atomistic simulations ($\approx$ 1 nsec). Thus we seek simpler energy functions that can encode the sequence-structure correlation. We are helped by the fact that evolution has already solved the problem of how to find different sequences $\{q_i\}$ compatible with a given structure $\{r_i\}$. For scientist trying to break the protein folding code, the problem is

- to use all the information that biologists and crystallographers have collected in their sequence and structural databases, $\{q_i\}^\mu$ and $\{r_i\}^\mu$, and
- to develop new energy functions that fold these proteins and can be generalized to fold the plethora of new sequences arising from the genome project.

Code breakers who wish to take this phenomenological route must understand both evolution and physics.

Friedrichs and Wolynes introduced in 1989 an *associative memory* Hamiltonian, $\mathcal{H}_{\mathrm{AM}}$, that encodes correlations between the sequence of the target protein whose structure is to be determined, and the sequences and structures of a set of memory proteins taken from the database. The associative memory Hamiltonian resembles the empirical energy functions used in conventional molecular dynamics, but its form was motivated by energy functions used in neural network theory to perform pattern recognition (Goldstein et al. (1992)).

$$\mathcal{H}_{\mathrm{AM}}(\{r_{ij}\}) = -\sum_{\mu}\sum_{i<j} \gamma_{ij}^{\mu}\ \theta(r_{ij} - r_{i'j'}^{\mu}) + \mathcal{H}_{\mathrm{o}}. \tag{25}$$

In its simplest form, the associative memory Hamiltonian is a function of the pairwise distances, r_{ij}, between the α-carbons of residues i and j. γ_{ij}^{μ} encodes a degree of similarity between residues i and j of the target protein and a corresponding pair in the memory protein, μ, and it may include information about the physicochemical properties of the residues, their probability of mutation, or context of the residues in the protein. $\theta(r_{ij} - r_{i'j'}^{\mu})$ is a Gaussian function of the difference between the pairwise distance in the target structure and the memory structure. The energy function has a minimum of varying depth for the structure of each memory protein. These minima are differentiated by the sequence "property" similarity weights, γ^{μ}. $\mathcal{H}_{\mathrm{o}}$ is a typical *chain molecule* Hamiltonian for the backbone atoms, and includes harmonic terms to induce backbone rigidity and the correct chirality, and to prevent the overlap of non-bonded α-carbons (Friedrichs and Wolynes (1990), Friedrichs et al. (1991)). In essence, each memory protein constructs a small folding funnel. If these funnels add coherently, or only a single one dominates, the Hamiltonian will not be very frustrated, and a single folding funnel to a structure consistent with empirical correlations will be formed. Achieving this requires a good choice of the γ-parameters and training proteins. The quantitative version of MFP helps us do this.

The energy parameters in the energy landscape analysis can be expressed in terms of the γ coefficients in the energy function:

$$\delta E_{\mathrm{s}} = \mathbf{A}\gamma\ , \tag{26}$$

$$\varDelta E^2 = \gamma\mathbf{B}\gamma\ . \tag{27}$$

The explicit values of $\mathbf{A}$ and $\mathbf{B}$ are obtained from a set of training proteins with known structures and simulated molten globule states. According to (16), $T_{\mathrm{f}}/T_{\mathrm{g}}$ is maximized when $\delta E_S/\sqrt{\varDelta E^2}$ is maximized. This maximization procedure leads to optimal values $\gamma = \mathbf{B}^{-1}\mathbf{A}$. Simulations with naive assignments of gamma, e.g. interaction energies between similar residues being $E_{\mathrm{sim}} = -3$, and between dissimilar residues $E_{\mathrm{dis}} = -1$, give rise to much smaller $T_{\mathrm{f}}/T_{\mathrm{g}}$ values.

Simulated annealing for the optimally encoded Hamiltonian generally leads to qualitatively correct structures when the target protein can be assigned to one of three broad classes of folding motifs: alpha, beta, or mixed alpha-beta proteins (Goldstein et al. (1992)). For a comparison code based only on hydrophobicity and proximity, the results of such a molecular dynamics run for the myoglobin shown earlier are presented in Fig. 9. The simulation begins with the protein in a random extended form with a radius of gyration typical of a random coil, $R_g = 60\,\text{Å}$. Collapse and compaction of the protein occurs quickly to a state that has roughly the correct topology or fold, but has incomplete secondary structure. The local Q score for helix-A is about 0.3 when the collapsed protein is first formed. Continued folding in this compact state completes the formation of the helices, and modifies the tertiary contacts. In the short trajectory shown here, roughly one-third of all the native contacts have been formed, but the local Q value is considerably higher. Preliminary studies of the local T_f/T_g ratios indicate that helix-A is the major part of a kinetically competent, quasi-independent folding unit or *foldon* (Panchenko et al. (1996)). The associative memory Hamiltonian

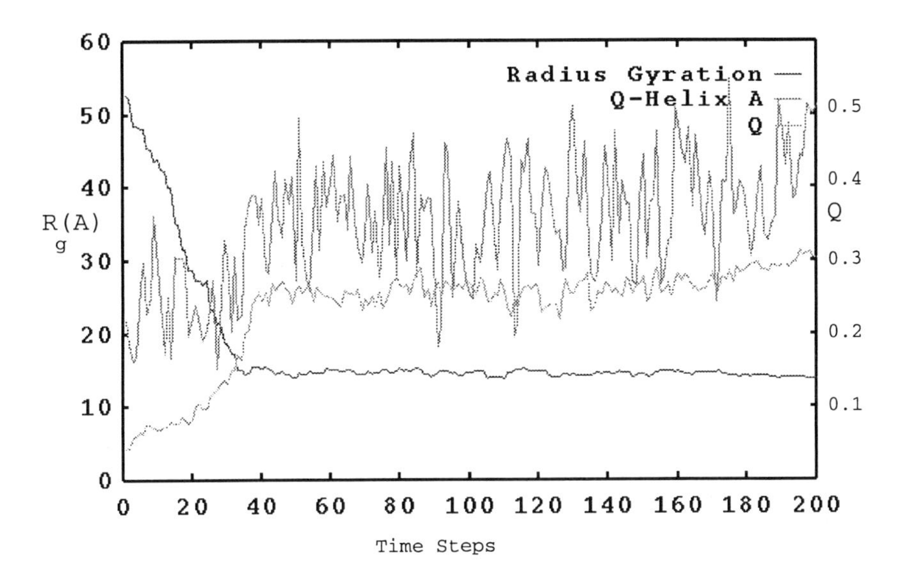

Fig. 9. Molecular dynamics simulation of the collapse and structure formation in myoglobin, using the $\mathcal{H}_{\text{AM}}$ energy function.

can function as a laboratory for similar studies with more realistic encodings and provides a powerful tool to determine the energy landscape parameters at various stages in the folding funnel.

The energy landscape perspective presented here is a simple, but powerful framework to view the complex nature of protein folding. Experimental techniques are just being developed to study the sub-millisecond stages of folding in which the protein is compact and partially folded. This region and the late stages of folding present challenges to statistical physicists because the partially folded protein is in a low entropy state. This low entropy state appears to be liquid-crystalline-like, i.e. a state in which subtle forces can cause subtle forms of partial ordering. The diverse routes leading to the native state of a protein can be better probed by adding at least one other reaction coordinate to the folding funnel in Fig. 4. For example, to further detail the folding process, the roles that substructure or foldon formation and and *micro-phase separation* of the hydrophobic residues to the core of the protein and the hydrophilic residues to the surface should be investigated. Many thermodynamic properties of proteins in these last stages of folding are non-self-averaging and can be important in determining whether a given sequence is foldable. The statistical physics view of the protein folding problem has already lead to partial successes in solving such practical problems in molecular biology as structure prediction and design. The energy landscape approach offers experimentalists a framework to analyze their folding experiments and guide the development of more precise probes of Nature's information highway.

References

Binder, K., Young, A. P. (1986): Rev. Mod. Phys. **58**, 801

Bryngelson, J., Wolynes, P. G. (1987): Spin glasses and the statistical mechanics of protein folding. Proc. Nat. Acad. Sci., U.S.A. **84**, 7524–7528

Bryngelson, J., Onuchic, J. N., Socci, N., Wolynes, P. G. (1995): Funnels, pathways and the energy landscape of protein folding: a synthesis. Proteins **21**, 167–195

Creighton, T. E. (1994): *In: Mechanisms of Protein Folding* (Oxford Univ., Oxford, Pain, R. H. editor), 1–25

Dill, K. A. et al. (1995): Principles of protein folding - A perspective from simple exact models. Protein Science **4**, 561–602

Dinner, A., Sali, A., Karplus, M., Shakhnovich, E. (1994): Phase Diagram of a model protein derived by exhaustive enumeration of the conformations. J. Chem. Phys **101**, 1444–1451

Flory, P. J. (1954): *Principles of Polymer Chemistry* (Cornell Univ., Ithaca), 523–530

Frauenfelder, H. et al. (1990): Proteins and pressure. J. Phys. Chem. **94**, 1024–37

Friedrichs, M., Wolynes, P. G. (1989): Toward protein tertiary structure recognition by means of associative memory hamiltonians. Science **246**, 371–373

Friedrichs, M., Wolynes, P. G. (1990): Molecular Dynamics of Associative Memory Hamiltonians for Protein Tertiary Structure Recognition. Tetrahedron Comp. Method. **3**, 175

Friedrichs, M., Goldstein, R., Wolynes, P. G. (1991): Generalized protein tertiary structure recognition using associative memory hamiltonians. J. Mol. Biol. **222**, 1013–1034

Garel, T., Orland, H., Thirumalai, D. (1995): *In: New Developments in Theoretical Studies of Proteins: Advanced Studies in Physical Chemistry* (Ron Elber, editor, World Scientific, Singapore)

Gierasch, L. M., King, J. (1990): *Protein Folding:Deciphering the Second Half of the Genetic Code* (AAAS, Washington), vii–vii

Goldstein, R., Luthey-Schulten, Z. Wolynes, P. G. (1992): Optimal protein-folding codes from spin-glass theory. Proc. Natl. Acad. USA **89**, 4918-4922

Huang, G. S., Oas, T. G. (1995): Proc. Natl. Acad. Sci. **92**, 6878

Luthey-Schulten, Z., Ramirez, B., Wolynes, P. (1994): Helix-coil, liquid-crystal and spin-glass transitions of a collapsed heteropolymer. J. Phys. Chem. **99**, 2177–2185

Onuchic, J. N., Wolynes, P. G., Luthey-Schulten, Z., Socci, N. D. (1995): Toward an outline of the topography of a realistic protein-folding funnel. Proc. Nat. Acad. Sci., U.S.A.**92**, 3626–3630

Panchenko, A., Luthey-Schulten, Z. Wolynes, P. G. (1996): Foldons, Protein Structural Modules and Exons. Proc. Nat. Acad. Sci., U.S.A. **93**, 2008–2013

Sasai, M., Wolynes, P. G. (1990): Molecular theory of associative memory hamiltonian models of protein folding. Phys. Rev. Lett. **65**, 2740–2743

Shakhnovich, E., Gutin, A. (1990): Implications of thermodynamics of protein folding for evolution of primary sequences. Nature **346**, 773–775

Socci, N., Onuchic, J. (1994): Folding kinetics of protein-like heteropolymers. J. Chem. Phys. **101**, 1519–1528

Socci, N., Onuchic, J. (1995): Kinetic and Thermodynamic Analysis of proteinlike heteropolmers: Monte Carlo histogram technique. J. Chem. Phys **103**, 4732–4744

Takada, S., Wolynes, P. G. (1996): Statics, metastable states and barriers in protein folding: A replica variational approach. Physical Review E (in press)

From Interatomic Interactions to Protein Structure

Joseph D. Bryngelson[1] and Eric M. Billings[2]

[1] Physical Sciences Laboratory,
[2] Laboratory of Structural Biology,

Division of Computer Research and Technology,
National Institutes of Health, Bethesda, MD 20892, USA

1 Introduction to Protein Structure

1.1 Introductory Remarks

The basic unit of life is the cell. Some organisms, e.g., bacteria and yeast, contain a single cell. Multicellular organisms such as humans and plants start as a single cell, but as this cell multiplies it differentiates to form distinct tissues and organs. The cell's flexibility to develop into distinct tissue types which perform different functions is the key to more complex organisms. The underlying similarity between living organisms results from the flexibility and specificity possible with the most common denominator of life—the cell.

An organisms genetic heritage is encoded in DNA, the molecules of heredity. The genetic code of DNA is translated directly into proteins. In a real sense, a cell's genetic heritage is reflected in the proteins which it is able to synthesize. A simple unicellular species requires about 350 different proteins to survive. More complex species, such as humans, have as many as $\sim 10^5$ different proteins. Each compound which is synthesized; each biochemical signal which is sent or received; each product synthesized for export to the rest of the organism is in part a result of the structure and function of some protein.

1.2 What Is the Chemical Composition of a Living Cell?

The predominant chemical component of most cells is **water**. A typical bacterial cell is about 70% water by weight. Of the remaining (dry) portion of a typical cell, the largest constituent, about 50%, is **protein**. The remaining dry weight consists mainly of **nucleic acids** (the molecules of heredity), **lipids** (the cell membrane), and **carbohydrates** (mainly used for energy storage and production). Proteins perform a great variety of functions in cells, as shown in Table 1. Each protein tends to perform a small number of functions (usually one) well; such **specificity** is important in biological systems.

Table 1. Some proteins and their functions

Functional Class	*Function*	*Examples*
Structural Proteins	Hold things together	Collagen, Anchorin, Integrins
Enzymes	Catalyze chemical reactions	Alcohol Dehydrogenase, Peptidases
Channels	Regulate cell interior	Sodium Channel, Porins
Repressors and **Activators**	Regulate gene expression	λ-Repressor
Receptors	Cell signalling	adrenergic receptors

1.3 What Are Proteins?

Proteins are polymers or long chains built from a basic set of **amino acids**. The 20 most common amino acids and their sequence within the protein chain are sufficient to create a wide variety of proteins, each suited to its unique function. The chemical structure of a single, generic amino acid is show in Fig. 1 below. More precisely, Fig. 1 shows the **zwitterion**, which is the most common molecular species in aqueous (water) solution at pH 7. More information on the zwitterion is contained in the caption to Fig. 1. The amino acids are distinguished by their sidechains or **R-groups**. These sidechains are shown in Fig. 3. The chemical structure of proteins is held together by **covalent** bonds. Crudely, each covalent bond may be thought of as a pair of electrons that the two connected atoms share. This sharing of electrons leads to a strong attractive force between the atoms. When two pairs of electrons are shared the bond is called a **double bond**. Covalent bonds are represented by cylinders in the figures. The amino acid's R-group is attached to the central carbon atom called the **alpha-carbon** and denoted by C^{α}. The C^{α} is in turn bonded to the nitrogen (N) of the NH_2 (amino) group and to the carbon (C') of the COOH (carboxylic acid) group. The (C') is often called the **carbonyl carbon** for reasons that will become clear shortly.

An amino acid **residue** is composed of an R-group and the backbone atoms which become part of the protein chain. A residue is added to a protein by removing a hydrogen atom from the new residue's NH_2 group and the OH unit from the terminal COOH and forming a bond between the N and C' atoms. The residues join in an unbranched chain to form a protein as shown in Fig. 2. The C'-N bond connecting two amino acid residues is called the **peptide bond**. [1] Polymers of amino acids are often called **polypeptides**. Proteins are large polypeptides. Once the peptide bond is formed, the C' has a double bond to an oxygen (O), called a **carbonyl** bond. The large dipole moment of this bond is important for protein structure. The NH_2 group of the first amino acid in the

[1] This bond is sometimes refered to as the **amide bond**.

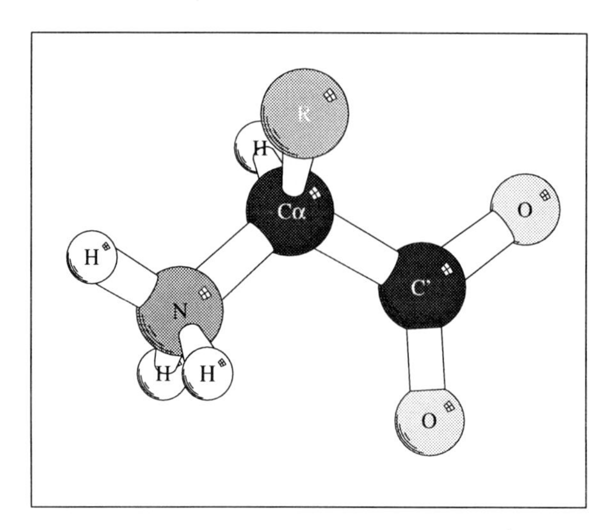

Fig. 1. A diagram representing the structure of a generic amino acid. The gray sphere labeled R represents the position of the amino acid sidechain. The structures of the 20 amino acid sidechains is shown in Fig. 3. The molecular species shown is the zwitterion, which is the dominant species at most biologically relavant values of pH. In the zwitterion the proton (hydrogen ion) from the acidic COOH group is transfered to the basic NH_2 group.

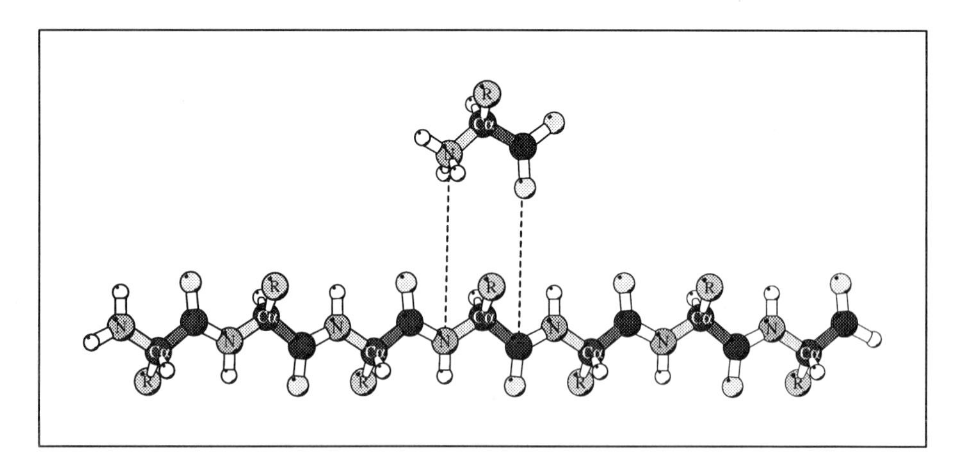

Fig. 2. The chemical structure of a protein. The unlabeled black spheres represent the carbonyl carbons (the C'). The white bonds between the C' atoms and the N atoms connect the residues. These bonds are peptide bonds. They have some double bond character and are flat. The light gray bonds between the C^{α} atoms and the N atoms are the axes of rotation for the ϕ angles and the dark gray bonds between the C^{α} atoms and the C' atoms are the axes of rotation for the ψ angles.

chain, the **N terminus**, and the COOH group of the last amino acid in the chain, the **C terminus**, remain intact.

The peptide bond is planar and is not free to rotate like the other bonds along the backbone. Therefore, rotations about the N-C^α and the C^α-C' bonds determine the **conformation**, that is, the three dimensional structure of the polymer **backbone**. Rotations about backbone angles are described by the **dihedral angles** which are formed by four consecutive atoms. This angle is determined by the angle between the first atom, the axis of the two central atoms, and the fourth atom. The dihedral angle that corresponds to rotation about the N-C^α bond is called the ϕ angle and the dihedral angle that corresponds to rotation about the C^α-C' bond is called the ψ angle. Listing of the ϕ and ψ angles for each amino acid residue provides a convenient, commonly used representation of the backbone of a protein.

1.4 What Determines Protein Structure?

An appreciation of the interactions that hold proteins together is required for a good understanding protein structures. In principle, all chemical forces are electromagnetic and may be understood as the behavior of charged particles obeying the laws of quantum electrodynamics. In practice, this approach to chemical forces is both impractical and unenlightening when applied to proteins, because of the large number of atoms involved. Alternatively, the chemical forces may be divided into different types, each with its own characteristics. This approach to chemical forces proves to be an effective method for understanding the behavior of biological macro-molecules. However, the reader should keep in mind that there is really only one interaction which has many manifestations.

Protein Energetics. The most commonly used energy unit in the molecular biophysics literature is the **kilocalorie per mole**, usually abbreviated **kcal/mole**. In terms of energy units that are more familiar to most physicists $1\,\mathrm{eV} = 23.06\,\mathrm{kcal/mole}$. The connection between calories and joules is given by the mechanical equivalent of heat: 1 calorie = 4.184 joules. The figure of merit for comparing energies in molecular biophysics is the thermal energy at room temperature, $k_\mathrm{B}T_\mathrm{room} = 0.617\,\mathrm{kcal/mole}$, where k_B denotes the Boltzmann constant and $T_\mathrm{room} = 298\,\mathrm{K}$ ($25\,°\mathrm{C}$) denotes room temperature in degrees Kelvin. (In this chapter T will always represent absolute temperature.)

Many biologically important interactions have energies of $\sim$2–6 kcal/mole and are often referred to as **weak interactions**. This range of energies allows for sufficient stability and flexibility for biological function. A clear, detailed discussion of the rôle of weak interactions in molecular biophysics is found in Chap. 5 of Watson et al. (1987).

Covalent Bonds. Covalent bonds are strong interactions; they typically have energies on the order $\sim$50–250 kcal/mole. These are the bonds that are represented as solid lines in chemical structure diagrams. Covalent bonds can be con-

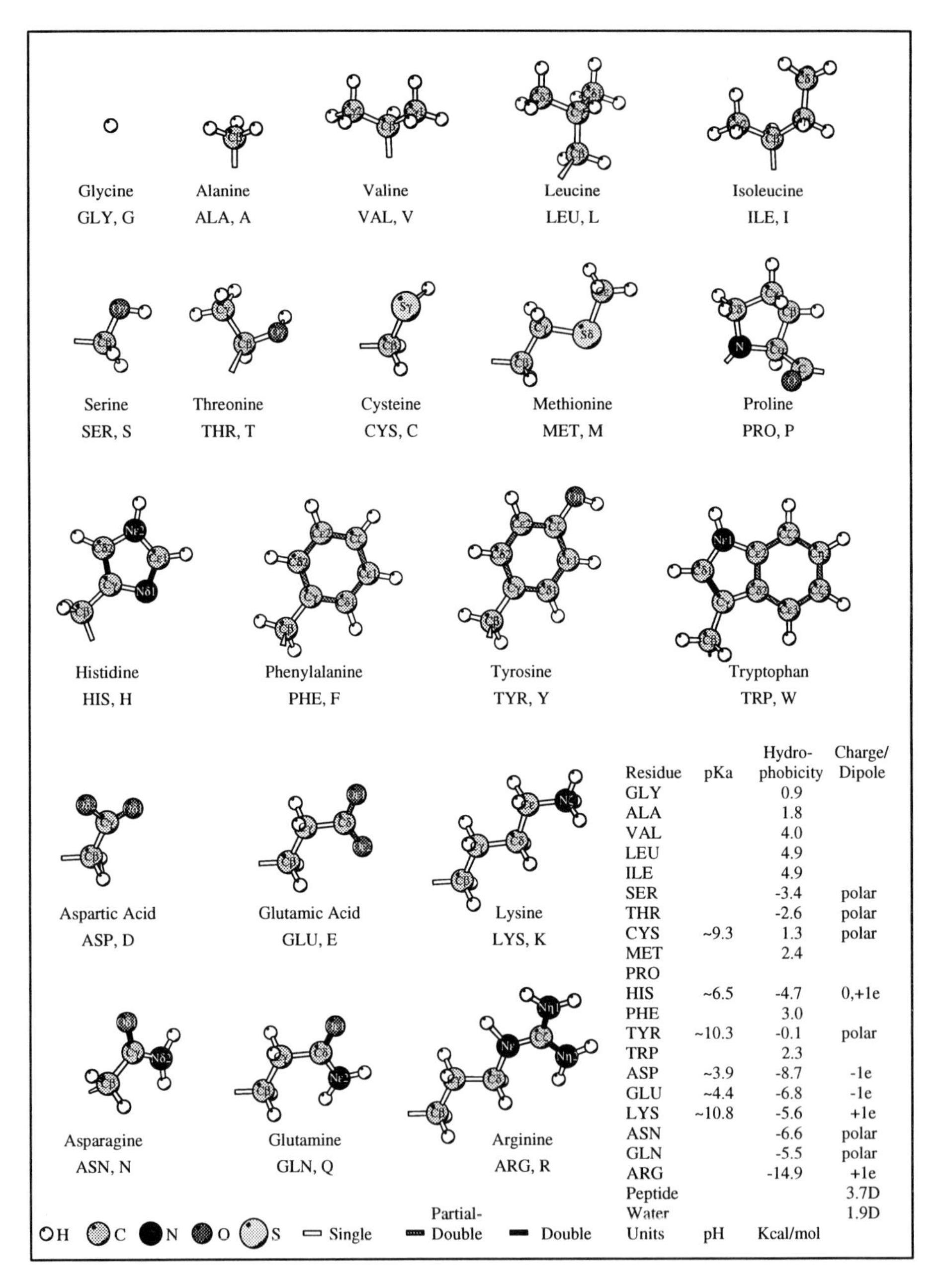

Residue	pKa	Hydro-phobicity	Charge/Dipole
GLY		0.9	
ALA		1.8	
VAL		4.0	
LEU		4.9	
ILE		4.9	
SER		-3.4	polar
THR		-2.6	polar
CYS	~9.3	1.3	polar
MET		2.4	
PRO			
HIS	~6.5	-4.7	0,+1e
PHE		3.0	
TYR	~10.3	-0.1	polar
TRP		2.3	
ASP	~3.9	-8.7	-1e
GLU	~4.4	-6.8	-1e
LYS	~10.8	-5.6	+1e
ASN		-6.6	polar
GLN		-5.5	polar
ARG		-14.9	+1e
Peptide			3.7D
Water			1.9D
Units	pH	Kcal/mol	

Fig. 3. Chemical structures of the 20 amino acid sidechains. Underneath the name of each amino acid is its standard three letter and one letter abbreviation. The labels on the carbon, nitrogen, oxygen, and sulfur atoms are standard abbreviations used in discussions of structure. We put them in the figure for easy reference. The table in the lower right hand corner lists some important properties of the sidechains. The meanings of the terms in this table are explained in the text, except for pH which is explained in elementary chemistry textbooks. The hydrophobicity data is from Radzicka and Wolfenden (1988) and is the energy required to move the R-group from a non-polar environment (cyclohexane) into water. This energy is thought to approximate the energy required to move the R-group from the protein's interior into water.

sidered essentially permanent on the time scale of a protein's lifetime. Biochemical reactions that involve the breaking of a covalent bond are usually catalyzed by enzymes. Covalent bonds hold together the protein backbone and sidechains. In addition, the side chain of the amino acid cysteine has a sulfur (S) atom that can form a covalent bond with the sulfur atom on another cysteine sidechain. This bond has an energy of $\sim 50\,\text{kcal/mole}$ and can be an important cross link between residues which are not close in the sequence but are close in the folded state. These disulfide bonds form in cells when a protein is folding into its active three-dimensional structure.

van der Waals Interactions. We use this term to refer to the combination of two interactions that operate between any two atoms, a weak long-ranged interaction and a strong short-ranged interaction. The long-ranged interaction results from the interaction between the quantum mechanical fluctuations of the dipole moments of the atoms; it is attractive and its energy varies as the inverse sixth power of the distance between the atoms. The short-ranged interaction results from a combination of the repulsion of the overlapping electron clouds, the Pauli exclusion principle, and the electrostatic repulsion of the nuclei as the atoms move closer; it is repulsive and increases much more rapidly than the attractive potential. The repulsive short-ranged interaction can be considered, for many purposes, to be an infinite-energy hard sphere interaction. This combination of interactions has a minimum which occurs at a distance called the **van der Waals radius**, which depends on the type of atom. Typical values of van der Waals radii are $1.2\,\text{Å}$ to $2.2\,\text{Å}$. The depth of the energy minimum is typically about the size of thermal energies. This energy can significantly stabilize structures if many close contacts are formed, i.e., if two molecular surfaces of complementary shape come into contact. The repulsive hard spheres of two or more atoms can not overlap. The results of this limitation are called **steric effects**; they sharply reduce the number of possible protein conformations. For example, steric effects limit the possible values of the ϕ and ψ angles. The overlap of atoms with adjacent residues and with atoms in side chains render many conformations improbable. Plots of the energy of interaction for different ϕ and ψ angles, have proven useful for thinking about protein structure; they are called **Ramachandran maps**. The Ramachandran map for an alanine dipeptide is shown in Fig. 4.

Electrostatic Effects. There are two sources of static electrical charges in proteins. First, five amino acids are charged under normal conditions (of pH and ionic strength). The two negatively charged amino acid residues are aspartic and glutamic acid. The three positively charged amino acids are lysine, arginine and histidine. These charged residues are often exposed to solvent or involved in electrostatic interactions with oppositely charged residues, often called **salt bridges**, at the core of a protein. The charge of a sidechain can change with pH. Some sidechains can gain a charge by losing a hydrogen ion (H^+), that is, a proton. The equilibrium for this ionization reaction depends on pH. The pH at

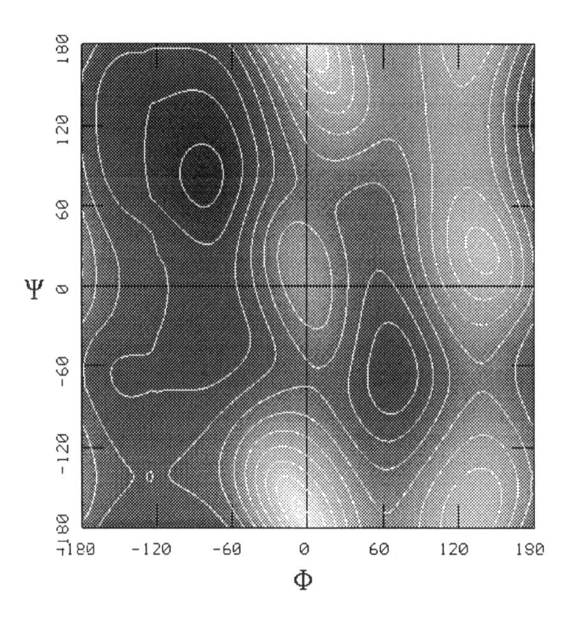

Fig. 4. Ramachandran map for an alanine dipeptide. The potential energy of the dipeptide is contoured for all combinations of ϕ and ψ and is plotted as a function of ψ (vertical axis) versus ϕ (horizontal axis). The dark regions have low potential energy and the light regions have high potential energy. The α-helix region is centered at $(-57°, -47°)$ and the β-sheet region is centered at $(-120°, +113°)$.

which the R-group is ionized 50% of the time is called the pK_a. At pHs below the pK_a (more acidic) the R-group is usually **protonated**, that is, the hydrogen ion is still bound to the R-group. The biologically relevant pK_a values are shown in Fig. 3.

The second source of charge comes from the unequal sharing of electrons in a covalent bond. Some elements are more **electronegative** than others, that is, they tend to attract electrons more strongly than their bond partner. For example, in a carbonyl bond an electronegative oxygen pulls the covalent electrons it shares with a carbon more strongly. This results in a negatively charged region around the oxygen due to the greater electron density. The informal, short-hand terminology is that the oxygen has a negative **partial charge** and the carbon has a positive partial charge. The induced partial charges create local dipoles which have strong effects on protein structure. Water has a dipole of 1.85 D, and the dipole of the peptide bond in the backbone is 3.7 D, where "D" is the abbreviation for **Debye**, a useful unit of molecular dipole moments. One Debye equals 10^{-18} electrostatic unit centimeters (esu cm); the dipole moment of an electron and a proton that are separated by 1 Å is approximately 4.8 D. The peptide dipole is one of the strongest contributors to the stability of local structures with proteins. In addition, six of the R-groups have significant dipole moments, namely, serine, threonine, cysteine, tyrosine, asparagine, and glutamine.

The energies associated with the electrostatic interactions of charges obey Coulomb's Law

$$E_{i,j} = \frac{q_i q_j}{\epsilon r_{i,j}} \tag{1}$$

where q_i and q_j represent charges at points i and j, $r_{i,j}$ represents the distance between points i and j, ϵ represents the dielectric constant of the medium, and $E_{i,j}$ represents the energy of the electrostatic interaction of the two charges. Having discussed the charges, we now turn to the other parameter in the expression (1) for electrostatic energy, the dielectric constant.

What is the dielectric constant of proteins? This seemingly simple question does not have a simple answer. The *static* macroscopic dielectric constant of the surrounding water is 78.5; the electrical properties of the interior of the protein probably resemble those of amide polymers, which typically have a dielectric constant of ~ 4. This difference in dielectric constants has a dramatic effect on electrostatic energies. For example, the potential energy of two singly ionized atoms placed 4 Å apart in water is approximately -1 kcal/mole, whereas in the protein interior this energy is on the order of -20 kcal/mole. The appropriate treatment of the dielectric constant remains a difficult technical problem in studies where highly accurate calculations of protein energies are desirable, e.g., molecular dynamics simulations of protein motion. Fortunately, for the purposes of this chapter it suffices to note that the dielectric constant of the protein interior is far smaller than that of the water which surrounds the protein.

Hydrogen Bonds. Hydrogen is the most common element in organic compounds. A glance at an R-group or the peptide backbone shows several hydrogens that are covalently bonded to strongly electronegative elements such as N, O and S. The hydrogen carries a partial positive charge in these bonds and its bond partner (**hydrogen donor**) carries a partial negative charge. This strongly polar bond creates a local dipole which readily interacts with nearby atoms. The interaction of the dipole with an atom with a negative partial charge (**hydrogen acceptor**) is referred to as a **hydrogen bond**. The hydrogen bond is predominantly an electrostatic effect.

The strength of a hydrogen bond can range from 1 to 7 kcal/mole. This bond energy is strongly dependent upon the geometry of the donor, hydrogen and acceptors atoms. A strong hydrogen bond requires that the distance between the donor and acceptor atoms be less than ~ 3.5 Å, and that the three atoms be nearly collinear.

The amide (NH) and carbonyl (CO) groups in the protein backbone make especially strong hydrogen donors and acceptors, respectively. The amide (NH) group nitrogen has one hydrogen to donate. The carbonyl (CO) group oxygen can accept two hydrogen bonds since it has two pairs of electrons which are not directly bonded. Hydrogen bonds in the backbone atoms are vitally important in stabilizing proteins structures. Since each residue's backbone atoms have three possible hydrogen bonds it is easy to see that a typical protein with 100 residues could be dominated by hydrogen bonding patterns.

Several R-groups are capable of forming hydrogen bonds. Any R-group with a strongly polar bond involving a hydrogen or unbonded electrons can participate in hydrogen bonds. This criterion includes all of the charged and polar R-groups. The **active site** of an enzyme is the subset of residues which interact directly with the substrate. Residues that can hydrogen bond frequently line the active site. These residues help to define the specificity of the enzyme for its substrate because effective hydrogen bonding requires that the substrate must match the spacing, geometry, and donor/acceptor pattern of the active site.

Digression on Water. Proteins function in a solvent, water. The unusual properties of water have a strong, often dominant effect on the energetics of protein structure. The dynamic structure of liquid water is not entirely understood, despite extensive experimental and theoretical analysis. Water consists of an electronegative oxygen and two hydrogens. The oxygen pulls the electrons that it shares with the hydrogens in covalent bonds towards itself. Consequently, in the water molecule the oxygen has a negative partial charge and the hydrogens have a positive partial charge resulting in a molecule with unusual electrostatic properties and a strong dipole moment of 1.85 D.

Water forms an extensive and stable hydrogen bonding network because each water molecule can participate in four hydrogen bonds. The oxygen can accept two hydrogens from adjacent molecules and each hydrogen can be donated in a hydrogen bond. Experimental data on *bulk* water (Thiessen and Narten (1982)) shows that water molecules form localized structures with tetrahedral coordination. These local structures last for 10^{-12} to 10^{-9} seconds before dissolving and forming again with different neighbors (Frank (1958)). However, water forms **clathrate cage**-like structures near proteins. A cage structure is formed around a solute and is the minimum number of waters necessary to construct a (nearly) spherical layer around the solute.

Hydrophobic Effect. The unique properties of water as a solvent may be analyzed from the protein's perspective. If the fluctuations of the water structure are averaged over time, the effect of the water on the protein may be thought of as constituting a **hydrophobic force**. A hydrophobic force is a way of describing the fact that water-water interactions tend to be more favorable than water-protein interactions; and that "forces" certain protein-protein interactions. Essentially, the water structure pushes the proteins together and causes them to aggregate in order to avoid water; hence the term "hydro-phobicity".

The precise physical nature of the hydrophobic effect is a subject of current research; a physically oriented review of the general effects of water on forces in biological molecules is Leikin et al. (1993). A few general insights concerning hydrophobicity have emerged from research and proven valuable for understanding protein structure. The hydrophobicity of a residue is determined by its polarity and size. The less polar a residue the more hydrophobic it is. Charged residues and those with strong dipoles are **hydrophilic**. Neutral residues which do not have polar functional groups are hydrophobic. In addition, the larger the non-polar R-group, the more hydrophobic it is. The hydrophobic effect is evident in the free energy required to move an amino acid from the gas phase into solvent. Among the eight hydrophobic residues, glycine, alanine, valine, leucine, isoleucine, methionine, phenylalanine, and tryptophan, this relative free energy is proportional to the surface area; the more it disrupts the water the less soluble it is. The hydrophobic residues therefore will aggregate to reduce their combined surface area. The energy associated with the **hydrophobic effect** is as much as $\sim 2\,\text{kcal/mole}$ per residue. As a result of the hydrophobic effect, the hydrophobic residues tend to be in the interior of the protein and the hydrophilic residues tend to be on the surface of the protein. Although the concept of hydrophobic R-groups being attracted to one another is helpful, it is important to keep in mind that this is really an entropy-driven hydrophobic effect.

1.5 Understanding Protein Structures

The reader may find it useful to keep a few generalities about proteins in mind in order to put the information in this chapter in context. Proteins can be either **globular**, i.e., shaped somewhat like a ball, or **fibrous**, i.e., shaped like a long, thin fiber. Enzymes are typically globular. Structural proteins are typically fibrous. Most discussion in this chapter is oriented towards globular proteins; however, the general principles are applicable to all proteins. The size range for individual globular proteins is from $\sim$ 50 to $\sim$ 500 amino acids and from 25 to 100 Å in diameter. The lower size limit depends on when one starts calling a molecule a protein instead of a small polypeptide. Fibrous proteins can have much longer dimensions. Collagen forms units which are 1800 Å long!

In a cell proteins are synthesized as linear chains at a rate of about one residue per second. During and after this synthesis the protein must self-assemble into its final, functional form. This self-assembly process is called **protein folding** From a thermodynamic perspective, we would expect the protein to adopt its minimum free energy structure. [2] This thermodynamic hypothesis has been confirmed for several proteins (Anfinsen (1973)). For example, a protein may be denatured (unfolded by heating or adding salts etc.) and then allowed to refold. In these experiments proteins typically adopt their naturally occurring, functional form after searching numerous conformations, thereby providing strong evidence that the information necessary to describe the active form is embedded in the amino acid sequence. The difference in free energy between the unfolded and correctly folded states is on the order of 10 to 15 kcal/mole. This energy difference is small for such a large molecule; its smallness illustrates the delicate balance of free energies involved in protein structure.

Primary Structure. The sequence of amino acids in a protein represents its **primary structure**. The primary sequence is often shown as a sequence of

[2] More precisely, it folds to the structure that minimizes the free energy of the protein plus solvent system.

either the one or three character codes of the amino acids (see Fig. 3). The residues are written in the order that they are synthesized, from the N (amino) to the C (carboxylic acid) terminus.

Proteins can be compared on the basis of their primary sequence. The **homology** between two proteins is defined by the percentage of identical or similar residues between the aligned sequences. Residues are considered similar if they have similar properties such as charge, hydrophobicity or size. For example, Aspartic acid and glutamic acid are similar because of their negative charge. The sequences of the sub-units in hemoglobin make a good example. There are regions of the proteins which have identical residues, that is, they are exactly homologous. However, other regions appear to have gaps and insertions. These gaps and insertions typically occur near residues which are at the surface of the protein or at the N or C termini.

Peptide	1									10										20										30
α	V	-	L	S	P	A	D	**K**	T	N	V	K	A	A	**W**	**G**	**K**	**V**	G	A	H	A	G	E	Y	**G**	A	**E**	A	**L**
β	V	H	L	T	P	V	E	**K**	S	A	V	T	A	L	**W**	**G**	**K**	**V**	N	-	-	V	D	E	V	**G**	G	**E**	A	**L**
γ	G	H	F	T	E	E	D	**K**	A	T	I	T	S	L	**W**	**G**	**K**	**V**	N	-	-	V	E	D	A	**G**	G	**E**	T	**L**

Fig. 5. The primary sequences of the first 30 residues in the α , β and γ peptides which can combine to form human hemoglobin. Adult hemoglobin consists of 2 α and 2 β peptide sub-units. Fetal hemoglobin contains the γ instead of the β peptide. Residues which are critical to protein function are either exactly conserved (bold) or are residues with similar properties. Seven residues are identical and several are similar. Similar mutations conserve the hydrophobic (G for A), charge (D for E), or hydrogen bonding (S for T) nature of the residue.

Secondary Structure. Proteins fold into local structures of short lengths displaying what is called **secondary structure**. This local structure appears to be dominated by two factors: hydrogen bonding patterns between nearby residues and favorable orientations of the ϕ and ψ angles. For residues that are very close in sequence, hydrogen bonding can occur only when these residues have specific combinations of ϕ and ψ. The number of possible secondary structures is reduced by this restriction. The most common secondary structures are helices and the β-strands which combine to form β sheets (see Fig. 6).

Helices. Helices are formed when a residue forms a hydrogen bond with a residue which is 3, 4 or 5 residues away (towards the C terminus). Three types of helices can be formed. They differ in the number of residues per turn, height of one turn (**pitch**), diameter, and the number of atoms from the oxygen to its hydrogen bond partner. All three of these structures are right-handed helices.

Left-handed helices are not formed because the R-groups can not be positioned along the internal diameter of the helix.

Table 2. Attributes of Secondary Structures

Type	ϕ	ψ	*Residues/Turn*	*Pitch (Å)*	*Atoms from O to H*
3-10-helix	−49	−26	3.0	6.0	10
α-helix	−57	−47	3.6	5.4	13
π-helix	−57	−70	4.4	5.1	16
β-sheet (parallel)	−119	+113	2.0	6.4	NA
β-sheet (anti-parallel)	−139	+135	2.0	6.8	NA

By far the most common helix in proteins is the α-helix. Roughly 25% of the residues in known protein structures are in an α-helix. An α-helix is formed when the carbonyl oxygen of residue i forms a hydrogen bond with the amide hydrogen of residue $i+4$. In the α-helix the dipoles formed by the amide bonds are close to parallel, thereby providing additional energetic stabilization. Other helices that sometimes appear in proteins are the 3-10 helix (named from the properties shown in second and fourth columns in Table 2), which is formed when the carbonyl of residue i bonds to residue $i+3$, and the uncommon π-helix, which may appear as the initial or final turn in an α-helix.

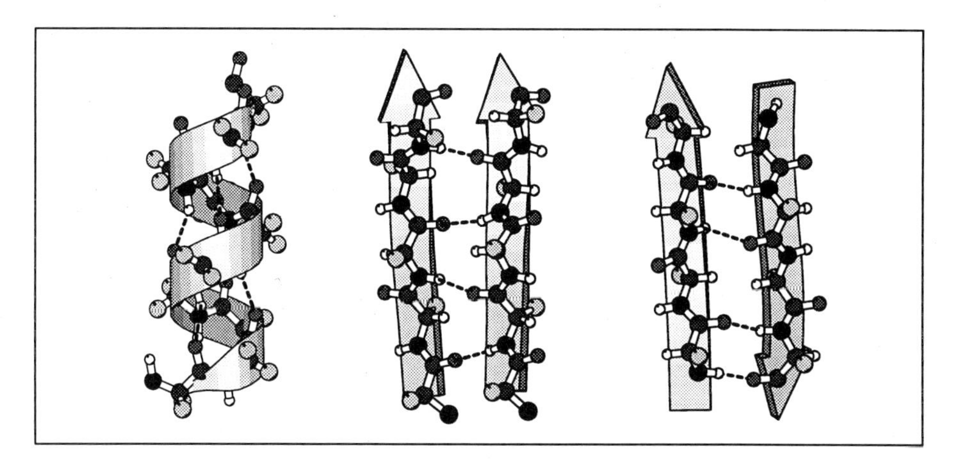

Fig. 6. Secondary Structural Elements. The three most common secondary structural elements are shown in the figure. They are, left to right, α-helix, parallel β-sheet, and anti-parallel β-sheet. The ball and stick figures show the position of the protein backbone with dashed lines representing hydrogen bonds. Superimposed on the ball and stick figures are the **ribbon diagram** symbols for the secondary structures. The ribbon diagram representation facilitates visual understanding of complex protein structures. It is often used to show tertiary structure, as in Fig. 7 and Fig. 8.

β-Strands and β-Sheets. The β-strand is the local conformation in which the backbone is as extended as possible. β-sheets are formed when a section of the peptide is in the β-strand conformation and is located next to another β-strand segment, thereby providing the opportunity for hydrogen bonding between the backbones of the two β-strands. The β-sheets may form with the β-strand segments aligned in a parallel or an anti-parallel fashion. The parallel β-sheets have the two segments with the direction of their N to C termini aligned whereas the anti-parallel β-sheets do not. The parallel and anti-parallel configurations result in distinct hydrogen bonding patterns and different orientations of the R-groups with respect to the β-sheet. Roughly 20% of residues in the database of protein structures are in some form of β-sheet.

Tertiary Structure. The secondary structural elements pack into a compact, native structure. This complete three dimensional structure of a single protein is called the **tertiary structure**. The tertiary structure is what is observed when viewing the three-dimensional structures as in Fig. 7. The tertiary structure of a protein, particularly of a protein with more than $\sim$ 200 amino acids, may be organized into several structural units called **domains**. The domain concept is somewhat subjective but useful. A domain is a typically globular group of residues which have many interactions with each other and relatively few interactions with residues outside the domain. Domains can often fold separately, without the rest of the protein. Tertiary structures may contain helical and β-sheet domains. Some simple geometric arrangements of secondary structure elements occur frequently in protein structures. These arrangements are called **motifs**. There are many common motifs names such as *leucine zippers*, *β-barrels*, and *greek keys*. Domains are typically build from one or more motifs.

Some proteins have small molecules bond to them. These small molecules are called **prosthetic groups** and they often play a vital role in the function of a protein. A good example of a prosthetic group is the heme group bound to hemoglobins. The structure of hemoglobin, including the bound heme groups, is shown in Fig. 8. The function of hemoglobin is to carry oxygen in blood. The prosthetic group is essential to this function because the transported oxygen binds on the iron atom in the heme group.

Quaternary Structure. Some functional proteins consist of more than one peptide chain. The way in which these peptide chains are arranged is called the **quaternary structure** These complexes may consist of two (**dimer**), three (**trimer**) four (**tetramer**) or more individual proteins. These multimers may be termed **homo-** or **hetero-**meric if they contain one or more type of protein. For example, hemoglobin is an excellent example of a hetero-tetramer. It consists of four distinct proteins: two α and two β chains. The structure of hemoglobin is shown in Fig. 8.

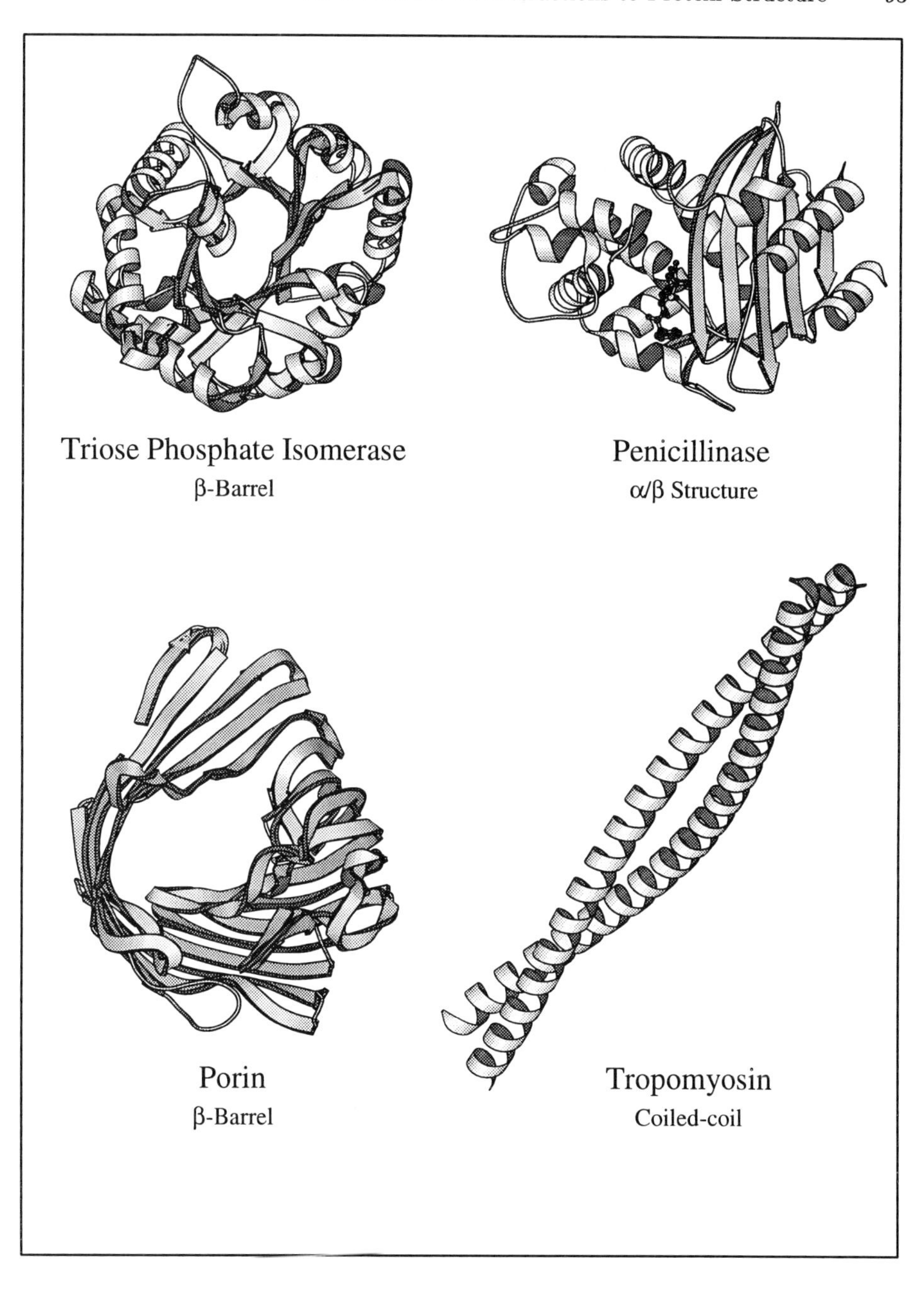

Fig. 7. Tertiary Structure. Secondary structures can be combined to form a variety of 3-dimensional structures. Triose phosphate isomerase is a combined α/β enzyme with a central core which is soluble in water. Penicillinase is an α/β enzyme with two domains. One domain has a β-sheet which is sandwiched between α-helices and the other domain is α-helical. The ball and stick figure represents Penicillin-G positioned in the active site between the two domains. Porins are β-barrel structural proteins which position themselves in the cell membrane to form regulatable channels through which various substrates can pass. Tropomyosin is a structural protein in the common **coiled-coil** motif. Its all α-helical content allows for long, sturdy structures.

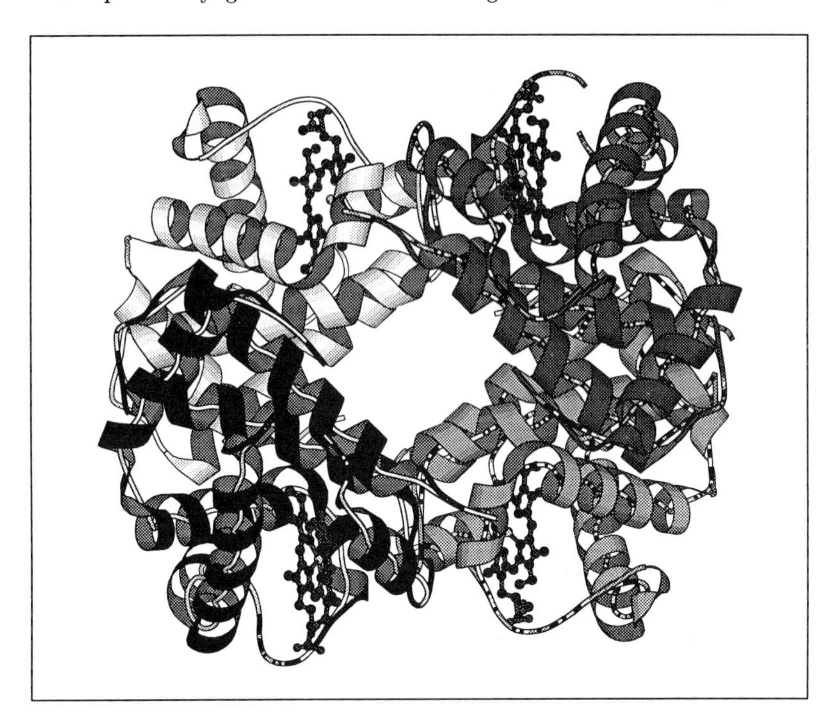

Fig. 8. Quaternary Structure. Four individual proteins, or sub-units comprise the functional form of hemoglobin. There is also a heme prosthetic group (represented by the ball and stick figures) embedded in each sub-unit which contains the iron that binds oxygen. There are two α and two β sub-units which comprise the hetero-tetramer.

2 Statistical Mechanics of Polymer Configurations

2.1 Introduction

Temperature and thermal interactions are essential to the description of any living thing, so statistical mechanics—the most comprehensive physical description of the thermal interactions—naturally finds application in the study of proteins. We will use statistical mechanics to develop tools for understanding proteins in this section and in the following one. Section 2 introduces the statistical mechanics of polymer configurations. Readers wishing to explore this subject further will find the very clear review by Chan and Dill (1991) accessible after reading Sect. 2. Section 3 introduces the helix-coil transition in proteins. The standard reference for this material is Poland and Scheraga (1970), which should be clear to readers who have read through Sect. 3.

Both of these sections make use of simple, mathematically tractable models which describe the essence of the phenomenon being studied. Since these models are simple, many quantities can be calculated directly and the origins and relations between different effects are transparent. The insights gained from these simple models can then be tested in experiments and used to build and analyze more realistic models. Furthermore, these insights also give a vocabulary for discussing the properties of the more complicated models.

2.2 Random Flight Chains

A polymer essentially consists of a large number of small molecular units, **monomers**, jointed by covalent bonds. Proteins are linear polymers, that is, polymers where each monomer bonds to only two other monomers (except for the two end monomers). For a linear polymer consisting of $n+1$ monomers and n bonds, we denote the position of monomer j by $\mathbf{r}_j$, where $j = 0, 1, 2, \ldots, n$, and the full set of these co-ordinate by $\{\mathbf{r}_j\}$. For convenience, we take the position of monomer 0 as the origin of the co-ordinate system, so $\mathbf{r}_0 \equiv \mathbf{0}$. We define the bond vectors by $\Delta\mathbf{r}_j = \mathbf{r}_j - \mathbf{r}_{j-1}$ and represent the full set of bond vectors by $\{\Delta\mathbf{r}_j\}$. The two coordinate sets $\{\mathbf{r}_j\}$ and $\{\Delta\mathbf{r}_j\}$ contain the same information because the position of the monomer 0 is fixed. We use these two representations interchangeably.

A simple way to represent the chain connectivity, the essential aspect of polymers, is through a function, $p(\Delta\mathbf{r})$, which represents the probability density of a bond vector $\Delta\mathbf{r}$. In physical terms this probability density is related to a potential energy function $u(\Delta\mathbf{r})$ through the Boltzmann distribution,

$$p(\Delta\mathbf{r}) = \exp(-\beta u(\Delta\mathbf{r})) \ , \tag{2}$$

where $\beta = 1/(k_B T)$. Traditionally, the zero of the energy function u is chosen so that $p(\Delta\mathbf{r})$ is normalized, that is, so that $\int p(\Delta\mathbf{r}) \mathrm{d}^3\Delta\mathbf{r} = 1$. This representation of the chain connectivity suggests a useful representation of the potential energy, $U(\{\mathbf{r}_j\})$,

$$U(\{\mathbf{r}_j\}) = \sum_{j=1}^{n} u(|\mathbf{r}_j - \mathbf{r}_{j-1}|) + W(\{\mathbf{r}_j\}) \ , \tag{3}$$

where the u_j energies only account for the chain connectivity and W contains all of the rest of the potential energy, such as solvent interactions, angular dependencies of backbone bond energies and steric effects. The probability of a polymer having the configuration with the co-ordinates $\{\Delta\mathbf{r}_j\}$, which we denote by $P_{\text{Config.}}(\{\Delta\mathbf{r}_j\})$, follows from the Boltzmann law,

$$P_{\text{Config.}}(\{\Delta\mathbf{r}_j\}) = Z^{-1} \left[\prod_{j=1}^{n} p(\Delta\mathbf{r}_j)\right] \exp(-\beta W(\{\Delta\mathbf{r}_j\}) \ . \tag{4}$$

The normalization factor in (4) is just the configurational partition function of the polymer

$$Z = \int \left[\prod_{j=1}^{n} p(\Delta\mathbf{r}_j)\right] \exp(-\beta W(\{\Delta\mathbf{r}_j\})) \prod_{j=1}^{n} \mathrm{d}^3\Delta\mathbf{r}_j \ . \tag{5}$$

In the simplest situation, $W = 0$. The remaining energy term, the first term in (3), enforces the chain connectivity. When $W = 0$ the probability of a set of bond vectors, $\{\Delta \mathbf{r}_j\}$, given by (4), simplifies to the product of the individual bond vectors. This factorization means that the bond vectors are independent, random variables, so the configuration of the polymer backbone is formally identical to a random walk of n independent steps, each of which has a mean of $\mathbf{0}$, starting at the origin. A polymer model with $W = 0$ is called a **random flight chain**.

We use $P_n(\mathbf{R})$ to denote the probability density of the end-to-end vector of a random walk after n steps, where

$$\mathbf{R} \equiv \sum_{j=1}^{n} \Delta \mathbf{r}_j = \mathbf{r}_n - \mathbf{r}_0 \ . \tag{6}$$

Here $\mathbf{r}_0 = \mathbf{0}$, so that $\mathbf{R} = \mathbf{r}_n$. The magnitude of $\mathbf{R}$, i.e., the distance between monomer 0 and monomer n, provides a crude, but frequently useful, measure of the spatial extent of a polymer chain called the **end-to-end distance**. The vector $\mathbf{R}$ equals the sum of n independent random variables so the probability density of $\mathbf{R}$ when n is large follows from the central limit theorem of probability theory (Weiss (1994)). This theorem states that if $\mathbf{R}$ is the sum of n independent random variables with probability density $p(\Delta \mathbf{r})$, then

$$P_n(\mathbf{R}) = \left(\frac{3}{2\pi n l^2}\right)^{3/2} \exp\left(-\frac{3|\mathbf{R}|^2}{2nl^2}\right) \tag{7}$$

in the limit as $n \to \infty$, where l^2 represents the second moment of p, $l^2 = \int |\Delta \mathbf{r}|^2 p(\Delta \mathbf{r}) \mathrm{d}^3 \Delta \mathbf{r}$. Equation (7) also follows from a saddle point calculation that leads to more accurate results. This calculation shows that the leading order corrections to the Gaussian behavior of $P_n(\mathbf{R})$ go as $1/n^{3/2}$. More details on the central limit theorem and the saddle point calculation may be found in Weiss (1994).

2.3 Gaussian Polymers

In general the expression (7) for the probability density of the end-to-end vector gives accurate results only when n is large. The mathematical expressions for this probability density can become quite complex when n is small. However, there is one polymer model in which (7) holds *exactly* for *all* values of n; this model is called the **Gaussian model**. This model captures the essence of the properties of random flight chains, yet mathematical manipulations with it are simple, so it has found wide use in the study of polymers. The Gaussian model consists of monomers joined to their sequence neighbors by harmonic bonds with a root mean squared length of l. The potential energy of this system of monomers joined by springs is

$$U_{\mathrm{Gauss}} = \frac{3k_{\mathrm{B}}T}{2l^2} \sum_{j=1}^{n} |\mathbf{r}_j - \mathbf{r}_{j-1}|^2 \ . \tag{8}$$

The explicit temperature dependence of the potential energy of the model results from the temperature *independence* of the bond length. For a system with potential energy given by (8) at a temperature T, the probability density of a configuration with bond vectors $\{\Delta\mathbf{r}_j\}$ is given by

$$P_{\text{config.}}(\{\Delta\mathbf{r}_j\}) = \left(\frac{3}{2\pi l^2}\right)^{3n/2} \exp\left(-\frac{3}{2l^2}\sum_{j=1}^{n}|\mathbf{r}_j - \mathbf{r}_{j-1}|^2\right) , \tag{9}$$

which is the product of a set of Gaussian densities. The validity of (7) for all values of n may be verified by substituting (9) for $P_{\text{config.}}$ into the identity

$$P_n(\mathbf{R}) = \int P_{\text{config.}}(\{\Delta\mathbf{r}_j\})\delta^3\left(\mathbf{R} - \sum_{j=1}^{n}\Delta\mathbf{r}_j\right)\prod_{j=1}^{n}\mathrm{d}^3\Delta\mathbf{r} , \tag{10}$$

using the Fourier representation of the three dimensional Dirac delta distribution, $\delta^3(\mathbf{r}) = (2\pi)^{-3}\int \exp(\mathrm{i}\mathbf{k}\cdot\mathbf{r})\mathrm{d}^3\mathbf{k}$, and evaluating the resulting Gaussian integrals. Results derived from the Gaussian model may be used to model the properties of any two monomers in any random flight chain, so long as they are sufficiently far apart, because the asymptotic result (7) is always accurate for large n.

The points that represent the positions of the monomers, $\mathbf{r}_j$, also may be thought of as labeling discrete points on a continuous curve, $\mathbf{r}_j = \mathbf{r}(s_j)$, where $s_j = j\Delta s$. Here s represents the distance along the contour of the continuous curve. Taking the limit as $\Delta s \to 0$ and $n \to \infty$ such that $n\Delta s = L$ (L denotes the fixed length of the polymer) turns the sum in the potential energy in (8) into an integral so the energy of the Gaussian model becomes a function of the *path* $\mathbf{r}(s)$ in the continuous representation. Likewise, the partition function becomes a path integral, so all of the tools developed to study path integrals in quantum mechanics and quantum field theory can be used in polymer physics (Freed (1972) and Doi and Edwards (1986)).

2.4 The Equivalent Chain

Although the Gaussian model has great elegance and intuitive appeal, our discussion of it was based on setting $W = 0$ in the expression (3) for the potential energy. This term accounts for all of the energies except for a crude chain connectivity term. If $W \neq 0$, then the probability density of a polymer configuration $\{\Delta\mathbf{r}_j\}$, given by (4), no longer equals the product of probability densities for each of the bond vectors separately, independent of the other bond vectors. The consequences of these new interactions depends on the **range** of the interaction along the sequence. Accordingly, we write W as

$$W(\{\Delta\mathbf{r}_j\}) = W_{\text{short}}(\{\Delta\mathbf{r}_j\}) + W_{\text{long}}(\{\Delta\mathbf{r}_j\}) \tag{11}$$

where W_{short} represents the **short ranged interactions**, those between monomers close in sequence, and W_{long} represents the **long ranged interactions**, those between monomers distant in sequence. This separation of interactions will prove useful in spite of its somewhat arbitrary character. A good illustration of a short ranged interaction between two adjoining monomers is provided by the restrictions on $\phi - \psi$ angles in Ramachandran plots. Conversely, an illustration of a long ranged interaction is provided by the hydrophobic attraction between two different α-helices.

What happens if we only consider short ranged interactions, setting $W_{\text{long}} = 0$? Suppose the short ranged interactions explicitly affect monomers up to m apart from one another on the sequence and consider two bond vectors, $\Delta\mathbf{r}_i$ and $\Delta\mathbf{r}_j$. Intuitively, if $|i-j| \lesssim m$, then the two bond vectors will be highly correlated; conversely, if $|i-j| \gg m$, then the bond vectors will have negligible correlation. Therefore, considering "chunks" of polymer of sufficient length leads back to independent bond vectors for the "chunks". The properties of the polymer with short ranged interactions and n "chunks" then become formally identical to those of a Gaussian polymer with n monomers. This intuitive notion of a "chunk" of sequence is quantified by the **effective segment**, which is defined below.

We illustrate the effective segment concept by studying a simple **lattice polymer model** which has been used in many computer simulations of polymers in general and proteins in particular. In this model all of the monomers are placed on the sites of a cubic lattice with lattice spacing ℓ. The length of each bond equals the lattice spacing. Two monomers can not occupy the same lattice site in most computer simulations; however, since we are only considering short ranged interactions at present, we will only restrict the bonds not to go back on themselves, i.e., monomer j can not occupy the same site as monomer $j+2$. With this short ranged restriction each bond can have one of 5 possible directions. This restriction may be thought of as a "steric" interaction of the sort that gives rise to Ramachandran plots. We take each of the 5 bond directions to be equally probable.

The average of the scalar product between two bond vectors, $\langle \Delta\mathbf{r}_i \cdot \Delta\mathbf{r}_j \rangle$, provides a measure of the correlations in the bond directions. To calculate these averages for the lattice model, take the axes of the lattice to define the x, y, and z axes of a co-ordinate system and consider a bond vector $\Delta\mathbf{r}_j = +\ell\hat{\mathbf{x}}$. The choice of the $\hat{\mathbf{x}}$ direction is arbitrary. The next bond vector, $\Delta\mathbf{r}_{j+1}$, equals either $+\ell\hat{\mathbf{x}}$, $+\ell\hat{\mathbf{y}}$, $-\ell\hat{\mathbf{y}}$, $+\ell\hat{\mathbf{z}}$, or $-\ell\hat{\mathbf{z}}$, each with probability $1/5$; therefore $\langle \Delta\mathbf{r}_j \cdot \Delta\mathbf{r}_{j+1} \rangle = \ell^2/5$. What about two non-adjacent bonds, say j and $j+2$? If $\Delta\mathbf{r}_{j+1} = +\ell\hat{\mathbf{y}}$, $-\ell\hat{\mathbf{y}}$, $+\ell\hat{\mathbf{z}}$, or $-\ell\hat{\mathbf{z}}$, then $\Delta\mathbf{r}_{j+2} = +\ell\hat{\mathbf{x}}$ or $-\ell\hat{\mathbf{x}}$ with equal probability, so the scalar product with $\Delta\mathbf{r}_j$ vanishes upon averaging. The only contribution to the average occurs when $\Delta\mathbf{r}_{j+1} = +\ell\hat{\mathbf{x}}$ *and* $\Delta\mathbf{r}_{j+2} = +\ell\hat{\mathbf{x}}$, which occurs with probability $(1/5)^2$. Thus, $\langle \Delta\mathbf{r}_j \cdot \Delta\mathbf{r}_{j+2} \rangle = \ell^2/25$. Continuing this line of reasoning leads to the general expression

$$\langle \Delta\mathbf{r}_j \cdot \Delta\mathbf{r}_{j+k} \rangle = \ell^2 \gamma^k \quad , \tag{12}$$

where $\gamma \equiv 1/5$. The essential point in deriving (12) is the need to explicitly consider correlations in the bond directions only over a small, *fixed* sequence length. Beyond this sequence length the correlations decay exponentially. As long as only short ranged interactions, W_{short}, are taken into account, then relations like that of (12) will be valid, though with different values of γ and perhaps a slightly more complex form.

The mean squared end-to-end distance of a polymer has a simple expression in terms of the averages we have just studied,

$$\langle R^2 \rangle = \langle | \sum_{j=1}^{n} \Delta \mathbf{r}_j |^2 \rangle = nl^2 + \sum_{i=1}^{n} \sum_{j=1}^{i-1} \langle \Delta \mathbf{r}_i \cdot \Delta \mathbf{r}_j \rangle \ , \tag{13}$$

where l^2 represents the second moment of $\Delta \mathbf{r}$, as above. The averages in the double sum vanish in the Gaussian model, and generally in random flight models, so $R^2 = nl^2$, as expected. These averages are given by (12) in the lattice model. Since they depend only on $i - j$, the expression (13) for the mean squared end-to-end distance simplifies to

$$\langle R^2 \rangle = n\ell^2 + \ell^2 \sum_{k=1}^{n-1} (n-k)\gamma^k \ , \tag{14}$$

because the second moment $l^2 = \ell^2$ and there are $n - k$ terms with $i - j = k$ in the double sum in (13). The summation over k in (14) equals the sum of a geometric series and the derivative of a geometric series which may be summed to yield $\langle R^2 \rangle = C_n n\ell^2$, where the coefficient

$$C_n = \frac{1+\gamma}{1-\gamma} - \frac{2\gamma(1-\gamma^n)}{n(1-\gamma^2)} \tag{15}$$

is called the **characteristic ratio**.

The characteristic ratio may approach a constant, C_∞, as $n \to \infty$. The rate at which C_n approaches C_∞ is a measure of the stiffness of the polymer. At one extreme, the characteristic ratio of a random flight polymer is $C_n = C_\infty = 1$ for all values of n, whereas, at the other extreme, the characteristic ratio of a rigid rod is $C_n = n$, which never approaches a constant. The characteristic ratio of a lattice polymer approaches a constant, $C_\infty = (1+\gamma)/(1-\gamma)$, when n becomes large because the correlations in the bond vectors fall off sufficiently quickly, in this case exponentially. As we noted above, this exponential decay occurs whenever the polymer has only short ranged interactions; therefore, the characteristic ratio will approach a constant for large n in these situations. This result suggests writing $\langle R^2 \rangle = C_\infty n\ell^2 = n_{\text{eff}} l_{\text{eff}}^2$ where n_{eff} represents the number of effective segments and l_{eff} represents the **effective segment length**. The length of an equivalent segment is also called the **Kuhn length** in the polymer physics literature.

For many purposes a polymer with only short ranged interactions and n bonds of length ℓ has the same statistical mechanical properties as a Gaussian model with n_{eff} bonds of length l_{eff}; this Gaussian model is called the **equivalent chain**. A second equation for the equivalent chain obtains from requiring that the polymer and the equivalent chain have the same length when maximally extended, i.e., $r_{\max} = fn\ell = n_{\text{eff}} l_{\text{eff}}$, where f accounts for restrictions on bond angles which may prohibit the bond vectors from becoming completely parallel to one another. Solving for n_{eff} and l_{eff} gives

$$l_{\text{eff}} = \left(\frac{C_\infty}{f}\right)\ell \tag{16}$$

$$n_{\text{eff}} = \left(\frac{f^2}{C_\infty}\right)n \ . \tag{17}$$

For the lattice polymer model $\gamma = 1/5$, so $C_\infty = 3/2$, and $f = 1$, so a lattice polymer with n bonds of length ℓ is, for large n, equivalent to a Gaussian polymer with $n_{\text{eff}} = 2n/3$ bonds of length $l_{\text{eff}} = 3\ell/2$. In this example the Kuhn length, l_{eff}, is temperature independent because there were no energies in our calculations. Usually l_{eff} is temperature dependent because the probabilities of bond vectors derive from a (temperature dependent) Boltzmann distribution.

2.5 Calculations for Proteins

The equivalent chain concept can be extended to proteins. The characteristic ratio for any polymer can be calculated from (13) for the mean squared end-to-end distance and a knowledge of $\langle \Delta\mathbf{r}_i \cdot \Delta\mathbf{r}_j \rangle$. The calculation of this average was easy for the lattice model because of its symmetry. This calculation is much more complex for proteins because they lack this symmetry. We only give the results; accounts of the methods can be found in Cantor and Schimmel (1980), Chap. 18 and especially in Flory (1969). Calculations using semi-empirical potential energy functions with sequences of natural globular proteins give a characteristic ratio of $C_\infty \approx 6$ (Miller and Goebel (1968)). Unlike the lattice model, where $f = 1$, the zig-zag of the (real) bonds in the backbone prevent the protein from being fully extendible; for a protein backbone $f \approx .95$. Therefore, the effective segment length of a typical globular protein $l_{\text{eff}} \approx 6.3$ amino acids. These calculations also show that the presence of glycine and proline residues in particular have a large effect on the characteristic ratio. The characteristic ratio of homopolymers of most of the twenty amino acids is about 8 to 9; however, the addition of a small number of glycines to the sequence diminishes the characteristic ratio substantially. These calculations, and the effects predicted by them, agree with measurements of the size of denatured proteins (Miller and Goebel (1968)).

The problem of interpreting computer simulations of proteins that use the lattice polymer model illustrates the utility of the effective segment concept. Such computer simulations have yielded valuable insights into protein behavior (Bryngelson et al. (1995) and Dill et al. (1995)). Interpreting these studies requires knowing the number of amino acids which are represented by one bond of the lattice model. One simple answer to this question is to require the protein and the lattice polymer to have the same number of effective segments in their equivalent chains. Recall that an effective segment of protein has ≈ 6.3 amino acids and that one bond in a lattice polymer represents two-thirds of an effective segment; therefore, one bond represents $\approx (2/3)(6.3) \approx 4.2$ amino acids. Care must be exercised in interpreting this number; it was derived by considering only the short range interactions and the equilibrium properties of the molecule. A different ratio of amino acids to lattice bonds may be appropriate when long range interactions are included or when dynamic effects are studied. However, the ratio derived above provides a conceptually straightforward starting point for thinking about these computer simulations.

2.6 Excluded Volume Effects

What happens when long range interactions are taken into account, so $W_{\text{long}} \neq 0$? This question is profoundly difficult; methods to answer it with rigor and in detail are still being developed. However, simple arguments can provide many significant insights. Here we will illustrate a simple, approximate argument of Flory to estimate the effect of long ranged interactions on the spatial extent, or size, of a polymer (Flory (1953) Chap. 14)

An important class of long range interactions in polymers result from **excluded volume** interactions. This name comes from the effect of the hard core repulsive interactions that we mentioned earlier in the discussion of van der Waals interactions. Two atoms can not occupy the same space, so the volume occupied by one atom is excluded from occupation by other atoms.

There are two major contributions to the free energy of a polymer with repulsive long range interactions, the entropy of swelling and the repulsive energy. A simple estimate of the entropy of swelling the polymer to a size R follows from using the end-to-end distance to measure the size of the polymer and taking the logarithm of the expression for the number of configurations in (7)

$$S(R) \approx k_{\mathrm{B}} \log P_n(R) = S(0) - \frac{3k_{\mathrm{B}}R^2}{2Nl^2} \ . \tag{18}$$

The quantity $-TS(R)$ is often called the elastic term in the free energy. A simple estimate of the repulsive energy follows from ignoring the correlations in the positions of the monomers; then the repulsive energy will be directly proportional to the number of monomers times the density of monomers,[3] yielding an estimate of the repulsive energy as a function of size,

$$E_{\text{rep}}(R) \approx \frac{k_{\mathrm{B}}Tan^2}{R^3} \ , \tag{19}$$

where a is a constant of proportionality representing the excluded volume. The appearance of $k_{\mathrm{B}}T$ in (19) is solely for the sake of convenience. Combining these estimates gives a free energy of

$$F(R) \approx E_{\text{rep}} - TS = k_{\mathrm{B}}T\left(\frac{an^2}{R^3} + \frac{3R^2}{2n}\right) \ , \tag{20}$$

[3] For the purposes of making these simple estimates we will ignore the surface terms and the difference between n and $n+1$ in the ideal polymer expressions.

where terms that are constant with respect to R have been ignored. Minimizing this free energy with respect to R by setting $\mathrm{d}F/\mathrm{d}R = 0$ yields a relation between the number of monomers in a polymer with long range repulsive interactions and its size, $R \propto n^{\nu}$, where $\nu = 3/5$. The exponent ν is called the Flory exponent. The proportionality relation with $\nu = 3/5$ agrees amazingly well with experiments and computer simulations.

2.7 Polymer Collapse

Functioning globular proteins are collapsed; yet so far in this section we have only discussed open, extended polymers. For a polymer to collapse, an attractive force must exist between monomers, that is, the excluded volume constant, a, must become negative. A free energy that included only an attractive energy and an elastic term would predict that the polymer would collapse to a point. Therefore, the free energy must include a repulsive three-body interaction term in order to prevent this catastrophe. At this point the reader may be somewhat confused. Doesn't a represent the *excluded* volume, that is, an intrinsically repulsive interaction? If fact, a really represents the first term in a perturbative expansion of the free energy, in the density of monomers, about the free energy of an ideal polymer, closely analogous to the virial expansion for the pressure of a gas. Just as the first coefficient of a virial expansion can become negative, so a can become negative. The next term in the expansion we denote by a_3; it represents the free energy contribution of three distant (along the sequence) monomers interacting simultaneously, just as the equivalent term in the virial expansion represents the contribution of three body collisions to the pressure. We take $a_3 > 0$. The free energy contribution of this new term scales as $E_3 \propto n\rho^2$, so the free energy (20) is modified to

$$F(R) \approx k_{\mathrm{B}}T\left(\frac{an^2}{R^3} + \frac{a_3 n^3}{R^6} + \frac{3R^2}{2n}\right) . \tag{21}$$

Minimizing F now yields two regimes separated by an intermediate state. (Bryngelson and Wolynes (1990)) For $a > 0$, the polymer swells and $R \propto n^{3/5}$ as before. For $a < 0$, the polymer collapses and $R \propto n^{1/3}$, that is, it becomes a compact globule. For $a = 0$, the polymer resembles an ideal polymer again and $R \propto n^{1/2}$. The point where $a = 0$ is called the **Θ-point** and is analogous to the Boyle point in gases, where the competition between the effects of the hard core repulsion and the effects of the attractive long-ranged force cause the second virial coefficient to vanish so the system behaves like an ideal gas. [4] This treatment of polymer collapse is not quite right because R, the end-to-end distance, is a poor measure of size for a collapsed polymer. A better measure of size is the **radius of gyration** , R_{G}, defined by $R_{\mathrm{G}}^2 \equiv (n+1)^{-1}\sum_j |\mathbf{r}_j - \mathbf{R}_{\mathrm{cm}}|^2$ for a polymer with $n+1$ monomers, where $\mathbf{R}_{\mathrm{cm}} \equiv (n+1)^{-1}\sum_j \mathbf{r}_j$, represents the polymer's center of mass. A treatment of polymer collapse using the radius of gyration is given by

[4] However, a polymer at the Θ-point is not quite ideal because the higher order terms in the virial expansion are non-zero.

Birshtein and Pryamitsyn (1991); the results are in qualitative agreement with the results above.

3 The Helix-Coil Transition

In Sect. 2 we did not mention any local structures. Yet, in Sect. 1.5 we emphasized the importance of such local structures as the α-helix and β-sheet in understanding the structure and stability of proteins. Perhaps the simplest of these local structures is the α-helix. Now we discuss the statistical mechanics of α-helix formation.

3.1 Alpha Helices and Cooperativity

A sketch of the hydrogen bond pattern of an α-helix is shown in Fig. 9. Notice

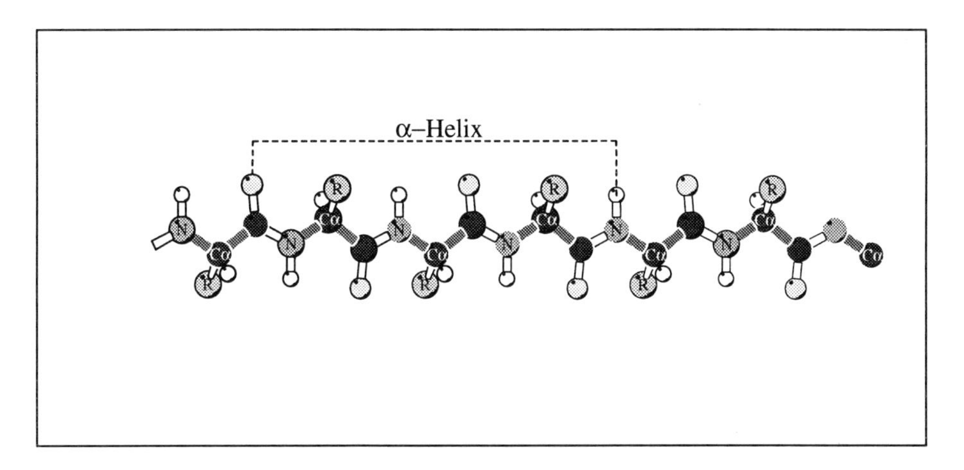

Fig. 9. Schematic drawing of the backbone hydrogen bonds in an α-helix. The dashed line represents a single α-helix hydrogen bond. The mobile bonds are shaded gray. The dihedral angles associated with the 6 mobile bonds that are under the dashed line must lie in the α-helical part of the Ramachandran plot in order for the indicated helix hydrogen bond to form.

that 3 sets of $\phi - \psi$ angles, 6 dihedral angles in all, must lie in the α-helical part of the Ramachandran plot in order for the *first* backbone hydrogen bond to form. However, forming an adjacent hydrogen bond requires only 1 more set of $\phi - \psi$ angles, 2 dihedral angles, to lie in the α-helical region. Thus, the entropy cost of forming the first hydrogen bond is three times greater than the cost of forming subsequent, adjacent hydrogen bonds. This difference in entropy cost means that α-helix formation is expected to be a co-operative transition and that very short helices should be rare.[5]

3.2 Lifson-Roig Model

A simple model, often called the **Lifson-Roig** model, captures the essential equilibrium effects of the co-operative hydrogen bonding in α-helices. Define a residue to be in a helix state if it resides in the α-helix portion of the Ramachandran plot, show in Fig. 4. Define a residue to be in a coil state if it resides in any other portion of the Ramachandran plot. Denote the free energies of residue j existing in a helix and a coil state by f_j^{h} and f_j^{c}, respectively. In terms of $E_j(\phi,\psi)$, the energy of residue j as a function of its backbone angles,

$$f_j^{\mathrm{h}} = -k_{\mathrm{B}}T \log \iint_{helix} \exp[-\beta E_j(\phi,\psi)] \mathrm{d}\phi \mathrm{d}\psi \tag{22}$$

$$f_j^{\mathrm{c}} = -k_{\mathrm{B}}T \log \iint_{coil} \exp[-\beta E_j(\phi,\psi)] \mathrm{d}\phi \mathrm{d}\psi \ , \tag{23}$$

where the subscripts on the integrals indicate integration over the helix and coil portions of ϕ-ψ space, respectively. The values of f_j^{h} and f_j^{c} will be different for each of the 20 amino acids. Finally, denote the average energy of the backbone hydrogen bonds in an α-helix by J. The first and last amino acid in a polypeptide chain do not participate in α-helical hydrogen bonds because the first amino acid has no peptide NH group and the last amino acid has no peptide CO group. We will specify the conformation of a polypeptide by specifying whether or not each residue is in a helix state. To this end, define the conformation variable μ_j by $\mu_j = 1$ if residue j is in a helix state and $\mu_j = 0$ if it is in a coil state. The Lifson-Roig theory studies an effective energy function $E(\{\mu_j\})$ equal to the difference between the free energy of the conformation $\{\mu_j\}$ and the all coil conformation for a polypeptide of N residues,

$$E_N(\{\mu_j\}) = \sum_{j=1}^{N} \epsilon_j \mu_j + J \sum_{j=2}^{N-3} \mu_j \mu_{j+1} \mu_{j+2} \ , \tag{24}$$

where $\epsilon_j \equiv f_j^{\mathrm{h}} - f_j^{\mathrm{c}}$.

The energy $E_N(\{\mu_j\})$ is really an effective free energy; all of the degrees of freedom that are not in the model have been thermally averaged in the manner of the expressions (22) and (23) for the free energies of the helix and coil states. In fact, the function $E_j(\phi,\psi)$ is also really an effective energy because it implicitly averages over the sidechain and solvent degrees of freedom. Such effective energy

[5] Although the helix-coil transition is sharp, it is not a true phase transition in the physicists sense because the system is one-dimensional and the interactions are short-ranged. In such situations a true phase transition is impossible (Landau and Lifshitz (1980) Sect. 163). However, a true one-dimensional phase transition may exist for the helix-coil transition in DNA, where some long-ranged effects come into play (Poland and Scheraga (1970) Chap. 10 D and papers 39 and 40).

functions form the substance of most models in condensed matter physics. These are often used without any comments about the implicit averaging over neglected degrees of freedom; we have commented on this averaging here because proteins have many levels of structure, so the degrees of freedom explicitly included in a useful model of one phenomenon may be averaged in an equally useful model of another phenomenon. The procedure of thermally averaging over the neglected degrees of freedom illustrated above is constructed so that the effective energy produces correct Boltzmann weights for the entire system when it is used as the energy function in a Boltzmann distribution.

3.3 Transfer Matrix Method of Calculating the Partition Function

The Lifson-Roig model is really just a thinly disguised Ising modeli. To see this connection, define $s_j = 2\mu_j - 1$; then s_j is an Ising variable, taking on only the values $+1$ and -1 and the effective energy in (24) becomes a one dimensional Ising Hamiltonian with 1, 2, and 3 spin interactions. This connection means that the thermal properties of the Lifson-Roig model can be found using methods developed for Ising models. In particular, the transfer matrix method has proven to be of great value in the study of the helix-coil transition.

We will use a slightly simplified version of the Lifson-Roig energy function to illustrate the basic idea of the transfer matrix method,

$$\tilde{E}_N(\{\mu_j\}) = \epsilon \sum_{j=1}^{N} \mu_j + J \sum_{j=1}^{N-1} \mu_j \mu_{j+1} \quad . \tag{25}$$

This model has only nearest neighbor interactions, has the same value of ϵ for all the residues, and has no special end effects. But it still contains the co-operativity that is the hallmark of the helix-coil transition. This model will allow us to focus on the main ideas of the helix-coil transition of the matrix method without unnecessary algebraic complications.

To begin we define Z_N to represent the partition function of a polypeptide of N residues, Z_N^{c} to represent the partition function of this polypeptide if the final residue is fixed in a coil state, and Z_N^{h} to represent the partition function if the final residue is fixed in a helix state. By definition, $Z_N = Z_N^{\mathrm{c}} + Z_N^{\mathrm{h}}$, which may be written as the inner product of two vectors,

$$Z_N = (Z_N^{\mathrm{c}}, Z_N^{\mathrm{h}}) \begin{pmatrix} 1 \\ 1 \end{pmatrix} \quad . \tag{26}$$

The essential idea of the matrix method is to find a 2×2 matrix $\mathbf{M}$ with the property $(Z_{k+1}^{\mathrm{c}}, Z_{k+1}^{\mathrm{h}}) = (Z_k^{\mathrm{c}}, Z_k^{\mathrm{h}})\mathbf{M}$; then calculating the partition function reduces to finding $(Z_1^{\mathrm{c}}, Z_1^{\mathrm{h}})$, multiplying it by a product of $\mathbf{M}$ matrices, and using (26) for Z_N. To find this matrix, we first define $\{\mu_j\}_k$ to represent $\{\mu_1, \mu_2, \ldots, \mu_k\}$ and note that $\tilde{E}_{k+1}(\{\mu_j\}_{k+1}) = \epsilon\mu_{k+1} + \epsilon\mu_k + J\mu_{k+1}\mu_k + J\mu_k\mu_{k-1} + \tilde{E}_{k-1}(\{\mu_j\}_{k-1})$. Therefore, the partial partition function Z_{k+1}^{h} can be written as

$$Z_{k+1}^{\mathrm{h}} = \sum_{\{\mu_j\}_k} \exp(-\beta\epsilon - \beta\epsilon\mu_k - \beta J\mu_k - \beta J\mu_k\mu_{k-1} - \beta\tilde{E}_{k-1}(\{\mu_j\}_{k-1})) \quad , \tag{27}$$

so breaking the above sum into two terms, the first with $\mu_k = 1$ and the second with $\mu_k = 0$ yields

$$Z_{k+1}^{\mathrm{h}} = \exp(-\beta\epsilon - \beta J) Z_k^{\mathrm{h}} + \exp(-\beta\epsilon) Z_k^{\mathrm{c}} \ . \tag{28}$$

The partial partition function Z_{k+1}^{c} can be analyzed in the same way, yielding

$$Z_{k+1}^{\mathrm{c}} = Z_k^{\mathrm{h}} + Z_k^{\mathrm{c}} \ . \tag{29}$$

The elements of the matrix $\mathbf{M}$ may be found by combining the recursive expressions (28) and (29) for the partial partition functions Z_{k+1}^{h} and Z_{k+1}^{c}.

A standard notation in the helix-coil transition literature defines $s = \exp(-\beta\epsilon -\beta J)$ and $\sigma = \exp(+\beta J)$. The quantity s is often called the **propagation parameter** because it represents the statistical weight of adding one more residue to a helix that already exists, while the quantity σ is often called the **nucleation parameter** because it represents the statistical weight of forming a single residue of helix in a coil region. Both of these weights are relative to the statistical weight of adding a residue in the coil state. In this notation

$$\mathbf{M} = \begin{pmatrix} 1 & \sigma s \\ 1 & s \end{pmatrix} \ . \tag{30}$$

For a single residue $Z_1^{\mathrm{c}} = 1$ and $Z_1^{\mathrm{h}} = \sigma s$, so the matrix expression $(1,0)\mathbf{M} = (1, \sigma s)$, can be used to start the recursive matrix calculation of Z_N, leading to the desired expression for the partition function in terms of a matrix product,

$$Z_N = (1,0)\mathbf{M}^N \begin{pmatrix} 1 \\ 1 \end{pmatrix} \ , \tag{31}$$

where the matrix $\mathbf{M}$ is defined above in (30).

Having obtained the partition function (31), we may now use the properties of matrices to calculate the properties of this simple helix-coil transition model. In particular, if Λ represents an $n \times n$ diagonal matrix with diagonal elements $\lambda_1, \lambda_2, \ldots, \lambda_n$, then the matrix Λ^m is also diagonal and has diagonal elements $\lambda_1^m, \lambda_2^m, \ldots, \lambda_n^m$. The 2×2 matrix $\mathbf{M}$ is not diagonal, but is easily transformed into a 2×2 diagonal matrix by a similarity transform, $\Lambda = \mathbf{T}^{-1}\mathbf{M}\mathbf{T}$, where $\mathbf{T}$ represents a 2×2 matrix, $\mathbf{T}^{-1}$ represents its inverse, and the diagonal elements of Λ equal the eigenvalues of $\mathbf{M}$. These eigenvalues are $\lambda_1 = (1/2)[(1+s) + \sqrt{(1-s)^2 - 4s\sigma}]$ and $\lambda_2 = (1/2)[(1+s) - \sqrt{(1-s)^2 - 4s\sigma}]$. The transform matrix $\mathbf{T}$ solves the underdetermined set of homogeneous linear equations $\mathbf{T}\Lambda = \mathbf{T}\mathbf{M}$.[6] A solution of this of equation is

$$\mathbf{T} = \begin{pmatrix} 1-\lambda_2 & 1-\lambda_1 \\ 1 & s \end{pmatrix} \tag{32}$$

[6] The requirement that this equation has a non-trivial solution, $\det(\mathbf{T}\Lambda - \mathbf{T}\mathbf{M}) = 0$, is equivalent to the requirement that the diagonal entries of Λ equal the eigenvalues of $\mathbf{M}$.

which has the inverse

$$\mathbf{T}^{-1} = \frac{1}{\lambda_1 - \lambda_2} \begin{pmatrix} 1 & \lambda_1 - 1 \\ -1 & 1 - \lambda_2 \end{pmatrix} \ . \tag{33}$$

With these relations, and noting that $\mathbf{M}^N = (\mathbf{T}\mathbf{T}^{-1}\mathbf{M}\mathbf{T}\mathbf{T}^{-1})^N = \mathbf{T}\Lambda^N\mathbf{T}^{-1}$, the expression for the partition function (31) may now be written as

$$\begin{aligned} Z_N &= (1,0)\mathbf{T}\Lambda^N\mathbf{T}^{-1}\begin{pmatrix} 1 \\ 1 \end{pmatrix} \\ &= (1-\lambda_2, 1-\lambda_1)\begin{pmatrix} \lambda_1^N & 0 \\ 0 & \lambda_2^N \end{pmatrix}\begin{pmatrix} \lambda_1/(\lambda_1-\lambda_2) \\ -\lambda_2/(\lambda_1-\lambda_2) \end{pmatrix} \\ &= \frac{1}{\lambda_1 - \lambda_2}[\lambda_1^{N+1}(1-\lambda_2) - \lambda_2^{N+1}(1-\lambda_1)] \ . \end{aligned} \tag{34}$$

Since $\lambda_1 > \lambda_2$, for $N \gg 1$, which is the relevant case for the helix-coil transition, $Z_N \approx \lambda_1^N$. The free energy per residue, $f = -k_{\mathrm{B}}T(1/N)\log Z_N$, for large N is given by

$$f = -k_{\mathrm{B}}T \log\{\frac{1}{2}[(1+s) + \sqrt{(1-s)^2 - 4s\sigma}] \ , \tag{35}$$

where we have substituted the expression for λ_1 in terms of s and σ into the above free energy (35).

The properties of this model may be calculated by analogy with the Ising model of magnetism. Perhaps the property of greatest interest is the average fraction of residues in helix, that is, the **helix fraction**, θ. The magnetic analog of the helix fraction is the average magnetization and, as is clear by examining the expressions (24) and (25) for the effective energy of a configuration[7], the parameter ϵ is the analog of the magnetic field. Continuing the analogy, the helix fraction is related to the free energy by $\theta = \partial f/\partial\epsilon$. Using this relation in conjunction with (35) for the free energy, the definitions of s and σ, and the chain rule for derivatives, yields a simple expression for the helix fraction as a function of the parameters s and σ,

$$\theta = \frac{s}{2\lambda_1(s,\sigma)}\left[1 - \frac{1 - s - 2\sigma}{\sqrt{(1-s)^2 + 4s\sigma}}\right] \ . \tag{36}$$

Similar calculations can be done with, e.g., the Lifson-Roig model. Even more importantly, these calculations lead to testable results.

This matrix method for calculating the partition function can be extended to include cases where ϵ and J vary from residue to residue. In these cases the matrix method becomes a technique for fast calculation of the partition function because the amount of computer time required is proportional to N, whereas the amount of computer time required for a brute force calculation of the partition function that sums the Boltzmann factors for all of the states is proportional to 2^N. See Bryngelson et al. (1990) for an application of the matrix method to rapid computation of partition functions on a parallel computer.

[7] These equations are the analog of the Hamiltonian in the Ising model of magnetism.

3.4 Comparison with Experiments

The theories of the helix-coil transition can be tested in experiments. Many of these tests involve the polarization of ultraviolet light. Most biological molecules are **optically active**, that is, they affect the polarization of light. When light goes through an optically active material the amount of light absorbed in the process depends on whether the light is right hand circularly polarized or left hand circularly polarized. This effect is called **circular dichroism** or **CD** for short (Cantor and Schimmel (1980) Chap. 8). The CD for light with wavelengths between about 1900 Å to about 2300 Å is very sensitive to the presence of α-helices and therefore it is used to measure the helix fraction of proteins and polypeptides in solution. The results of CD measurements agree quite will with the theories of the helix-coil discussed above (Cantor and Schimmel (1980) Chap. 20).

4 Application to Protein Structure Prediction

4.1 Introduction

Sections 2 and 3 of this chapter stand in sharp contrast with the Sect. 1. Section 1 discussed detailed interatomic interactions and protein structures. Sections 2 and 3 discussed the statistical mechanics of simple models of polymers and helices. This section brings together these two approaches to understanding proteins and their structures; it discusses an application of statistical mechanics to understanding the prediction of protein tertiary structure from protein primary structure, i.e., amino acid sequence.

In general, the determination of the amino acid sequence of a protein, i.e., the primary structure, is relatively easy, whereas the determination of the three dimensional spatial structure of a protein, i.e., the secondary and tertiary structure, is extremely difficult. The spatial structure of a protein, plus its dynamics, determines what a protein *does*, so this situation is unfortunate. The structure contains far more information than the sequence alone and much of this information is useful.[8] Knowing the structure is vital to many practical applications. An example of a biomedical application of structural knowledge occurs in the problem of drug design. Most drug targets are proteins and most drugs bind to a specific pocket in the target molecule to render it inactive. The more specific the binding, the fewer the side effects caused by the drug. Therefore, knowing the structure of a protein drug target can help in the design of an effective drug with few side effects. Similarly, knowledge of the protein structure is necessary for a fundamental understanding of life at a molecular level.

4.2 Potential Functions and the Potential Accuracy Problem

However, the one dimensional protein sequence generally *determines* the three dimensional protein structure, as we mentioned in Sect. 1.5. All of the information needed to construct the structure must somehow be contained in the

[8] By structure, we mean the placement of all of the atoms, except possibly the hydrogens, so that the structure includes the sequence information.

sequence. Therefore, there must exist some algorithm for determining the structure from the sequence. The search for this algorithm is the **protein structure prediction problem**, often called the **protein folding problem**. This problem is one of the most important problems in molecular biophysics. Its importance derives from the power of the knowledge available in protein structures and the relative ease of sequence determination. It will be discussed more fully in the chapter by P. G. Wolynes in this book, so we will only give a brief introduction here.

The general strategy used for predicting protein structures consists of two steps. First, one obtains an effective energy function, usually referred to as a **potential function**, which represents the free energy of the protein plus solvent system as a function of protein conformation. This potential function could come from an analysis of the physical forces involved in protein structure (we discussed these forces in Sect. 1.4), or from a statistical analysis of the database of know protein structures. Second, one finds the structure that minimizes this potential function for the desired sequence. This structure is the predicted structure. Unfortunately, the minimization is difficult, because the potential function always has many minima, and therefore requires large amounts of computer time and very intelligent algorithms. The search for good methods for carrying out the minimization step is a subject of current research interest.

However, this minimization calculation is useless if the potential function is bad. Since proteins *do* fold in nature, a potential function that was a perfectly accurate representation of the potential function used by nature would lead to good structures once the minimization problem was solved. This perfectly accurate potential function is unknown, so we must content ourselves using some approximate potential function. Therefore, we must ask: How accurate must the potential be to obtain reliable predictions? This question is the **potential accuracy problem**. The importance of this problem derives from the extreme importance of the structure prediction problem.

4.3 A Simple Potential Accuracy Calculation

Consider the simplest possible case: one has two alternative structures, labeled 0 and 1, with calculated energies E_0' and E_1' respectively. The real energies of these states, as determined by nature's potential function, we call E_0 and E_1 respectively. We take $E_1' > E_0'$ and wish to know $R(E_0', E_1')$, the probability that $E_1 > E_0$, that is, the probability that the prediction that state 0 has a lower energy is correct. The relation between the real and calculated energies can be written as $E_0' = E_0 + \delta E_0$ and $E_1' = E_1 + \delta E_1$, where δE_0 and δE_1 represent the errors in the calculated energies of structures 0 and 1. The requirement for a correct prediction, $E_1 > E_0$, may now be written as $E_1' - E_0' > \Delta E$, where we haved defined $\Delta E \equiv \delta E_1 - \delta E_0$. If we denote the probability density of ΔE, by $P(\Delta E)$, and the energy gap between the two calculated energies by

$E'_{\text{gap}} \equiv E'_1 - E'_0$, then the probability of predicting the correct structure is

$$R(E'_0, E'_1) = \int_{-\infty}^{E'_{\text{gap}}} P(\Delta E) \mathrm{d}(\Delta E) \ . \tag{37}$$

To complete the calculation of the probability of predicting a correct structure for this simple case, we must express $P(\Delta E)$ in terms of the probability density of the errors in the energies of the individual structures, $p(\delta E)$. No general connection exists between these two probability densities. For example, if the errors in E'_0 and E'_1 are highly correlated, then $\delta E_0 \approx \delta E_1$ and $P(\Delta E)$ will be sharply peaked at $\Delta E = 0$, regardless of the form of $p(\delta E)$. If the errors in the two calculated energies are independent then

$$P(\Delta E) = \int_{-\infty}^{+\infty} p(\delta E) p(\Delta E + \delta E) \mathrm{d}(\delta E) \ . \tag{38}$$

We call this simplest case the **independent error assumption** . The independent error assumption is a worst case because correlations in errors tend to narrow $P(\Delta E)$, as noted above. The most important correlations in errors result from correlations in structure. Similar structures have many interactions in common. The errors in the calculated energies of these common interactions will be identical and therefore not contribute to the difference in errors $\Delta E = \delta E_1 - \delta E_0$. Only parts of the structures which differ will contribute to the difference between the errors in the energies.

4.4 The Random Heteropolymer Model

The random heteropolymer model is the simplest model of a protein. It was first proposed, with some important additions, as a model for proteins in Bryngelson and Wolynes (1987). In the random heteropolymer model the conformation is specified by the set of contacts $\{\Delta(i,j)\}$, where $\Delta(i,j) = 1$ if amino acid residues i and j are in contact and 0 otherwise. This specification of conformation establishes the positions of the C^αs with a precision of about 1 Å for a typical protein (Bryngelson (1994)). The energy, E, of this conformation is given by

$$E(\{\Delta(i,j)\}) = \sum_{i<j} B_{i,j} \Delta(i,j) \ , \tag{39}$$

where the $B_{i,j}$ are random variables whose probability density is

$$p_{\text{contact}}(B_{i,j}) = (1/\sqrt{2\pi B^2}) \exp(-B_{i,j}^2 / 2B^2) \ . \tag{40}$$

The standard deviation, B, in (40) sets the scale of contact energies.[9]

[9] In principle, the probability density in (40) should be centered about some $B_0 < 0$, to induce collapse, and the energy expression (39) should contain a three body term, $C \sum \Delta(i,j)\Delta(j,k)$, with $C > 0$, to prevent collapse to a single point. The terms containing B_0 and C would add a constant energy to each of the collapsed conformations. However, here (Sect. 4) we will be calculating only the *relative* energies of different *collapsed* conformations, so we will consider only collapsed configurations with excluded volume in our calculations and set $B_0 = C = 0$ for simplicity.

The properties of the random heteropolymer model were first discussed by Bryngelson and Wolynes (1987) within a random energy approximation which assumes that the energies of the random heteropolymer are statistically independent. The later work of Shakhnovich and Gutin (1989) and others have related this approximation to more conventional mean field approximations. A particularly simple and attractive derivation of the random energy approximation as a high dimension limit is given in Garel et al. (1995). We will use four properties of the random energy approximation for the random heteropolymer:

1. the number of compact structures scales exponentially with the number of residues, i.e., there are $\Omega = \nu^N$ compact structures, where ν is approximately independent of N;
2. the energies of different structures are independent random variables;
3. the probability density that a structure has energy E is

$$\rho(E) = (1/\sqrt{\pi N z B^2}) \exp(-E^2/NzB^2) \ ; \tag{41}$$

4. the low energy structures have few contacts, and therefore few interactions, in common.

Errors in the potential function are modeled by adding random noise to each of the contact energies, so

$$E'(\{\Delta(i,j)\}) = \sum_{i<j} B'_{i,j} \Delta(i,j) \ , \tag{42}$$

where $B'_{i,j} = B_{i,j} + \eta_{i,j}$ and $\eta_{i,j}$, the noise, is a random variable with mean 0 and standard deviation η. The error in calculating the energy of the conformation $\{\Delta(i,j)\}$ is

$$\delta E(\{\Delta(i,j)\}) = \sum_{i<j} \eta_{i,j} \Delta(i,j) \ . \tag{43}$$

Denoting by z the average number of contacts each amino acid residue has with other residues, the quantity δE consists of the sum of $(1/2)zN$ independent random variables, so the central limit theorem, (7), implies that δE is a random variable whose probability density is $p(\delta E) = (1/\sqrt{\pi N z \eta^2}) \exp(-\delta E^2/N z \eta^2)$. The expression for the total error in (43) has the same form as the random heteropolymer model, so by property 2 above the errors (the δE) are random independent variables. Therefore, (38) can be used to calculate $P(\Delta E)$, the probability density of the difference in the error in two structures, yielding

$$P(\Delta E) = 1/(\sqrt{2\pi N z \eta^2}) \exp(-\delta E^2/2Nz\eta^2) \ . \tag{44}$$

4.5 Accuracy Calculation for Random Heteropolymers

In Sect. 4.3 we calculated the probability of predicting a correct structure from two alternatives as a function of the energy gap between these structures. In this section we calculate this probability for the random heteropolymer model in the limit of small inaccuracies. For small inaccuracies in the potential function the correct structure will, with high probability, be one of the two structures with the lowest calculated energies. Therefore, for sufficiently accurate potential functions the results of Sect. 4.3 for two structures can be applied to the random heteropolymer once the energy gap between the two lowest energy structures is known.

Consider a random heteropolymer with an energy gap of E_{gap} between its two lowest energy structures. To first order in η, the inaccuracy of the potential function, what is the probability of a correct prediction? Since, in this limit, the probability of an incorrect prediction is dominated by the probability of confusing the lowest energy structure with the next lowest energy structure, we can use (37) for $R(E_{\mathrm{gap}})$ for the probability of a correct prediction with two structures. Substituting (44) for the distribution of ΔE into (37) gives

$$R(E_{\mathrm{gap}}) = (1/2)[1 + \mathrm{erf}(E_{\mathrm{gap}}/\sqrt{2Nz\eta^2})] \ , \tag{45}$$

where $\mathrm{erf}(x)$ represents the error function

$$\mathrm{erf}(x) = \frac{2}{\sqrt{\pi}} \int_0^x e^{-t^2} \mathrm{d}t \ . \tag{46}$$

However, calculating E_{gap} exactly requires calculating the energies of all the structures, which may not be feasible if Ω is large, and for protein models is in general infeasible. Similarly, one may wish to investigate the accuracy requirements for predicting the structure of a large class of molecules, such as globular proteins. The energy gap E_{gap} is a random variable in the random heteropolymer model. We denote the probability density of the energy gap by $f(E_{\mathrm{gap}})$. If $f(E_{\mathrm{gap}})$ is known, then we can calculate the average of the probability of predicting a correct structure, $\overline{R} = \int f(E_{\mathrm{gap}})R(E_{\mathrm{gap}})dE_{\mathrm{gap}}$.

To calculate $f(E_{\mathrm{gap}})$ we first consider the probability density that a realization of the model has an energy gap larger than E_{gap}, which we denote $F(E_{\mathrm{gap}})$. Since the energies of the model are independent random variables (property 2 above),

$$F(E_{\mathrm{gap}}) = \Omega \int_{-\infty}^{+\infty} \rho(E_0 - E_{\mathrm{gap}}) \left[\int_{E_0}^{+\infty} \rho(E)\mathrm{d}E \right]^{\Omega-1} \mathrm{d}E_0 \ . \tag{47}$$

The factor of Ω in front of the integral in (47) occurs because the labeling of the states is arbitrary so any of the Ω structures could have the lowest energy (denoted E_0 in (47)). Notice that $F(0) = 1$, as it should. To calculate $F(E_{\mathrm{gap}})$ for the case of large N, we define $\gamma(E_0, E_{\mathrm{gap}}) \equiv \rho(E_0 - E_{\mathrm{gap}})/\rho(E)$ and

$$\mathcal{P}(E_0) \equiv \Omega\rho(E_0) \left[\int_{E_0}^{+\infty} \rho(E)\mathrm{d}E \right]^{\Omega-1} \ . \tag{48}$$

Then the expression (47) for $F(E_{\mathrm{gap}})$ can be written as

$$F(E_{\mathrm{gap}}) = \Omega \int_{-\infty}^{+\infty} \gamma(E_0, E_{\mathrm{gap}})\mathcal{P}(E_0)\mathrm{d}E_0 \ . \tag{49}$$

The function $\mathcal{P}(E_0)$ is precisely the probability that the lowest energy structure of a realization of the random heteropolymer has energy E_0. As such, $\mathcal{P}(E_0)$ is normalized so that its integral over all possible energies equals 1. Furthermore, $\mathcal{P}(E_0)$ is sharply peaked at its maximum value, E_0^*, when N—and therefore $\Omega = \nu^N$—is large. Consequently, $F(E_{\text{gap}}) \approx \gamma(E_0^*, E_{\text{gap}})$ for large N. An estimate of E_0^* follows from a simple argument: given the number of states, Ω, and the probability of a state having energy E, $\rho(E)$, we expect that the lowest energy state has energy $\rho(E_0^*) \approx 1/\Omega$. This equation gives $E_0^* \approx -NB\sqrt{z \log \nu}$ as an estimate of the typical value of E_0. The same result is obtained from a straightforward calculation of the maximum of $\mathcal{P}(E_0)$: by taking its derivative and setting it equal to 0, one obtains the same result up to some finite size corrections. Substituting $E_0^* = -NB\sqrt{z \log \nu}$ into $F(E_{\text{gap}}) \approx \gamma(E_0^*, E_{\text{gap}}) = \rho(E_0^* - E_{\text{gap}})/\rho(E)$ and taking the derivative with respect to E_{gap} gives $f(E_{\text{gap}})$, the probability of a random heteropolymer having an energy gap E_{gap},

$$f(E_{\text{gap}}) = -\frac{\mathrm{d}F}{\mathrm{d}E_{\text{gap}}} = \left(\frac{2\sqrt{\log \nu}}{\sqrt{z}B} + \frac{2E_{\text{gap}}}{NzB^2}\right) \exp\left(-\frac{2\sqrt{\log \nu}E_{\text{gap}}}{\sqrt{z}B} - \frac{E_{\text{gap}}^2}{NzB^2}\right) . \tag{50}$$

The expression for $f(E_{\text{gap}})$ in (50) is well approximated by

$$f(E_{\text{gap}}) = (2\sqrt{\log \nu}/\sqrt{z}B) \exp(-2\sqrt{\log \nu}E_{\text{gap}}/\sqrt{z}B) \tag{51}$$

for large N because typical gaps are of the order $E_{\text{gap}} \sim B$,[10] so the other terms in (50) are negligible in comparison.

Using the expression (51) for $f(E_{\text{gap}})$, the probability of having an energy gap E_{gap}, and the expression (45) for $R(E_{\text{gap}})$ yields

$$\overline{R} = \frac{\sqrt{\log \nu}}{\sqrt{z}B} \int_0^\infty \exp\left(-\frac{2\sqrt{\log \nu}E_{\text{gap}}}{\sqrt{z}B}\right) \left[1 + \operatorname{erf}\left(\frac{E_{\text{gap}}^2}{\sqrt{2Nz\eta^2}}\right)\right] \mathrm{d}E_{\text{gap}} . \tag{52}$$

The easiest way to do the integral in (52) is to notice that it is essentially a Laplace transform of $R(E_{\text{gap}})$, and therefore may be found in standard tables. In any case, the result is

$$\overline{R} = (1/2)[1 + \exp(2N\eta^2 \log \nu/B^2)\operatorname{erfc}(\sqrt{2N \log \nu}\eta/B)] , \tag{53}$$

where $\operatorname{erfc}(x) = 1 - \operatorname{erf}(x)$. Of course, this result is valid only for accurate potential functions, so we expand (53) to first order in η to obtain

$$\overline{R} = 1 - \sqrt{\frac{2 \log \nu}{\pi}} \left(\frac{\sqrt{N}\eta}{B}\right) . \tag{54}$$

This result for small η (or, more precisely, for small $\sqrt{2N \log \nu}\eta/B$) is identical to the small η result of a more rigorous calculation that takes into account all $\nu^N - 1$ possible incorrect predictions in the random heteropolymer model (Bryngelson (1994)). Also, note that since (54) is only the first term in an expansion, it does not imply that $\overline{R}$ is a linear function of η.

[10] This statement may be checked explicitly by calculating the average of E_{gap}, $\overline{E}_{\text{gap}} = (1/2)\sqrt{\pi NzB\nu^N}\operatorname{erfc}(\sqrt{N \log \nu})$. For large N, $\overline{E}_{\text{gap}} = (1/2)\sqrt{z/\log \nu}B + O(B/N)$.

4.6 Implications for Protein Structure Prediction

What does this result (54) mean? The quantities B and η represent the typical magnitudes of the monomer-monomer interactions and the inaccuracies in those interactions, respectively, so their ratio, η/B, represents the relative inaccuracy of the monomer-monomer interactions. The presence of the factor of $\sqrt{N}$ in (54) is not surprising; the number of interactions scale as $\sim N$ so the total error in the energy scales as $\sim \sqrt{N}\eta$. The constant $K \equiv \sqrt{2\log\nu/\pi}$ has only a very weak dependence on ν. For reasonable estimates of ν the constant K is of order 1. Therefore, to predict the structure of a 100 residue random polypeptide with 80% confidence would require a potential function in which the errors in the interactions between residues was less than the order of 2%. Naïvely, this result seems to rule out the possibility of predicting protein structure from sequence in the near future. Remember, the error represented by η is the error in the entire contact energy of two amino acids, an energy which will include all of the forces discussed in the Sect. 1.5 of this chapter. A potential function that could include all of these forces calculated with such accuracy so that the residue-residue interaction were accurate to 1% is well beyond current scientific and technological capabilities.

However, if one steps back and re-examines the meaning of the calculation, then the result (54) is not nearly so pessimistic. First, in our calculations a correct structure means that all of the contacts are correct. Perfect prediction of the contact map is a stringent requirement for a protein structure prediction to meet. As we noted above, this requirement alone fixes the positions of the C^{α}s to about 1Å. If two protein structures have 60% of their contacts in common they are often judged to be structurally similar. Thus, (54) applies to a quality of prediction that no one is hoping for. Second, our calculations assumed a *random* amino acid sequence. Amino acid sequences are obviously not random in the sense that a random sequence would have little probability of performing a useful biological function; however, requirements of folding and stability may also influence amino acid sequence and these influences would affect our result (54).

The issue of random versus non-random sequences deserves more comments. For the random heteropolymer, at least in the random energy approximation, if a structure prediction is wrong at all, then it is totally wrong. By this statement we mean that if we just miss predicting the true lowest energy structure and instead predict the true second lowest energy structure, then the prediction is useless because the low energy structures are completely different, from property 4 above. However, if the low energy structures are structurally similar, as is the case when the **principle of minimal frustration** (see Bryngelson and Wolynes (1987), and Bryngelson et al. (1995)) is applicable (this principle is fully explained in chapter devoted to protein folding), then the accuracy requirement for the potential function is less severe because one does not need to find the single lowest energy structure for an approximate prediction. Unlike the random heteropolymer, a structure with an energy that is slightly higher than the correct structure would be similar to the correct structure. Thus, we end this chapter on a provocatives note by proposing that *any* real success in protein tertiary structure prediction is evidence for non-randomness in the amino acid sequence.

Acknowledgements

We would like to thank Dr. William A. Eaton, Dr. Silvo Franz, Dr. V. Adrian Parsegian, Dr. Sri Sastry, and Dr. George H. Weiss for discussions and assistance, Dr. John A. Hertz for inviting one of us (JDB) to give the lectures at the European-Nordic Summer School in Humlebæk, Denmark, upon which this chapter is based, and Dr. Henrik Flyvbjerg for infinite patience as editor of this book.

References

Anfinsen, C. B. (1973): Principles that Govern the Folding of Protein Chains. Science **181**, 223-230.

Birshtein, T. M., Pryamitsyn, V. A. (1991): Coil-Globule Type Transitions in Polymers. 2. Theory of Coil-Globule Type Transition in Linear Macromolecules. Macromolecules **24**, 1554-1560.

Bryngelson, J. D. (1994): When is a Potential Function Accurate Enough for Structure Prediction? Theory and Application to a Random Heteropolymer Model of Protein Folding. J. Chem. Phys. **100**, 6038–6045

Bryngelson, J. D., Wolynes, P. G. (1987): Spin Glasses and the Statistical Mechanics of Protein Folding. Proc. Natl. Acad. Sci. U.S.A. **84**, 7524–7528

Bryngelson, J. D., and Wolynes, P. G. (1990): A Simple Statistical Field Theory of Heteropolymer Collapse with Application to Protein Folding. Biopolymers **30**, 177–188

Bryngelson, J. D., Hopfield, J. J., Southard, S. N., Jr. (1990): A Protein Structure Predictor Based on an Energy Model with Learned Parameters. Tetrahedron Comput. Methodol. **3**, 129–144

Bryngelson, J. D., Onuchic, J. N., Socci, N. D., Wolynes, P. G. (1995): Funnels, Pathways, and the Energy Landscape of Protein Folding: A Synthesis. Proteins Struct Funct Genet **21**, 167–195

Cantor, C. R., Schimmel, P. R., *Biophysical Chemistry,* Vol. I-III (W. H. Freeman, San Francisco, California)

Chan, H. S. and Dill, K. A. (1991): Polymer Principles in Protein Structure and Stability. Annu. Rev. Biophys. Biophys. Chem. **20**, 447–490

Dill, K. A., Bromberg, S., Yue, K., Fieberg, K. M., Yee, D. P., Thomas, P. D., and Chan, H. S. (1995): Principles of Protein Folding—A Perspective from Simple Exact Models. Protein Science **4**, 561–602

Doi, M., and Edwards, S. F. (1986): *The Theory of Polymer Dynamics* (Oxford University Press, Oxford)

Flory, P. J. (1953): *Principles of Polymer Chemistry* (Cornell University Press, Ithaca, New York)

Flory, P. J. (1969): *Statistical Mechanics of Chain Molecules* (John Wiley, New York, New York)

Frank, H. S. (1958): Covalency in the Hydrogen bond and the Properties of Water and Ice. Proc. R. Soc. London, Ser. A **247**, 481–492

Freed, K. F. (1972): Functional Integrals and Polymer Statistics. Adv. Chem. Phys. **22**, 1–129

Garel, T., Orland, H., and Thirumalai, D. (1995): Analytical Theories of Protein Folding, in R. Elber (ed.), *New Developments in Theoretical Studies of Proteins*, (World Scientific, Singapore)

Landau, L. D. and Lifshitz, E. M. (1980): *Course of Theoretical Physics,* Vol. 5, *Statistical Physics,* Part I 3rd ed., J. B. Sykes and M. J. Kearsley trans., (Pergamon Press, Oxford)

Leikin, S., Parsegian, V. A., and Rau, D. C. (1993): Hydration Forces. Annu. Rev. Phys. Chem. **44**, 369–395.

Miller, W. G., and Goebel, C. V. (1968): Dimensions of Protein Random Coils. Biochemistry **7**, 3925–3935

Poland, D. and Scheraga, H. A. (1970): *Theory of the Helix-Coil Transitions in Biopolymers* (Academic Press, New York, New York)

Radzicka, A. and Wolfenden, R. (1988): Comparing the Polarities of the Amino Acids: Side-Chain Distribution Coefficients Between the Vapor Phase, Cyclohexane, 1-Octanol, and Neutral Aqueous Solution. Biochemistry **27**, 1664–1670.

Shakhnovich, E. I., and Gutin, A. M. (1989): Formation of Unique Structure in Polypeptide Chains: Theoretical Investigation with the Aid of a Replica Approach. Biophys. Chem. **34**, 187-199.

Thiessen, W. E., and Narten, A. H. (1982): Neutron Diffraction of Light and Heavy Water Mixtures as 25 °C.
J. Chem. Phys. **77**, 2656–2662

Watson, J. D., Hopkins, N. H., Roberts, J. W., Steitz, J. A., Weiner, A. M. (1987): *Molecular Biology of the Gene,* 4th ed. (Benjamin-Cummings, Menlo Park, California)

Weiss, G. H. (1994): *Aspects and Applications of the Random Walk* (North-Holland, Amsterdam)

Probing Protein Motion Through Temperature Echoes

Klaus Schulten[1,2], Hui Lu[1,3], and Linsen Bai[1,2]

[1] Beckman Institute, University of Illinois at Urbana-Champaign, Urbana, IL 61801, USA
[2] Department of Physics, University of Illinois at Urbana-Champaign
[3] Department of Nuclear Engineering, University of Illinois at Urbana-Champaign

1 A Word to Theoretical Physicists Contemplating Research in Biology

The intellectual mastering of life at the molecular level naturally attracts physicists whose discipline has been so eminently successful in carrying the torch of mathematics into the natural sciences [1, 2]. But physicists ought to be aware that there are differences as well as similarities in the quests towards physical and biological theory. The most important difference for a theoretical physicist to keep in mind in the pursuit of biological theory is that the latter is about advancing biology and not about advancing physics. A theoretical physicist must also realize that she or he is not as welcome by experimentalists in biology as she or he is in physics, since successes of theory in biology have been few, the field of evolutionary and hereditary biology being the notable exception. In fact, ever since the discovery of the structure of the double helix by Watson and Crick, which many molecular biologists considered a discovery snapped away from a deserving experimentalist who could have achieved it at her own pace, molecular biology has been skeptical about the role of theory. Recently, though, the standing of theory has dramatically improved, both due to challenges in the biological sciences which beg for theoretical approaches as well as due to the increasing and extremely supportive role of computation, e.g., for sequence analysis, structure determination and neural network modelling.

Challenges for theoretical work are many. Most pressing are needs for concepts and algorithms to handle and analyze the rapidly increasing genetic data bases. The complete genomes of several biological species, including *homo sapiens*, have recently become available or will soon become available. For many proteins, variants for numerous biological species are known with rich and still unearthed information underlying the conservation and variability of amino acids. The problem of predicting the structure of proteins from genetic sequences, the so-called protein folding problem, begs a solution with immeasurable opportunities once a solution is at hand. A further challenge is the structure–function relationship of proteins, a problem with extremely wide scope and characterized both through universality and diversity. An important class of proteins, so-called regulatory proteins, control the expression of the cell's genetic information through their capacity to recognize DNA sequences and alter local DNA packing [3]. Of particular interest to physicists are biomolecular systems for which aggregates of biopolymers are the smallest functional units; examples of such systems are biological membranes and their complexes with proteins [4, 5], motor proteins polymerizing and depolymerizing into strands controlling cell movement and intracellular transport, and the coats of viruses, spherical shells of hundreds of interlocking protein units which exist in a metastable state to disintegrate during the infection process.

The present chapter focuses on proteins, actually only on a single one, bacteriorhodopsin [6]. Theoretical studies of proteins began with the advent of sufficiently powerful computers to simulate large particle systems and with the explosive increase of atomic resolution structures of proteins. The latter structures, though necessary prerequisites, in themselves are not sufficient for any physical theory of protein function; the motions in a protein play an equal role. This chapter provides a narrow slice of the theory of protein motion. This narrow field of view allows us to complement the other chapters on proteins in this book by giving more details than those chapters could do with the broader points of view. So the reader will bear in mind that the agenda of the field of protein dynamics is much broader than the issues presented in the present chapter. On the methodological side the field is concerned with providing accurate, yet simple force fields which govern the atomic motion of proteins. Ultimately, force fields will be determined in combined classical (for the nuclear motion)/quantum chemical (for the valence electrons) calculations employing, e.g., the Carr-Parinello method [7]. Researchers are investing currently strong efforts in developing efficient computational methods for classical dynamics of proteins involving tens to hundred thousands of atoms; a serious hurdle, for example, is the calculation of Coulomb forces since they need to be evaluated for all pairs of atoms for all time steps of the classical motion. Suitable integration schemes can economize the costly update of forces, in particular the Coulomb forces, with a resulting boost in computational efficiency [8]. At present a practitioner of the theory of proteins needs to be extremely competent in scientific computing with an understanding of massively parallel computing holding a particular promise for further success.

On the conceptual side an entry to protein dynamics is provided by studying normal modes. Since the classical Hamiltonian describing atomic motion is significantly non-harmonic and also extremely heterogenous, conventional normal mode analysis as applied, e.g., for crystals, is not suitable. A quasi-harmonic description derives normal modes from a principal component analysis, i.e., from a diagonalization of the covariance matrix of all atomic positions, averaged over time. However, a gliding average with a, say, 100 ps window, reveals that modes derived in such way vary in time due to significant conformational transitions and disorder in proteins [9]. Protein motion needs to be characterized also on spatial scales longer than atomic resolution, i.e., involving multi-atom segments of proteins. In fact, many proteins exhibit conformational changes which can be described as rotations of segments around hinges or as motions of flaps formed by secondary structure elements, e.g., α-helices or loops between α-helices [10].

The abstraction of functional dynamics from molecular dynamics simulations still remains an important challenge. Following established approaches one can try to identify correlation functions and susceptibilities which provide essential characteristics of protein dynamics and relate to observations and function: the dynamic structure function provides the Fourier transform of the motion of a protein's constituents and is observable through neutron diffraction or Mössbauer spectroscopy; the correlation function of the energy difference between two quantum states with diagonal coupling to the protein matrix accounts for the transition rate between the two states and, hence, for spectral or thermal transfer rates, e.g., the rate of electron transfer [11]; the dielectric susceptibility and thermal susceptibility, determined through monitoring dipolar or energy fluctuations account for dielectric properties, e.g., at membrane surfaces [12], or specific heats, e.g., of ordered and disordered water.

The theory of proteins, as a relatively young field, can benefit tremendously from related and already established fields. The closest relative is the theory of liquids since solvent molecules, though not connected into a polymer and much more homogeneous in structure, are subject to similar forces and disorder phenomena (see [13]). On larger length and longer time scales, molecular hydrodynamics [14] can provide much guidance to gain understanding of low frequency motion of protein segments encompassing many atoms. Condensed matter theory of disordered materials [15] likewise deals with systems, e.g., glasses, of great conceptual similarity. Condensed matter theory can also serve as a reminder that the primary role of theory is not quantitative description, but rather qualitative understanding; anybody suspecting that not much useful can come of such a role should have a close look at the triumphs of condensed matter theory.

The beauty of theoretical protein science stems from its rapidly increasing treasure of new structures and functions; one could hardly imagine a science with more relevance to the existence and well-being of humans, and with greater riches in new discoveries and, consequently, new challenges.

2 Periodic Motions in Proteins

In this tutorial chapter we demonstrate the existence of protein normal modes which maintain phase coherence for about one picosecond. We also study the dissipation of energy in proteins. The analysis is based on numerical experiments in which motional coherence, generated through two reassignments of Cartesian atomic velocities, induces *echoes* in the kinetic energy (temperature) and potential energy of proteins. Various echoes are produced in the case of the membrane protein *bacteriorhodopsin.* The echo phenomenon is then explained through a description of protein motion as an ensemble of harmonic oscillators. This description reveals that the echo can be expressed in terms of the temperature-temperature correlation function. A description in terms of *Langevin oscillators* allows one to account for decoherence effects. Finally, we consider echoes arising in an analytically tractable linear harmonic chain.

Motions in proteins have been studied by observation and by molecular

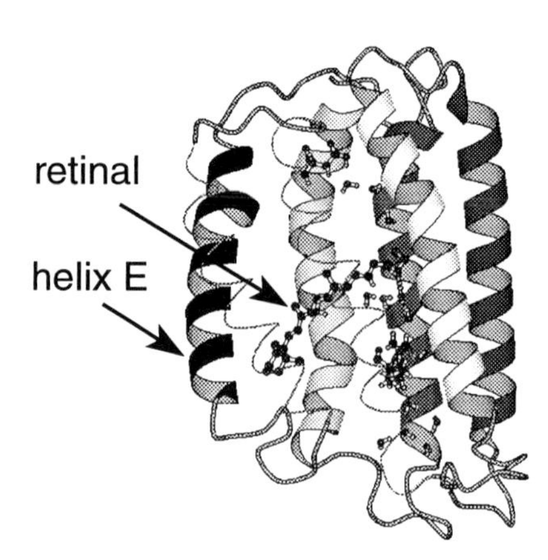

Fig. 1. Structure of the trans-membrane protein *bacteriorhodopsin* (from [27]). The protein has seven membrane-spanning helices A, B, C, ... G which form an elliptical cylinder around the *chromophore* retinal. Helices C and D are shown as thin lines to allow a view of the retinal. This protein serves to illustrate the various temperature echoes discussed in this chapter.

dynamics simulation for many years, but are still only poorly understood. In the case of solids and small molecules our understanding of characteristic motions is in a vastly better state, owing much to investigations of periodic motions in these systems [16, 17]. Following this example researchers have studied normal modes in proteins. Normal mode analysis has been used, for example, to describe the fluctuations and to display concerted motions of proteins [9, 18, 19, 20, 21, 22, 23, 24]. Normal modes have also been invoked to model slow motions between protein domains, for example, the hinge-bending motion of *lysozyme* [25, 26].

The widely adopted method to obtain normal modes for proteins is to calculate the second derivative (*Hessian*) matrix of the potential energy with respect to Cartesian coordinates or with respect to internal coordinates, and to diagonalize this matrix [18]. Due to anharmonic effects, normal modes defined through this method are not unique, but rather depend on the conformation of a protein. One would like to know how these modes, which are defined strictly only at $T = 0$, manifest themselves at higher temperatures. The significant anharmonic contributions of force fields in proteins, such as *torsional potentials, electrostatic interactions,* and *van der Waals interactions,* call into question the existence of protein normal modes and even the existence of periodic or coherent motions.

In this chapter we apply a computational experiment to describe normal modes in proteins by temperature echoes. Such echoes were first observed in simulations of disordered solids by Grest *et al.* [28, 29, 30, 31] and had been studied recently in the protein *bovine pancreatic trypsin inhibitor* (BPTI) in [32, 33]. The theory of echoes can also be applied to describe dipole relaxation and hysteresis in proteins [34]. In these first studies echoes were generated through two quenches of the kinetic energy of the protein, i.e., through reassignment of zero velocities to all atoms of the protein at times $t = 0$ and $t = \tau$. After the second quench one observes at time 2τ a dip in the temperature $T(t)$ which is defined through the total kinetic energy $E_k(t)$

$$T(t) = \frac{2}{3k_B N} E_k(t) \tag{1}$$

$$E_k(t) = \sum_{i=1}^{N} \frac{1}{2} m_i \mathbf{v}_i^2(t). \tag{2}$$

Here N denotes the number of atoms of the protein.

The chapter familiarizes the reader with several approaches to the study of protein motion. First, *molecular dynamics simulations* are utilized to induce and observe temperature echoes. Such simulations are by far the most frequently used and most accurate tool for theoretical investigations of proteins. We will explain how molecular dynamics simulations produce echoes and will provide a few examples of echoes. Second, we model proteins as an ensemble of harmonic modes with a given frequency distribution and derive a representation which averages over initial thermal conditions. This normal mode analysis of temperature echoes is very idealistic, but has the benefit of a mostly analytical mathematical description which encapsulates the relationship between thermal fluctuations of protein motion and the echo phenomenon. We will recover the well-known result of linear response theory, which describes relaxation immediately after a perturbation through correlation functions of thermal fluctuations; however, we will express also echoes which, due to a build-up of motional coherence through two consecutive perturbations, occur long after the perturbations.

Normal mode analysis overestimates coherence and, hence, we introduce a representation of protein motion through an ensemble of Langevin oscillators, again for a given frequency distribution. This representation is still amenable to an analytical description which properly carries out averages over initial thermal conditions. We finally consider an extremely idealized model of protein modes in which the frequency distribution actually results from the dynamics, namely a chain of linearly coupled harmonic oscillators. Temperature echoes exemplify the combination of numerical simulation and statistical mechanical analysis which is invoked in the theory of protein dynamics. It may give the reader a view, however narrow, of the conceptual approaches utilized in the physics of proteins.

This chapter adopts the common style of theoretical physics textbooks in that all analytical results are derived in detail, starting from a basic knowledge in classical and statistical mechanics; in this respect the chapter should be particularly useful for students.

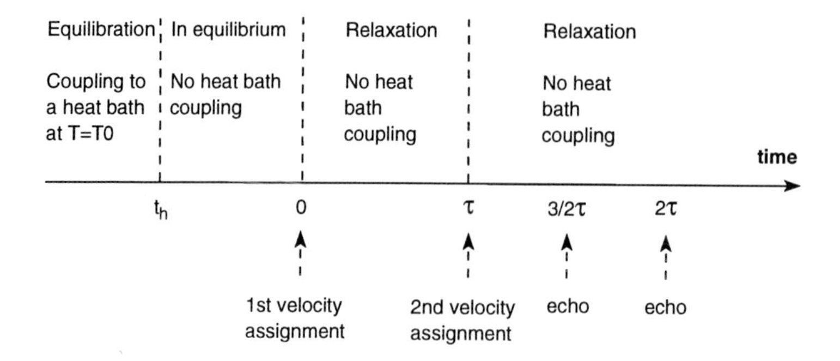

Fig. 2. Procedure of producing velocity reassignment echoes. No simulation is carried out until a coupling to a heat bath to achieve a desired temperature T_0 at $t = t_h < 0$. The simulations are then continued in the microcanonical ensemble, beginning with equilibration, the reassignment of Cartesian velocities to all (or to a selected group of) protein atoms at $t = 0$ using the set of velocities (4) followed by molecular dynamics simulation. At time $t = \tau$, atoms of the protein (all or a select group) are again reassigned Cartesian velocities, chosen from the set of velocities (7). Echoes arise, as described in the text, at times $\frac{3}{2}\tau$ and 2τ.

3 Generating Echoes in the Protein Bacteriorhodopsin

In order to produce coherent motion in a protein, we follow the procedure suggested in [35]. The protein to be probed is bacteriorhodopsin, a seven helix trans-membrane protein shown in Fig. 1. The choice of the protein structure [27] was dictated by a desire to demonstrate that the echo phenomenon does not depend on the nature of the protein: in [35] a water soluble protein had been chosen, here we study a membrane protein with mainly α-helical content. The protein was described through molecular dynamics (MD) simulations employing the program XPLOR [36] with the CHARMm force field [37]; all simulations employed the standard X-PLOR protein topology file `topallh6x.pro` and parameter file `parmallh3x.pro` to model bacteriorhodopsin. A dielectric constant of $\epsilon = 1$ was assumed and an integration time step of 0.5 fs was selected.

The echo procedure is presented schematically in Fig. 2. One first equilibrates the protein in an MD simulation at a desired temperature T_0 by coupling to a heat bath through repeated rescaling of the velocities such that the temperature T of the system, defined through Eqs. (1) and (2), assumes on average the value T_0. From then on all MD simulations are carried out in the microcanonical ensemble, i.e., energy is strictly conserved. This is a very important condition since motional coherence in a protein can be neither achieved nor maintained over any length of time if frictional and fluctuating forces are applied in any form.

Much of our attention in this chapter will be focussed on the temperature T as defined in Eqs. (1) and (2) and, hence, we characterize the quantity a bit further.

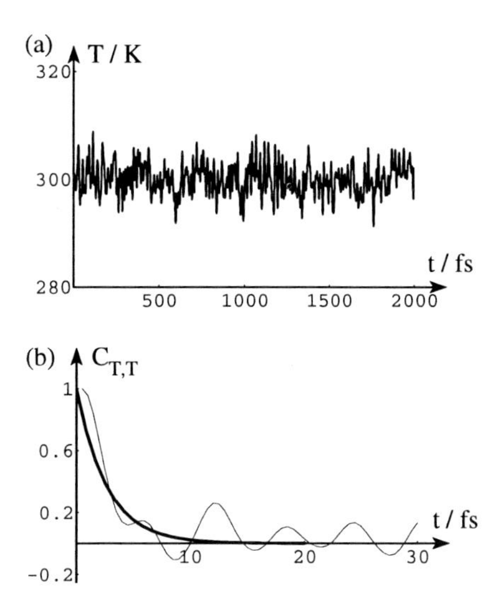

Fig. 3. (a) Temperature fluctuations in bacteriorhodopsin. The protein had been equilibrated at $T = 300\,\mathrm{K}$ and $T(t)$ was evaluated according to Eqs. (1) and (2) and plotted. The fluctuations reflect dynamic properties of the protein. (b) Temperature-temperature correlation function $C_{T,T}(t)$ evaluated according to Eq. (3) from the trace of $T(t)$ as shown in (a) (thin solid line) and compared to a matched exponential $\exp(-t/\tau_0)$ for $\tau_0 = 2.67$ fs (thick solid line).

T is actually fluctuating in time around its average value T_0 as demonstrated in Fig. 3. The fluctuations show a correlation in time which is conventionally characterized through the so-called temperature-temperature correlation function

$$C_{T,T}(t) = \frac{\langle T(t)\,T(0)\,\rangle - \langle\,T(0)\,\rangle^2}{\langle\,[\,T(t)\,]^2\,\rangle - \langle\,T(0)\,\rangle^2} \tag{3}$$

where $\langle\cdots\rangle$ denotes an ensemble average. One can recognize that $C_{T,T}$ decays to small values within about 5 fs, but then exhibits a slow decay of systematic oscillations. The behaviour of the protein hidden in these systematic oscillations is actually the main focus of this chapter. We seek to determine the dephasing time of these oscillations and want to inquire into the nature of the protein motions which remain coherent over periods actually much longer than shown in Fig. 3b.

In order to probe long time coherence in proteins, one prepares an initial state in two steps, as shown in Fig. 2. In a first step, at some instance defined through $t = 0$, one prepares the system by assigning to the protein's N atoms Cartesian velocities

$$\mathcal{V}^{(1)} = \{v_1^{(1)}, v_2^{(1)}, v_3^{(1)}, \ldots, v_{3N}^{(1)}\}\ . \tag{4}$$

This can be realized readily by most MD simulation programs simply by restarting a simulation with the 'old' positions of a protein's atoms and with velocities assigned from the set (4). The velocities in (4) are chosen according to the Maxwell distribution at a temperature $T = T_1$, e.g., according to

$$f(v_i^{(1)}) = \sqrt{\frac{m_i}{2\pi k_B T_1}} \exp\left[-\frac{m_i (v_i^{(1)})^2}{2 k_B T_1} \right] . \tag{5}$$

T_1 can differ from the initial temperature T_0. For our first example we chose $T_0 = 300\,\mathrm{K}$ and $T_1 = 50\,\mathrm{K}$. After such assignment the protein has a potential energy content, determined by the 'old' atomic positions, of temperature T_0, and a kinetic energy content, determined by the 'new' velocities, of T_1. The subsequent dynamics will mix potential and kinetic energy such that the new temperature, measured again through Eqs. (1) and (2), will relax to a value of about $\frac{1}{2}(T_0 + T_1) = 175\,\mathrm{K}$. This relaxation is seen in Fig. 4a during the time period $0 < t < 100$ fs, i.e., after the first velocity reassignment. Apparently, the protein exhibits random behaviour a few femtoseconds after the reassignment. One may ask what has been gained through the reassignment. The answer is: by assigning the velocities one knows at this point the phases of all protein modes; the phases are randomly distributed, but they are known through the set of velocities (4). This knowledge can be exploited in the second step of the procedure.

In fact, one stores the set (4) of velocities in memory and uses this information when, as before in the first step, a second set of random velocities

$$\mathcal{V}^{(2)} = \{v_1^{(2)}, v_2^{(2)}, v_3^{(2)}, \ldots, v_{3N}^{(2)}\} \tag{6}$$

is assigned to the protein's atoms at time $t = \tau$. This second set of velocities is selected strongly correlated to the velocities in (4), choosing

$$\mathcal{V}^{(2)} = \{\lambda\, v_1^{(1)}, \lambda\, v_2^{(1)}, \lambda\, v_3^{(1)}, \ldots, \lambda\, v_{3N}^{(1)}\} = \lambda\, \mathcal{V}^{(1)} . \tag{7}$$

The new velocities obey again the Maxwell distribution (5), however, for a temperature $T_2 = \lambda^2 T_1$, as can be readily verified. For our first example we adopted $\lambda = \sqrt{2}$, corresponding to $T_2 = 100\,\mathrm{K}$ and $\tau = 100$ fs. The choice of velocities (7) indeed induces strong coherence in the protein's motion. This is born out by the trace of the temperature $T(t)$ evaluated through Eqs. (1) and (2) and presented in Fig. 4a. After the reassignment of velocities the protein's kinetic energy corresponds to $T = 100\,\mathrm{K}$ while the potential energy corresponds to 175 K. Resurgence of the simulation leads to a rapid equilibration at a new temperature value of about $\frac{1}{2}(T_1 + T_2) = 137.5\,\mathrm{K}$. However, the systematic oscillations in $T(t)$ include now a hidden phase coherence which surfaces through two distinct features in the trace of $T(t)$, a dip at $t = 3\tau/2$ and a dip at $t = 2\tau$, both of which can be clearly recognized in Fig. 4a. These dips will be referred to as *temperature echoes*.

Through the choice of the temperatures T_1 and T_2 and variation of the time τ one can alter the depths of the temperature echoes in Fig. 4a, but one cannot

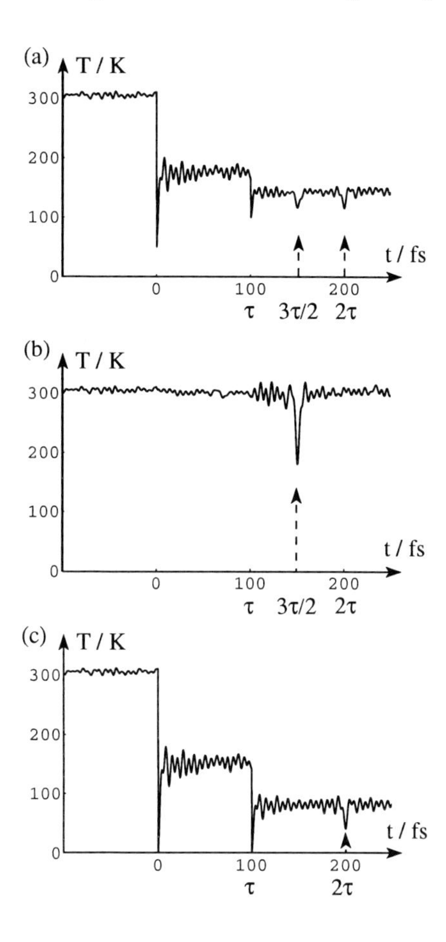

Fig. 4. Examples of temperature echoes in bacteriorhodopsin for $\tau = 100$ fs. (a) Velocity reassignment echoes for $T_0 = 300\,\mathrm{K}$, $T_1 = 50\,\mathrm{K}$, $T_2 = 100\,\mathrm{K}$; the velocities assigned twice at $t = 0$ and $t = \tau$ are strongly correlated according to (7). (b) Constant velocity reassignment echo, with $T_0 = T_1 = T_2 = 300\,\mathrm{K}$; identical velocities are assigned at $t = 0$ and at $t = \tau$; a sole echo appears at $t = 3\tau/2$. (c) Temperature quench ccho for $T_0 = 300\,\mathrm{K}$; in this case the kinetic energy of the protein is quenched (all atomic velocities set to zero) at $t = 0$ and at $t = \tau$; a sole echo appears at $t = 2\tau$.

alter the instances at which the echoes arise. For example, choosing $T_1 = T_2 = T_0$ the protein does not alter its temperature during the echo procedure. In this case an echo arises solely at $t = 3\tau/2$ as shown in Fig. 4b; this echo is referred to as *the constant temperature echo.* Choosing $T_1 = T_2 = 0$, a procedure called *temperature quench*, an echo arises solely at $t = 2\tau$; such an echo is demonstrated in Fig. 4c.

Conservation of energy implies $E_k(t) + V(t) = \text{const}$, where $V(t)$ is the total potential energy. As a result, one can monitor echoes not only through

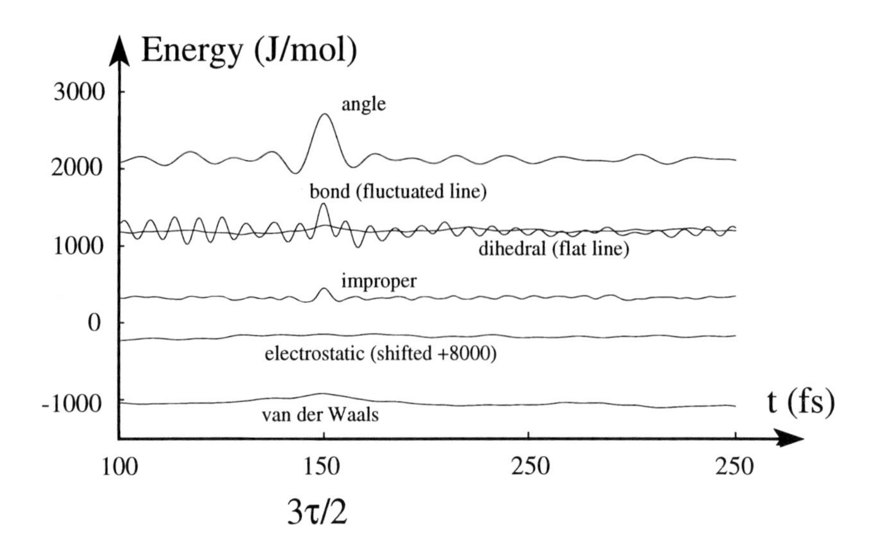

Fig. 5. Echoes reflected in the traces of the various potential energy contributions for bacteriorhodopsin prepared as in Fig. 4b. The potential energy contributions are defined in Eqs. (8–11). The dominant echo feature in the bond angle potential indicates that the normal modes involved in the echoes are skeletal motions.

the temperature as defined in Eqs. (1) and (2), but also through the potential energy. On first sight this is unappealing since the total potential energy is more difficult to evaluate than the kinetic energy. However, this approach permits one to dissect the $V(t)$-echo into its various contributions, e. g., contributions from bond, van der Waals and electrostatic energies which allows one then to conclude which type of motions participate significantly in the normal modes underlying the echo effect. In fact, the potential energy of a protein, as employed in molecular dynamics simulations, is partitioned into so-called bond, angle, dihedral, improper, electrostatic and van der Waals contributions [37, 38]:

$$V(t) = V_{\text{bond}}(t) + V_{\text{angle}}(t) + V_{\text{dihe}}(t) + V_{\text{impr}}(t) + V_{\text{elec}}(t) + V_{\text{vdw}}(t) \ . \quad (8)$$

The bond energy and angle energy are described by quadratic functions

$$V_{\text{bond}} = \sum_{\text{bonds}} k_b(|\mathbf{r}| - r_0)^2 \ ; \ V_{\text{angle}} = \sum_{\text{angles}} k_a(\theta - \theta_0)^2 \ . \quad (9)$$

The dihedral and improper terms can be expressed in a common functional form

$$V_{\text{torsion}} = \begin{cases} \sum_{\text{torsions}} k_\phi \left[1 + \cos(n\phi + \phi_0)\right] & (n = 1, 2, 3, \ldots) \\ \sum_{\text{torsions}} k_\phi \, (\phi - \phi_0)^2 \ . \end{cases} \quad (10)$$

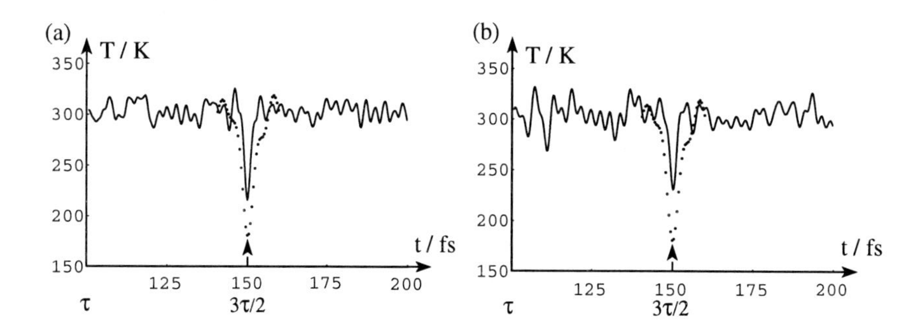

Fig. 6. Local velocity reassignment echoes (solid lines) in bacteriorhodopsin equilibrated at $T_0 = 300\,\mathrm{K}$ compared to all atom echoes (dotted lines). The protein has been prepared as for a constant temperature echo, i.e., with $T_0 = T_1 = T_2$. (a) Local echo involving all atoms of helix E of the protein (see Fig. 1); (b) Local echo involving all atoms of the chromophore retinal in the protein.

The electrostatic and van der Waals energies are given by the functions

$$V_{\text{elec}} = \sum_{\text{pairs}} \frac{q_1 q_2}{\epsilon r} \; ; \; V_{\text{vdw}} = \sum_{\text{pairs}} \left(\frac{A}{r^{12}} - \frac{B}{r^6} \right) . \tag{11}$$

Figure 5 presents the traces of the various potential energy contributions in (8) as they arise after two constant temperature velocity reassignments for $\tau = 100$ fs. One can see that $V_{\text{angle}}(t)$ exhibits the largest contribution to the $3\tau/2$-echo. The second largest contribution arises from $V_{\text{bond}}(t)$. The remaining potential energy terms contribute insignificantly. In particular, the electrostatic energy term shows no discernible contribution. One can interpret the results in Fig. 5 as a proof that the normal modes participating in the echoes for $\tau = 100$ fs are skeletal motions involving bending vibrations (which affect bond angles) and, to a lesser degree, stretch vibrations.

One can also generate local echoes in proteins. To demonstrate this we have carried out a constant temperature velocity reassignment echo at $T = 300\,\mathrm{K}$ and for $\tau = 100$ fs by replacing twice only atomic velocities of subsegments of bacteriorhodopsin, leaving the velocities outside of those subsegments unaltered. Two examples are shown in Fig. 6. In one case we have produced an echo in the helix E of the protein (see Fig. 1) and in the second case in the chromophore retinal. The echo depth measures only about 80 K which is significantly less than that of an all-atom echo for the same τ (see Fig. 7).

The delay time τ between velocity reassignments determines the time scale over which $3\tau/2$- and 2τ-echoes arise. Naturally, dephasing of protein modes described by a time scale τ_{dephase}, competes with the appearance of echoes, since the latter rely on coherence of the motion. One expects that for $\tau > \tau_{\text{dephase}}$ echoes are weak. For this purpose, constant temperature all-atom echoes have been systematically generated for τ-values increasing from a few femtoseconds

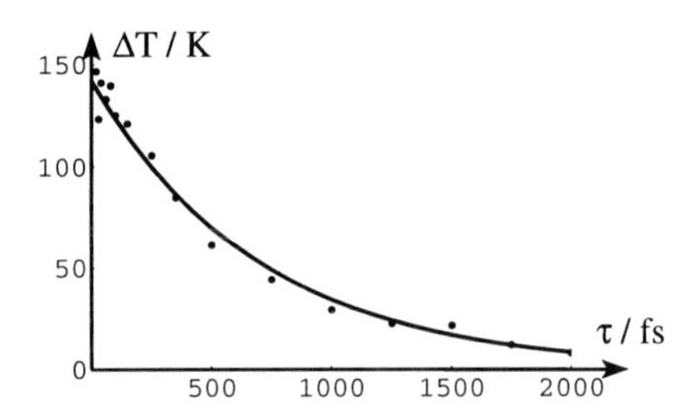

Fig. 7. Dependence of the depth of echo generated in bacteriorhodopsin for $T_0 = T_1 = T_2 = 300\,\mathrm{K}$ on the delay time τ between velocity reassignments. The simulation data are presented as dots and are matched to $\exp(-\tau/\tau_{\mathrm{dephase}})$ for $\tau_{\mathrm{dephase}} = 700$ fs (solid line).

to two picoseconds. The resulting depths of the echoes, presented in Fig. 7, show an exponential dependence on τ which apparently reflects an exponential decay of phase coherence. Remarkably, the results show that coherence of oscillations in proteins are maintained to a significant degree over a time of about a picosecond, a time which amounts to hundreds of periods of most bond stretch vibrations in proteins. Matching the τ-dependence of the echo depth in Fig. 7 to an exponential $\exp(-\tau/\tau_{\mathrm{dephase}})$ yields $\tau_{\mathrm{dephase}} = 700$ fs.

The temperature and potential energy echoes presented above are a signature of the motion of proteins on a 1 ps time scale. The value of the echoes hinges, however, on a suitable interpretation of the echo phenomena. This interpretation is furnished in the following two sections.

4 Harmonic Oscillator Description of Echoes

The echoes exhibited by the kinetic energy (temperature) and potential energy of a protein, after preparation of a coherent initial state, will be analyzed now in the framework of *normal mode analysis.* For this purpose the protein is considered an ensemble of uncoupled harmonic oscillators described by the Hamiltonians

$$H_\alpha = \frac{p_\alpha^2}{2m_\alpha} + \frac{1}{2} m_\alpha \omega_\alpha^2 q_\alpha^2 \,, \quad \alpha = 1, 2, \ldots \tag{12}$$

Consequently, we can solve the system's equations of motion, and find

$$q_\alpha(t) = A_\alpha \cos(\omega_\alpha t + \theta_\alpha) \tag{13}$$

$$p_\alpha(t) = m_\alpha \frac{\mathrm{d}q_\alpha(t)}{\mathrm{d}t} = -m_\alpha A_\alpha \omega_\alpha \sin(\omega_\alpha t + \theta_\alpha) \,. \tag{14}$$

This representation allows one to derive an analytical description of temperature echoes, as produced numerically in the previous section; in particular, one can deduce the relationship of echoes to other protein properties. However, normal mode analysis neglects decoherence effects.

Simple Explanation of Temperature Echoes

Before providing a systematic analysis we want to furnish a simple geometrical picture of the echo effect. For this purpose we notice that in a system with coordinates

$$\tilde{p}_\alpha = p_\alpha / \sqrt{m_\alpha} \tag{15}$$

$$\tilde{q}_\alpha = \sqrt{m_\alpha} \omega_\alpha q_\alpha \tag{16}$$

the trajectory $(\tilde{q}_\alpha(t), \tilde{p}_\alpha(t))$ of each oscillator α circumscribes a circle in counter-clockwise direction with angular frequency ω_α and radius $\sqrt{2E_\alpha}$ where E_α is the energy of the oscillator; for an ensemble of oscillators in thermal equilibrium, the energy assumes the average value $\langle E_\alpha \rangle = k_B T$. Figure 8a provides a snapshot of the oscillators in this presentation, indicating the circular trajectories for two oscillators. The reader should note that the individual oscillators precess with different angular velocities ω_α.

First velocity reassignment. We want to describe now the behaviour of the ensemble of oscillators. We consider the motion of the ensemble in the case of a temperature quench echo. Such an echo is prepared by setting in the initial preparation step, i.e., at $t = 0$, all momenta $\tilde{p}_\alpha$ to zero. In the $\tilde{q}$, $\tilde{p}$-diagram the vectors characterizing the oscillators come to lie then all on the $\tilde{q}$-axis. Since there are many oscillators, we can divide them into subsets with the same energy. For the sake of simplicity we pick, as shown in Fig. 8b, that one subset of the oscillators, numbering 8 in this case, which have the same energy $E = E_\alpha$: all circumscribe the same circle in the $\tilde{q}$, $\tilde{p}$-diagram, i.e., their trajectories are $(\tilde{q}_\alpha(t), \tilde{p}_\alpha(t)) = (\sqrt{2E}\cos\omega_\alpha t, \sqrt{2E}\sin\omega_\alpha t)$, $\alpha = 1, 2, \ldots 8$. From now on we will follow only these 8 oscillators and show how the quench echo occurs for these oscillators. Other subsets can be treated using the same procedure as from Fig. 8b to Fig. 8f, but for different radii, i.e., different energies. It then follows for the same reason that the temperature of the ensemble, which is the total kinetic energy of all oscillators, will have an echo at time 2τ.

Second velocity reassignment. The oscillators precess subsequently with different angular frequencies $\omega_\alpha, \alpha = 1, 2, \ldots 8$. A snapshot of the motion at time $t = \tau$ is shown in Fig. 8c; the isotropic spread of the vectors indicates that the motion is incoherent due to the random distribution of ω_α values; the kinetic energy of the ensemble is $\frac{1}{2}\tilde{p}_\alpha^2 = E\sin^2(\omega_\alpha \tau)$, the average of which is $\frac{1}{2}E$, as expected. At this instance, i.e., at $t = \tau$, the ensemble of oscillators is quenched a second time by resetting $\tilde{p}_\alpha = 0$, $\alpha = 1, 2, \ldots 8$, as shown in Fig. 8d. This realigns all vectors $(\tilde{q}_\alpha, \tilde{p}_\alpha)$, $\alpha = 1, 2, \ldots 8$, with the $\tilde{q}$-axis, as they were initially, at $t = 0$. Now, however, roughly half the vectors point to the right and half to the left—in fact, we have that $(\tilde{q}_\alpha(\tau), \tilde{p}_\alpha(\tau)) = (\sqrt{2E}\cos(\omega_\alpha \tau), 0)$. The energies of the ensemble are $E\cos^2(\omega_\alpha \tau)$, and one expects the kinetic energy for the ensuing ensemble to be $\frac{1}{2}E\langle \cos^2 \omega_\alpha \tau \rangle_\alpha = \frac{1}{4}E$. The reduction of the average kinetic energy from $\frac{1}{2}E$ to $\frac{1}{4}E$ reflects the energy drained from the ensemble at $t = \tau$.

After the second velocity reassignment. At times $t > \tau$ the vectors $(\tilde{q}_\alpha, \tilde{p}_\alpha)$ spread uniformly again in both the $\tilde{q}$- and $\tilde{p}$-directions as shown in Fig. 8e. The kinetic energies of the ensemble measure $E\cos^2(\omega_\alpha \tau)\sin^2(\omega_\alpha(t-\tau))$, $\alpha = 1, 2, \ldots 8$, or $\frac{1}{4}E[\sin(\omega_\alpha(t-2\tau)) + \sin(\omega_\alpha t)]^2$. The average over all ω_α yields, indeed, a value $\frac{1}{4}E$, except near $t = 2\tau$ where the average reduces to $\frac{1}{8}E$. The latter feature corresponds to the echo shown in Fig. 4b. The $(\tilde{q}_\alpha, p_\alpha)$ vectors at time $t = 2\tau$ are shown in Fig. 8f. One can recognize that the vectors in Fig. 8f "concentrate" more closely around the $\tilde{q}$-axis than in case of Fig. 8e, such that the kinetic energies are significantly smaller at $t = 2\tau$. In fact, the vectors at $t = 2\tau$ are given by $(\tilde{q}_\alpha(2\tau), \tilde{p}_\alpha(2\tau)) = (\sqrt{2E}\cos^2(\omega_\alpha\tau), \sqrt{2E}\cos(\omega_\alpha\tau)\sin(\omega_\alpha\tau))$, $\alpha = 1, 2, \ldots 8$, in which all $\tilde{q}$-components are now positive, as in the initial case (see Fig. 8b). This deviation from an isotropic symmetry originates from the orientation in Fig. 8b, such that a better sample in case of Fig. 8e would include the mirror image at the $\tilde{p}$-axis. However, even with that complement included, the kinetic energy of the ensemble is given by $\frac{1}{4}E\sin^2(2\omega_\alpha\tau)$, $\alpha = 1, 2, \ldots 8$, the average of which measures only $\frac{1}{8}E$ at the $t = 2\tau$ instance. This implies an echo depth of $\frac{1}{4}E - \frac{1}{8}E = \frac{1}{8}E$. Another way to see the echo arising at the stage depicted in Fig. 8f is to notice that at $t = 2\tau$ every single mode goes back to the same phase or to same phase plus π as in Fig. 8c. And in Fig. 8d, the modes near $\tilde{p}$-axis, such as modes 3 and 7, lost more energy in the second quench than those modes near the $\tilde{q}$-axis, such as modes 1 and 6. As a result the total kinetic energy, which is the sum of $\frac{1}{2}\tilde{p}_\alpha^2$, is lower in Fig. 8f, i.e., at $t = 2\tau$, than in Fig. 8e, i.e., at any arbitrary $t > \tau$.

We would like to note the following two points which emerge from the discussion of Fig. 8. First, the quench echo results from an average behavior of all the normal modes; every mode contributes to the echo feature. The echo depth, for an ensemble of harmonic oscillators, measures half of the current temperature. Second, the initial energies do not need to obey a Boltzmann distribution, rather, it is the distribution of oscillator frequencies that matters.

Temperature Echo in the Harmonic Approximation

We want to provide now a systematic derivation of velocity reassignment echoes which will provide us with an expression for the shape of the echoes as shown in Fig. 4. For this purpose we repeat the simple derivation above, carrying out, however, proper thermal averages and averages over the ensemble of frequencies ω_α. The present description assumes that the motion of a protein can be decomposed into a set of uncoupled normal modes described through the Hamiltonian in (12). The description follows closely the one given in [35]. For a protein with N atoms there exist $3N - 6$ different internal normal modes. Six degrees of freedom, which describe overall translation and rotation, are not counted as normal modes. We denote the frequency of the α-th mode by ω_α, the corresponding effective mass by m_α, and the associated vibrational coordinate by q_α, where $\alpha = 1, 2, \ldots, 3N - 6$.

The position of the α-th normal mode can be expressed as

$$q_\alpha(t) = A_\alpha \cos(\omega_\alpha t + \theta_\alpha) \tag{17}$$

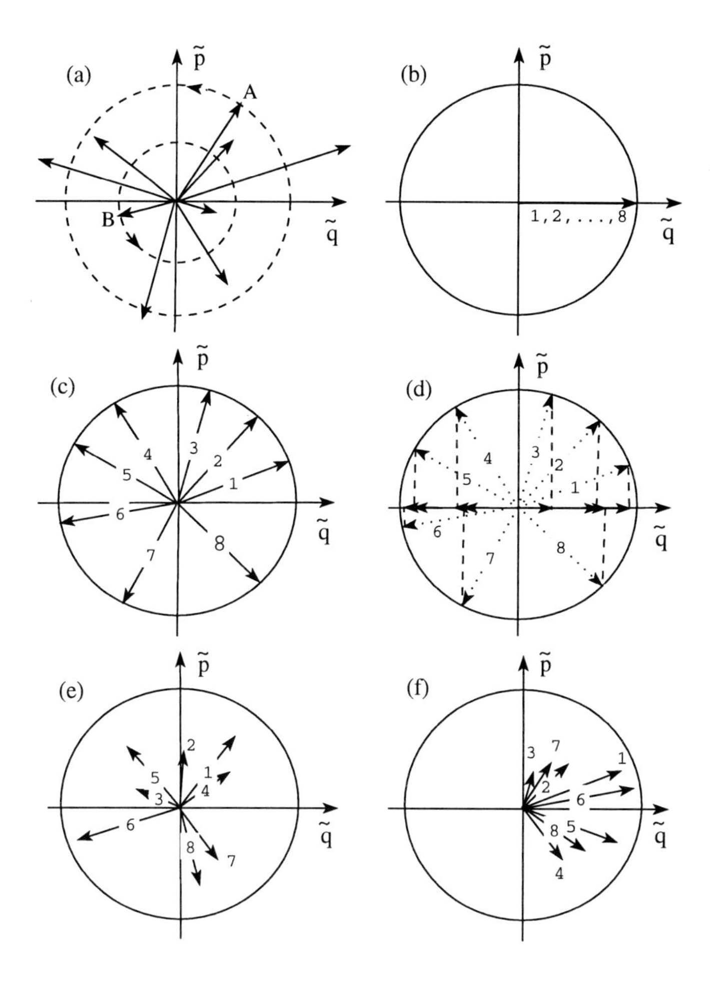

Fig. 8. Trajectories $(\tilde{q}_\alpha(t), \tilde{p}_\alpha(t))$ of an ensemble of modes to demonstrate the temperature echo. The coordinates $(\tilde{q}_\alpha(t), \tilde{p}_\alpha(t))$ are defined in Eqs. (15) and (16); (a) shows normal modes with different energies at $t < 0$; for modes A and B the circles circumscribed by $(\tilde{q}_\alpha(t), \tilde{p}_\alpha(t))$ are indicated by dashed lines; the modes precess around the origin with angular frequencies ω_α. (b) represents the state of eight normal modes at $t = 0+$, after the initial quench of the kinetic energy at $t = 0$; all momenta $\tilde{p}_\alpha(0)$ are set to zero; for the sake of simplicity we choose eight modes which now have the same initial energy, i.e., $(\tilde{q}_\alpha(t), \tilde{p}_\alpha(t))$ begin to move on the same circle in a counterclockwise direction. (c) presents the normal modes at $t = \tau$ -, the vectors $(\tilde{q}_\alpha(t), \tilde{p}_\alpha(t))$ are distributed more or less homogeneously since the modes precess with different frequencies ω_α. At the instance $t = \tau$ the kinetic energy is quenched a second time, i.e., all momenta $\tilde{p}_\alpha$ are set to zero; the resulting states at $t = \tau +$ are shown in (d). (e) shows the modes at some arbitrary time $t > \tau$; the mode vectors $(\tilde{q}_\alpha(t), \tilde{p}_\alpha(t))$ are again homogeneously distributed. However, at $t = 2\tau$ the vectors $(\tilde{q}_\alpha(2\tau), \tilde{p}_\alpha(2\tau)), \alpha = 1, 2, \ldots 8$ all point in the direction of the positive $\tilde{q}$ axis and are "focussed" near this axis, i.e., the momenta $\tilde{p}_\alpha(2\tau)$ are small as shown in (f).

with corresponding velocity

$$u_\alpha(t) = \frac{\mathrm{d}q_\alpha(t)}{\mathrm{d}t} = -A_\alpha \, \omega_\alpha \sin(\omega_\alpha t + \theta_\alpha) \ . \tag{18}$$

Here θ_α are initial phases which are homogeneously distributed in the interval $[0, 2\pi]$. A_α are the initial amplitudes which are randomly distributed according to the Rayleigh distribution [39]

$$P(A_\alpha) = \frac{m_\alpha \omega_\alpha^2 A_\alpha}{k_\mathrm{B} T_0} \exp\left(- \frac{m_\alpha \omega_\alpha^2 A_\alpha^2}{2 k_\mathrm{B} T_0} \right) \tag{19}$$

where T_0 is the equilibrium temperature of the system. This distribution can be easily derived from the Boltzmann distribution of the energy. The total kinetic energy is

$$E_k(t) = \sum_\alpha \frac{1}{2} m_\alpha \omega_\alpha^2 A_\alpha^2 \sin^2(\omega_\alpha t + \theta_\alpha) \ . \tag{20}$$

The kinetic energy averaged over the phases θ_α is

$$\langle E_k(t) \rangle_\theta = \left\langle \sum_\alpha \frac{1}{2} m_\alpha \omega_\alpha^2 A_\alpha^2 \sin^2(\omega_\alpha t + \theta_\alpha) \right\rangle_\theta = \sum_\alpha \frac{1}{4} m_\alpha \omega_\alpha^2 A_\alpha^2 \ . \tag{21}$$

Temperature-Temperature Correlation Function. Before calculating the echo we want to consider the thermal fluctuations of the temperature, defined through Eqs. (1) and (2) and discernable in the $T(t)$ traces in Figs. 4 and 6. An enlarged view of T(t) for bacteriorhodopsin at $T = 300\,\mathrm{K}$ is presented in Fig. 3a. The time dependence of $T(t)$ can be characterized through the temperature-temperature correlation function (3). We want to determine $C_{T,T}(t)$ for the ensemble of thermal oscillators described through Eqs. (17) and (18). For this purpose we express $C_{T,T}(t)$ through the kinetic energy using Eqs. (1) and (2)

$$C_{T,T}(t) = \frac{\langle E_k(t) \, E_k(0) \rangle_\theta - \langle E_k(0) \rangle_\theta^2}{\langle [E_k(0)]^2 \rangle_\theta - \langle E_k(0) \rangle_\theta^2} \ . \tag{22}$$

For the kinetic energy correlation function it holds that

$$\langle E_k(t) \, E_k(0) \rangle_\theta = \left\langle \left(\sum_\alpha \frac{1}{2} m_\alpha \omega_\alpha^2 A_\alpha^2 \sin^2(\omega_\alpha t + \theta_\alpha) \right) \left(\sum_\lambda \frac{1}{2} m_\lambda \omega_\lambda^2 A_\lambda^2 \sin^2 \theta_\lambda \right) \right\rangle_\theta \tag{23}$$

where the summation over α and λ is from 1 to $3N-6$ and where $\langle \ldots \rangle_\theta$ denotes the average over the random phases θ_α and θ_λ. Replacing $\sin\theta$ by $\frac{1}{2i}[\exp(i\theta) - \exp(-i\theta)]$ and employing the averaging technique proposed by Rayleigh [39, 40],

$$\langle \exp[\pm i(\theta_\alpha + \theta_\lambda)] \rangle_\theta = 0 \, ; \quad \langle \exp[\pm i(\theta_\alpha - \theta_\lambda)] \rangle_\theta = \delta_{\alpha\lambda} \ , \tag{24}$$

one obtains

$$\langle E_k(t)\, E_k(0) \rangle_\theta = \left(\sum_\alpha \frac{1}{4} m_\alpha \omega_\alpha^2 A_\alpha^2 \right)^2 + \sum_\alpha \frac{1}{32} m_\alpha^2 \omega_\alpha^4 A_\alpha^4 \cos(2\omega_\alpha t) \; . \quad (25)$$

Combining this result with (21) yields for (22)

$$C_{T,T}(t) = \frac{\sum_\alpha m_\alpha^2 \omega_\alpha^4 A_\alpha^4 \cos(2\omega_\alpha t)}{\sum_\alpha m_\alpha^2 \omega_\alpha^4 A_\alpha^4} \; . \quad (26)$$

For each molecular dynamics trajectory, the amplitudes A_α are constants of motion. However, as shown in [41], the correlation function $C_{T,T}(t)$ is almost identical for different trajectories. One may assume, therefore, that $C_{T,T}(t)$ calculated from one trajectory represents the temperature-temperature correlation function evaluated from the average over many trajectories with amplitudes A_α distributed according to Eq. (19). One may replace then in Eq. (26) the quantities $m_\alpha^2 \omega_\alpha^4 A_\alpha^4$ by their average values

$$\langle m_\alpha^2 \omega_\alpha^4 A_\alpha^4 \rangle_A = m_\alpha^2 \omega_\alpha^4 \int_0^\infty dA_\alpha \, A_\alpha^4 \, p(A_\alpha) = 8(k_\mathrm{B} T_0)^2 \; . \quad (27)$$

Consequently, one obtains

$$C_{T,T}(t) = \langle \cos(2\omega_\alpha t) \rangle_\alpha \quad (28)$$

where $\langle \ldots \rangle_A$ denotes the average over the amplitudes of the oscillators determined by use of Eq. (19), and where $\langle \ldots \rangle_\alpha$ is an average over all the normal modes, i.e.,

$$\langle f(\omega_\alpha) \rangle_\alpha = \frac{1}{3N-6} \sum_\alpha f(\omega_\alpha) = \int_0^\infty \mathrm{d}\omega D(\omega) \, f(\omega) \; . \quad (29)$$

$D(\omega)$ denotes the normalized density of states.

The correlation function $C_{T,T}(t)$ could be evaluated from Eqs. (28) and (29) if the density of states $D(\omega)$ were known. Here we determine $C_{T,T}(t)$ according to Eq. (3) from molecular dynamics simulations, which do not necessarily satisfy the harmonic approximation. $C_{T,T}(t)$ was calculated from a 20 ps interval in the simulation described in Section 3. The result is shown in Fig. 3. The correlation function can be matched to

$$C_{T,T}(t) \approx e^{-t/\tau_0} \; , \quad \tau_0 = 2.67\,\mathrm{fs} \; . \quad (30)$$

In the following, we will assume at various occasions

$$C_{T,T}(t) \approx 0 \;\; (t \gg \tau_0) \; . \quad (31)$$

As is evident from Fig. 3b, the exponential decay suggested in Eq. (30) is a rather poor approximation. The correlation function $C_{T,T}(t)$ has a long-time oscillatory behavior, which contains essential information on the density of states [33].

Temperature-Temperature Correlation Function and Temperature Quench Response Function[1]

We want to demonstrate now that, in linear response theory, the response of a protein to a brief temperature pulse is related to the temperature-temperature correlation function. To demonstrate this relationship, we consider a single mode described by the Langevin equation

$$m\,\ddot{x} \;=\; f(x) - \gamma\dot{x} + \sigma\,\xi \;. \tag{32}$$

To this stochastic differential equation corresponds a Fokker-Planck equation $\partial_t p(x,v,t) \;=\; \mathcal{L}_0(x,v)\,p(x,v,t)$ where

$$\mathcal{L}_0(x,v) \;=\; \frac{k_B T_0 \gamma}{m^2}\partial_v^2 \;+\; \frac{1}{m}\partial_v[\gamma v - f(x)] \;-\; \partial_x\, v \;. \tag{33}$$

We denote by $p_0(x,v)$ the stationary position-velocity distribution for this mode for which holds $p_0(x,v) \;\sim\; \exp\left(-\frac{mv^2}{2k_B T_0}\right)$, suppressing the x-dependence since it is immaterial in the following.

We consider now a perturbation applied to the mode at $t = 0$ and described through the operator $\ell \;=\; \partial_v^2$, i.e., we consider the perturbation $\delta(t)\,\epsilon\,\partial_v^2$ for small ϵ. Adding this perturbation to the Fokker-Planck operator (33) corresponds to a sudden temperature pulse $\Delta T\,\delta(t)$ in the system, where

$$\Delta T \;=\; \frac{m^2 \epsilon}{k_B \gamma} \;. \tag{34}$$

Application of ℓ, for small ϵ, induces a response in the kinetic energy E_k described by [44]

$$R_{E_k,\ell}(t) \;=\; \langle\, E_k(t)\, A(0)\, \rangle \;, \tag{35}$$

the so-called response function. Here A is

$$A \;=\; p_0^{-1}(x,v)\,\ell\,p_0(x,v) \;. \tag{36}$$

and, in the present case, is given by

$$A \;=\; -\frac{m}{k_B T_0} \;+\; \frac{2m}{(k_B T_0)^2}\,(\frac{1}{2}\,mv^2) \;. \tag{37}$$

Hence, one can express the response to a brief temperature pulse through

$$\begin{aligned} R_{E_k,\ell}(t) \;&=\; -\frac{m}{k_B T_0}\,\langle\, E_k(t)\,\rangle \;+\; \frac{2m}{(k_B T_0)^2}\langle\, E_k(t)\,E_k(0)\,\rangle \\ &=\; -\frac{m}{2} \;+\; \frac{2m}{(k_B T_0)^2}\langle\, E_k(t)\,E_k(0)\,\rangle \;. \end{aligned} \tag{38}$$

[1] As an introduction to the Langevin equations, Fokker-Planck equations and linear response theory employed here, we recommend to the reader the relevant chapters in the monographs [42, 43].

Following the derivation of Eq. (28) one can state

$$C_{T,T}(t) = \frac{\langle E_k(t)\, E_k(0) \rangle}{8(k_B T_0)^2} - \frac{1}{32} . \tag{39}$$

Comparing Eq. (38) and Eq. (39), one arrives at

$$R_{E_k,\ell}(t) = 16 m C_{T,T}(t) . \tag{40}$$

Hence, we have demonstrated that the response of the protein mode to a temperature pulse $\Delta T\, \delta(t)$, described by $R_{E_k,l}(t)$, is equal to the temperature-temperature correlation function $C_{T,T}(t)$. We expect, then, that the relaxation of the temperature of the system after velocity reassignments is described by the temperature-temperature correlation function (28). We will show now that the temperature echo can also be expressed in terms of $C_{T,T}(t)$ within the harmonic approximation.

Three stages of normal mode dynamics. As shown in Figs. 4 and 6, echoes can be observed for a protein prepared through two velocity reassignments by monitoring $T(t)$. Obviously, one needs to evaluate the total kinetic energy of the protein. For the purpose of this evaluation, we consider in turn the three stages of the echo dynamics, (0) before the first velocity reassignment, (1) between the first and the second reassignment, and (2) after the second reassignment, as discussed in connection with Fig. 8.

Normal modes imply concerted motions in which many protein atoms participate. The modes are described through a linear transformation from atomic coordinates X_j, $j = 1\,2, \ldots, 3N$ to normal mode coordinates $q_\alpha, \alpha = 1, 2, \ldots 3N-6$

$$q_\alpha = \sum_{j=1}^{3N} S_{\alpha j}(t)\, X_j . \tag{41}$$

We have indicated through a time-dependence of the transformation matrix $\mathbf{S}(t)$ that the normal modes in a non-harmonic system, like a protein, are not invariant in time. In fact, one expects that the modes in proteins vary in time and, consequently, that the matrix $\mathbf{S}(t)$ experiences significant changes while a protein moves across conformational substates [45]. Since the evolution of the transformation matrix $\mathbf{S}(t)$ is unknown it is, strictly speaking, impossible to carry the velocity correlation expressed in Eq. (7) over to a normal mode analysis. In fact, defining the normal mode velocities at $t = 0$

$$u_\alpha^{(1)} = \sum_{j=1}^{3N} S_{\alpha j}(0)\, v_j^{(1)} , \tag{42}$$

and at $t = \tau$

$$\tilde{u}_\alpha^{(2)} = \sum_{j=1}^{3N} S_{\alpha j}(\tau)\, v_j^{(2)} , \tag{43}$$

the sets of velocities, which should be assigned to the normal modes, are

$$\mathcal{U}^{(1)} = \{u_1^{(1)}, u_2^{(1)}, u_3^{(1)}, \ldots, u_{3N-6}^{(1)}\} \tag{44}$$

and

$$\tilde{\mathcal{U}}^{(2)} = \{\tilde{u}_1^{(2)}, \tilde{u}_2^{(2)}, \tilde{u}_3^{(2)}, \ldots, \tilde{u}_{3N-6}^{(2)}\} \tag{45}$$

corresponding to the velocity sets (4) and (6), respectively. However, the lack of knowledge of $\mathbf{S}(\tau)$ forces us to rather employ, at time $t = \tau$, the velocities transformed by $\mathbf{S}(0)$

$$u_\alpha^{(2)} = \sum_{j=1}^{3N} S_{\alpha j}(0)\, v_j^{(2)} \tag{46}$$

and, hence, the set

$$\mathcal{U}^{(2)} = \{u_1^{(2)}, u_2^{(2)}, u_3^{(2)}, \ldots, u_{3N-6}^{(2)}\} \tag{47}$$

can be written, according to Eq. (7),

$$\mathcal{U}^{(2)} = \{\lambda\, u_1^{(1)}, \lambda\, u_2^{(1)}, \lambda\, u_3^{(1)}, \ldots, \lambda\, u_{3N-6}^{(1)}\} = \lambda \mathcal{U}^{(1)} \ . \tag{48}$$

For the statistical characteristics of the velocities $u_\alpha^{(1)}$ and $u_\alpha^{(2)}$ and their correlation, the transformation $\mathbf{S}(0)$ is immaterial. Since the reassigned velocities are characterized only through their average properties, the transformation matrix $\mathbf{S}(0)$ is not required; one can apply the statistical characteristics directly to $u_\alpha^{(1)}$ and $u_\alpha^{(2)}$ without knowing the Cartesian velocities $v_j^{(1)}$ and $v_j^{(2)}$. However, the replacement $\mathbf{S}(\tau) \to \mathbf{S}(0)$ implies an error for the correlation of velocities, as described, for example, by Eq. (7). The correlation of two velocities $u_\alpha^{(1)}$ and $\tilde{u}_\alpha^{(2)}$ can be written, using (48),

$$\langle\, u_\alpha^{(1)}\, \tilde{u}_\alpha^{(2)} \,\rangle_u = g_\alpha(\tau, T)\, \langle\, u_\alpha^{(1)}\, u_\alpha^{(2)} \,\rangle_u = g_\alpha(\tau, T)\, \lambda\, \langle\, [u_\alpha^{(1)}]^2 \,\rangle_u \ . \tag{49}$$

Here $g_\alpha(\tau, T)$ is a factor accounting for the difference between $\mathbf{S}(\tau)$ and $\mathbf{S}(0)$ and is a function of the time interval τ and temperature T. In the harmonic case, which is assumed in this section, $g_\alpha(\tau, T) = 1$. But for proteins, due to anharmonic effects, one expects $0 < g_\alpha(\tau, T) < 1$, with $g_\alpha(\tau, T)$ differing more from unity for longer τ.

(0) Before the first reassignment. During this stage, the position of the α-th normal mode at $t < 0$ can be expressed as

$$q_\alpha^{(0)}(t) = A_\alpha \cos\left(\, \omega_\alpha\, t + \theta_\alpha \,\right) \tag{50}$$

where A_α denotes the amplitude of the mode, distributed according to Eq. (19), and where θ_α denotes the phase of the mode which is homogeneously distributed in $[0, 2\pi]$. Obviously, averaging over the phases yields

$$\langle\, \cos(n\theta_\alpha) \,\rangle_\theta = 0\ ;\ n = 1, 2, 3, \ldots\ , \tag{51}$$

a result needed further below. The velocities corresponding to Eq. (50) are

$$\frac{\mathrm{d}q_\alpha^{(0)}(t)}{\mathrm{d}t} = -A_\alpha\, \omega_\alpha \sin\left(\, \omega_\alpha\, t + \theta_\alpha \,\right) \ . \tag{52}$$

The temperature correlation function $C_{T,T}(t)$ during this stage is given by (28).

(1) After the first reassignment, but before the second. During this stage, i. e., for $0 \leq t < \tau$, the position for the α-th normal mode can be expressed as

$$q_\alpha^{(1)}(t) \;=\; A_\alpha^{(1)} \cos\left(\omega_\alpha t + \theta_\alpha^{(1)}\right) \;. \tag{53}$$

In this case, the amplitudes $A_\alpha^{(1)}$ and phases $\theta_\alpha^{(1)}$ are determined through two sets of conditions, namely, that the positions (53) at $t = 0$ must match the corresponding expression (50), from which follows

$$A_\alpha^{(1)} \cos\theta_\alpha^{(1)} \;=\; A_\alpha \cos\theta_\alpha \;, \tag{54}$$

and that the velocities for all modes α are assigned values from the set $\mathcal{U}^{(1)}$ [see Eq. (4)], which implies

$$\frac{\mathrm{d}q_\alpha^{(1)}(0)}{\mathrm{d}t} \;=\; -\,A_\alpha^{(1)}\,\omega_\alpha \sin\theta_\alpha^{(1)} \;=\; u_\alpha^{(1)} \;. \tag{55}$$

(2) After the second reassignment. During this stage, i. e., at $t \geq \tau$, the position of the α-th normal mode can be expressed as

$$q_\alpha^{(2)}(t) \;=\; A_\alpha^{(2)} \cos\left[\omega_\alpha (t-\tau) + \theta_\alpha^{(2)}\right] \;, \tag{56}$$

where amplitudes $A_\alpha^{(2)}$ and phases $\theta_\alpha^{(2)}$ follow from two conditions, a match of Eq. (56) with Eq. (53) at $t = \tau$, i. e., from

$$A_\alpha^{(2)} \cos\theta_\alpha^{(2)} \;=\; A_\alpha^{(1)} \cos\left(\omega_\alpha \tau + \theta_\alpha^{(1)}\right) \;, \tag{57}$$

and from the reassignment of the velocities, i.e., from matching the velocities corresponding to (56) to new velocities [c.f. (45, 48)]

$$\frac{\mathrm{d}q_\alpha^{(2)}(\tau)}{\mathrm{d}t} \;=\; -\,A_\alpha^{(2)}\,\omega_\alpha \sin\theta_\alpha^{(2)} \;=\; \lambda\, u_\alpha^{(1)} \;. \tag{58}$$

Defining

$$u_\alpha \;=\; u_\alpha^{(1)}/\lambda_1 \;, \tag{59}$$

where $\lambda_1 = \sqrt{T_1/T_0}$, one can restate Eq. (58)

$$-\,A_\alpha^{(2)}\,\omega_\alpha \sin\theta_\alpha^{(2)} \;=\; \lambda_2\, u_\alpha \;, \tag{60}$$

where $\lambda_2 = \lambda\,\lambda_1 = \sqrt{T_2/T_0}$. We note that the velocities u_α satisfy

$$\left\langle m_\alpha u_\alpha^2 \right\rangle_u \;=\; k_\mathrm{B} T_0 \;. \tag{61}$$

Equations (57) and (60) allow one to determine $A_\alpha^{(2)}$ and $\theta_\alpha^{(2)}$ and to describe the motion for $t \geq \tau$ according to (56).

Expression for the Temperature Echoes

One can now determine the kinetic energy after the velocity reassignments. The resulting kinetic energy is

$$E_k(t) = \sum_\alpha \frac{1}{2} m_\alpha \omega_\alpha^2 [A_\alpha^{(2)}]^2 \sin^2 [\omega_\alpha (t-\tau) + \theta_\alpha^{(2)}] \tag{62}$$

$$= \sum_\alpha \frac{1}{2} m_\alpha \omega_\alpha^2 \left\{ A_\alpha^{(2)} \sin \theta_\alpha^{(2)} \cos [\omega_\alpha (t-\tau)] + A_\alpha^{(2)} \cos \theta_\alpha^{(2)} \sin [\omega_\alpha (t-\tau)] \right\}^2 . \tag{63}$$

Using Eqs. (57) and (60), and then (54) and (55), one can express the amplitudes $A_\alpha^{(2)}$ and phases $\theta_\alpha^{(2)}$ in terms of the initial amplitudes A_α and phases θ_α in Eq. (50) as well as through the velocities u_α introduced in Eq. (59). One obtains

$$E_k(t) = \sum_\alpha \frac{1}{2} m_\alpha \omega_\alpha^2 \left\{ -\frac{\lambda_2 u_\alpha}{\omega_\alpha} \cos [\omega_\alpha (t-\tau)] + A_\alpha^{(1)} \cos [\omega_\alpha \tau + \theta_\alpha^{(1)}] \sin [\omega_\alpha (t-\tau)] \right\}^2$$

$$= \sum_\alpha \frac{1}{2} m_\alpha \omega_\alpha^2 \left\{ -\frac{\lambda_2 u_\alpha}{\omega_\alpha} \cos [\omega_\alpha (t-\tau)] + \sin [\omega_\alpha (t-\tau)] \left(\frac{\lambda_1 u_\alpha}{\omega_\alpha} \sin \omega_\alpha \tau + A_\alpha \cos \theta_\alpha \cos \omega_\alpha \tau \right) \right\}^2 . \tag{64}$$

This expression needs to be averaged over A_α, θ_α, and u_α. Employing Eqs. (19), (51) and (61) one can carry out the necessary averages which results in

$$\langle E_k(t) \rangle = \left[\frac{(3N-6)\, k_B T_0}{2} \right] \left\langle \frac{1+\lambda_1^2+2\lambda_2^2}{4} + \frac{1-\lambda_1^2}{4} \cos (2\omega_\alpha \tau) - \frac{1-\lambda_1^2}{8} \cos (2\omega_\alpha t) - \frac{1+\lambda_1^2-2\lambda_2^2}{4} \cos [2\omega_\alpha (t-\tau)] - \frac{1-\lambda_1^2}{8} \cos [2\omega_\alpha (t-2\tau)] + \frac{\lambda_1 \lambda_2}{2} \cos \left[2\omega_\alpha (t - \frac{\tau}{2}) \right] - \frac{\lambda_1 \lambda_2}{2} \cos \left[2\omega_\alpha (t - \frac{3\tau}{2}) \right] \right\rangle_\alpha \tag{65}$$

where $\langle \ldots \rangle_\alpha$ denotes the remaining average over all normal modes as described by Eq. (29). In the above derivation we have employed the property that the average of $u_\alpha A_\alpha$ over A_α and u_α vanishes.

At this point one can introduce the temperature-temperature correlation function expressed through Eq. (28). According to (28), one can replace all occurrences of $\langle \cos [2\omega_\alpha (t-t')] \rangle_\alpha$ by $C_{T,T}(t-t')$. Using (65), one obtains then for $t \geq \tau$

$$T(t) = T_0 \left[\frac{1+\lambda_1^2+2\lambda_2^2}{4} + \frac{1-\lambda_1^2}{4} C_{T,T}(\tau) - \frac{1-\lambda_1^2}{8} C_{T,T}(t) - \frac{1+\lambda_1^2-2\lambda_2^2}{4} C_{T,T}(t-\tau) - \frac{1-\lambda_1^2}{8} C_{T,T}(|t-2\tau|) + \frac{\lambda_1\lambda_2}{2} C_{T,T}\left(t-\frac{\tau}{2}\right) - \frac{\lambda_1\lambda_2}{2} C_{T,T}\left(\left|t-\frac{3\tau}{2}\right|\right) \right] . \quad (66)$$

$C_{T,T}(t)$ decays on a time scale of τ_0 [c.f. Eqs. (30) and (31) and Fig. 3] and, hence, we can note $C_{T,T}(\tau) \approx 0$, $C_{T,T}(t) \approx 0$ and $C_{T,T}(|t-\frac{\tau}{2}|) \approx 0$ when $t \geq \tau \gg \tau_0$. This leads to the expression

$$T(t) \approx T_0 \Bigg[\underbrace{\frac{1+\lambda_1^2+2\lambda_2^2}{4}}_{\text{new temperature}} - \underbrace{\frac{1+\lambda_1^2-2\lambda_2^2}{4} C_{T,T}(t-\tau)}_{\text{temperature recovery}} - \underbrace{\frac{\lambda_1\lambda_2}{2} C_{T,T}\left(\left|t-\frac{3\tau}{2}\right|\right)}_{\frac{3}{2}\tau\text{-pulse}} - \underbrace{\frac{1-\lambda_1^2}{8} C_{T,T}(|t-2\tau|)}_{2\tau\text{-pulse}} \Bigg] . \quad (67)$$

The terms in this expression can be interpreted in a straightforward way. We first note that the second, third, and fourth term do not contribute, except for $t \approx \tau, 3\tau/2, 2\tau$. The first term, accordingly, describes the average temperature after the second velocity reassignment. The second term describes the recovery of the temperature immediately after the second reassignment, i. e., at $t = \tau$. Since $C_{T,T}(0) = 1$, the temperature at $t = \tau$, given by the first two terms in Eq. (67), is $T_2 = \lambda_2^2 T_0$, the expected result. The second term in Eq. (67) describes the relaxation of the temperature from this initial value to the average temperature $\frac{1}{4}T_0(1 + \lambda_1^2 + 2\lambda_2^2)$. This term conforms with Eq. (40) which states that the temperature recovery can be described by $C_{T,T}(t)$.

The third term in (67) describes the $\frac{3}{2}\tau$-pulse; the prefactor of $C_{T,T}(t)$ in this term is the depth $\widetilde{\Delta T}(3\tau/2)$ of this pulse, i. e.,

$$\widetilde{\Delta T}\left(\frac{3\tau}{2}\right) = \frac{T_0}{2}\lambda_1\lambda_2 . \quad (68)$$

The fourth term describes the 2τ-pulse; the depth of this pulse is

$$\widetilde{\Delta T}(2\tau) = \frac{T_0}{8}(1 - \lambda_1^2) . \quad (69)$$

Expression (67) not only correctly predicts the existence of echoes at $t = 3\tau/2$ and at $t = 2\tau$, but also provides a description of the detailed time dependence of the temperature recovery at $t = \tau$ and of the two echoes. Figure 9 demonstrates that Eq. (67) describes the temperature recovery and the echoes well. However, Eq. (67) predicts that the echo depths are independent of the

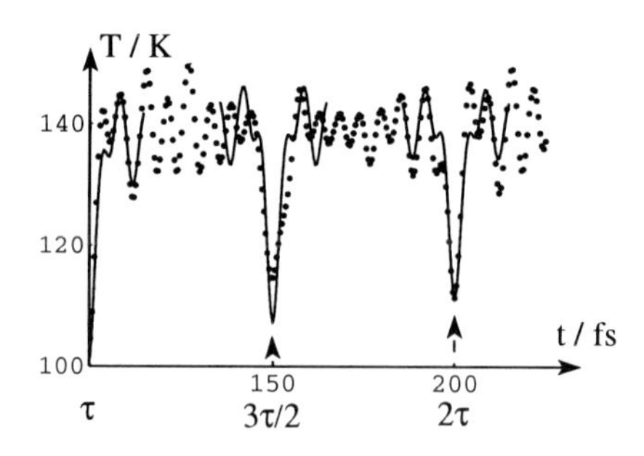

Fig. 9. Comparison of the kinetic energy of bacteriorhodopsin after velocity reassignments at $t = 0$ and $t = \tau$ for $\tau = 100$ fs and $T_0 = 300\,\mathrm{K}$, $T_1 = 50\,\mathrm{K}$, $T_2 = 100\,\mathrm{K}$ (same as Fig. 4) with the expression (67) (thin line); the simulated values $T(t)$ are shown as dots.

delay time τ between velocity reassignments. This contradicts the results shown in Fig. 7, the deficiency stemming from the neglect of interactions between vibrational modes which give rise to dephasing. A theory which accounts for dephasing and correctly describes the τ-dependence of the temperature echoes will be provided in the next section.

5 Langevin Oscillator Description of Echoes

To account for dephasing as an explanation for a decrease of the echo depth with increasing τ we adopt in this section a description of protein motion in terms of an ensemble of Langevin oscillators. Such an ensemble is provided by the set of stochastic differential equations [46, 47]

$$\ddot{x}_\alpha + b_\alpha \dot{x}_\alpha + \omega_\alpha^2 x_\alpha = \eta_\alpha(t) , \quad \alpha = 1, 2, \ldots, 3N - 6 . \tag{70}$$

In this model residual (anharmonic) interactions between normal modes are accounted for through dissipative (friction) and fluctuating forces $-b_\alpha \dot{x}_\alpha$ and $\eta_\alpha(t)$, respectively. These two terms are related through (fluctuation-dissipation theorem [42])

$$\langle \eta_\alpha(0)\, \eta_\alpha(t) \rangle = 2 k_\mathrm{B} T_0 b_\alpha \delta(t)/m_\alpha , \tag{71}$$

a property which ensures thermal equilibrium at temperature T_0. We assume that all modes are under-damped, i.e., $\omega_\alpha > \frac{1}{2}\, b_\alpha$, an assumption which we will justify below. Accordingly, we assume that the quantity

$$\Omega_\alpha = \sqrt{\omega_\alpha^2 - \frac{1}{4} b_\alpha^2} \tag{72}$$

is real.

Time-dependent averages for a Langevin oscillator. To describe the temperature echo, one needs to determine the average kinetic energy of the ensemble of Langevin oscillators (70) after two consecutive reassignments of velocities. For this purpose the average quantities $\langle x \rangle$, $\langle x^2 \rangle$, $\langle v \rangle$ and $\langle v^2 \rangle$ for a given initial position $x_\alpha(0)$ and velocity $v_\alpha(0)$ of Langevin oscillators will be required. We will demonstrate here the well known [48] evaluation of $\langle x \rangle$, leaving the evaluation of the remaining quantities as an exercise to the reader.

Ensemble averaging of Eq. (70) results in the equation

$$\langle \ddot{x}_\alpha \rangle + b_\alpha \langle \dot{x}_\alpha \rangle + \omega_\alpha^2 \langle x_\alpha \rangle = 0 . \tag{73}$$

We seek the corresponding equation for the Laplace transform

$$\langle \hat{x}_\alpha \rangle = \int_0^\infty dt\, e^{-\omega t} \langle x_\alpha \rangle . \tag{74}$$

In the case that the initial position of the oscillator is $x(0) = x_o$ and the initial velocity is $\dot{x}(0) = v_o$, it holds that

$$\left\langle \widehat{\dot{x}_\alpha} \right\rangle = \omega \langle \hat{x}_\alpha \rangle - x_o \tag{75}$$

$$\left\langle \widehat{\ddot{x}_\alpha} \right\rangle = \omega^2 \langle \hat{x}_\alpha \rangle - \omega x_o - v_o . \tag{76}$$

Accordingly, Laplace transformation of Eq. (73) results in

$$\omega^2 \langle \hat{x}_\alpha \rangle - \omega x_o - v_o + b\omega \langle \hat{x}_\alpha \rangle - b x_o + \omega_\alpha^2 \langle \hat{x}_\alpha \rangle = 0 \tag{77}$$

which can be written, using Eq. (72),

$$\langle \hat{x}_\alpha \rangle = \frac{\left(\omega + \frac{b}{2}\right) x_o}{\left(\omega + \frac{b}{2}\right)^2 + \Omega^2} + \frac{\frac{b}{2} x_o + v_o}{\left(\omega + \frac{b}{2}\right)^2 + \Omega^2} . \tag{78}$$

Inverse Laplace transform yields for real Ω_α

$$\langle x_\alpha(t) \rangle = x_\alpha(0)\, e^{-b_\alpha t/2} A(\Omega_\alpha t) + \frac{v_\alpha(0)}{\Omega_\alpha} e^{-b_\alpha t/2} \sin(\Omega_\alpha t) \tag{79}$$

where we defined

$$A(\Omega_\alpha t) = \cos(\Omega_\alpha t) + \frac{b_\alpha}{2\Omega_\alpha} \sin(\Omega_\alpha t) . \tag{80}$$

The remaining averages can be derived similarly and are given by (see Eq. (214) in [48])

$$\langle x_\alpha^2(t) \rangle = \langle x_\alpha(t) \rangle^2 + \frac{k_B T_0}{m_\alpha \omega_\alpha^2} (1 - e^{-b_\alpha t} B(\Omega_\alpha t)) \tag{81}$$

$$\langle v_\alpha(t) \rangle = -\frac{x_\alpha(0)\omega_\alpha^2}{\Omega_\alpha} e^{-b_\alpha t/2} \sin(\Omega_\alpha t) + v_\alpha(0)\, e^{-b_\alpha t/2} A(\Omega_\alpha t) \tag{82}$$

$$\langle v_\alpha^2(t) \rangle = \langle v_\alpha(t) \rangle^2 + \frac{k_B T_0}{m_\alpha} (1 - e^{-b_\alpha t} B(\Omega_\alpha t)) \tag{83}$$

where

$$B(\Omega_\alpha t) = \frac{b_\alpha^2}{2\Omega_\alpha^2}\sin^2(\Omega_\alpha t) + \frac{b_\alpha}{2\Omega_\alpha}\sin(2\Omega_\alpha t) + 1 \ . \quad (84)$$

We will consider in the following the case of a constant temperature velocity reassignment echo for which one needs to assign twice the same velocities u_α to all modes. In order to determine the kinetic energy we consider again, in sequence, three periods: (0) before the first velocity reassignment ($t < 0$), (1) after the first and before the second velocity reassignment ($0 \le t < \tau$), and (2) after the second velocity reassignment ($t \ge \tau$). The derivation below follows again closely [35].

(0) Before the first velocity reassignment. This period does not need to be described in detail. All that is required is information on the positions at the end of this period, namely, $x_\alpha^{(0)}(0)$. In fact, we solely need the average values

$$\left\langle x_\alpha^{(0)}(0) \right\rangle_{(0)} = 0 \ , \quad \left\langle \left[x_\alpha^{(0)}(0)\right]^2 \right\rangle_{(0)} = \frac{k_\mathrm{B} T_0}{m_\alpha \omega_\alpha^2} \ , \quad (85)$$

as becomes evident below. Here $\langle \ldots \rangle_{(0)}$ denotes the average for the system before the first reassignment. Since the velocities are reassigned at $t = 0$, no information on the velocities $v_\alpha(0)$ is required.

(1) After the first reassignment but before the second. From this period of the dynamics again only the positions at time $t = \tau$, i.e., $x_\alpha^{(1)}(t)$, are required since the velocities will be reassigned. In fact, one needs solely the averages of $x_\alpha^{(1)}(t)$ and of $[x_\alpha^{(1)}(t)]^2$. Using Eqs. (79) and (81) one obtains for initial positions $x_\alpha^{(1)}(0) = x_\alpha^{(0)}(0)$ and the assigned velocities $v_\alpha^{(1)}(0) = u_\alpha$

$$\begin{aligned} \left\langle x_\alpha^{(1)}(\tau) \right\rangle_{(1)} &= x_\alpha^{(0)}(0)\, e^{-b_\alpha \tau/2} \left[\cos(\Omega_\alpha \tau) + \frac{b_\alpha}{2\Omega_\alpha}\sin(\Omega_\alpha \tau)\right] \\ &\quad + \frac{u_\alpha}{\Omega_\alpha}\, e^{-b_\alpha \tau/2} \sin(\Omega_\alpha \tau) \end{aligned} \quad (86)$$

$$\begin{aligned} \left\langle [x_\alpha^{(1)}(\tau)]^2 \right\rangle_{(1)} &= \left\langle x_\alpha^{(1)}(\tau) \right\rangle_{(1)}^2 + \frac{k_\mathrm{B} T_0}{m_\alpha \omega_\alpha^2} \{1 - \\ &\quad e^{-b_\alpha \tau}\left[\frac{b_\alpha^2}{2\Omega_\alpha^2}\sin^2(\Omega_\alpha \tau) + \frac{b_\alpha}{2\Omega_\alpha}\sin(2\Omega_\alpha \tau) + 1\right]\Big\} \end{aligned} \quad (87)$$

In using these quantities below one needs to carry out the averages over all positions $x_\alpha^{(0)}(0)$, denoted by $\langle \cdots \rangle_{(0)}$, and over all velocities u_α, denoted by $\langle \cdots \rangle_u$.

(2) After the second reassignment. From this period one seeks solely information on the velocities in order to determine the average kinetic energy. For specific initial positions $x_\alpha^{(2)}(\tau) = x_\alpha^{(1)}(\tau)$ and assigned velocities $v_\alpha^{(2)}(\tau) = u_\alpha$ it follows, using Eqs. (82) and (83), that

$$\begin{aligned}\left\langle v_\alpha^{(2)}(t) \right\rangle_{(2)} &= -\frac{x_\alpha^{(1)}(\tau)\omega_\alpha^2}{\Omega_\alpha}\, e^{-b_\alpha(t-\tau)/2} \sin \Omega_\alpha(t-\tau) \\ &\quad + u_\alpha\, e^{-b_\alpha(t-\tau)/2} \left[\cos \Omega_\alpha(t-\tau) - \frac{b_\alpha}{2\Omega_\alpha} \sin \Omega_\alpha(t-\tau) \right] \qquad (88)\end{aligned}$$

$$\begin{aligned}\left\langle [v_\alpha^{(2)}(t)]^2 \right\rangle_{(2)} &= \left\langle v_\alpha^{(2)}(t) \right\rangle_{(2)}^2 + \frac{k_B T_0}{m_\alpha} \left\{ 1 - e^{-b_\alpha(t-\tau)} \left[\frac{b_\alpha^2}{2\Omega_\alpha^2} \sin^2 \Omega_\alpha(t-\tau) \right.\right. \\ &\quad \left.\left. - \frac{b_\alpha}{2\Omega_\alpha} \sin 2\Omega_\alpha(t-\tau) + 1 \right] \right\} . \qquad (89)\end{aligned}$$

In using these quantities below, one needs to carry out the averages over all positions $x_\alpha^{(1)}(\tau)$, denoted by $\langle \cdots \rangle_{(1)}$, and over all velocities u_α, denoted by $\langle \cdots \rangle_u$. Note that the velocities u_α must be averaged simultaneously for Eqs. (86–89) since the same velocities are assigned at $t = 0$ and at $t = \tau$.

Evaluation of the Temperature Echoes

The average kinetic energy and, hence, the depth of the temperature echo, can be determined from Eqs. (88) and (89) after the following additional averages are taken:

1. average $\langle \cdots \rangle_{(1)}$ over all initial positions $x_\alpha^{(1)}(\tau)$, employing Eqs. (86) (87);
2. average $\langle \cdots \rangle_{(0)}$ over all initial positions $x_\alpha^{(1)}(0)$, employing Eq. (85);
3. average $\langle \cdots \rangle_u$ over all reassigned velocities u_α using $\langle u_\alpha \rangle = 0$ and Eq. (61);
4. average $\langle \cdots \rangle_\alpha$ over all modes α.

One obtains in this way for the echo depth

$$\Delta T(3\tau/2) = T_0 - \left\langle \left\langle \left\langle [m_\alpha v_\alpha^{(2)}(3\tau/2)]^2 / k_B \right\rangle_{(2),(1),(0)} \right\rangle_u \right\rangle_\alpha \qquad (90)$$

or

$$\begin{aligned}\Delta T(3\tau/2) &= \frac{T_0}{2} \left\langle e^{-b_\alpha \tau} \frac{\omega_\alpha^2}{\Omega_\alpha^2} \left[1 - \cos(2\Omega_\alpha \tau) - \frac{b_\alpha}{\Omega_\alpha} \sin(\Omega_\alpha \tau) + \right.\right. \\ &\quad \left.\left. \frac{b_\alpha}{2\Omega_\alpha} \sin(2\Omega_\alpha \tau) \right] \right\rangle_\alpha . \qquad (91)\end{aligned}$$

For $b_\alpha = 0$ for all α, i. e., for oscillators without friction and fluctuating forces, (91) yields a τ-independent echo depth

$$\Delta T(3\tau/2) = \frac{T_0}{2} [1 - C_{T,T}(\tau)] \approx \frac{T_0}{2} \qquad (92)$$

which reproduces the result (68) derived for the harmonic model for $\lambda_1 = \lambda_2 = 1$.

In the case $b_\alpha \neq 0$, it holds that [49]

$$\frac{\langle m_\alpha\, x_\alpha(0)\, x_\alpha(2\tau)\rangle}{\langle m_\alpha\, x_\alpha^2\rangle} = e^{-b_\alpha \tau} \left[\cos(2\Omega_\alpha \tau) + \frac{b_\alpha}{2\Omega_\alpha} \sin(2\Omega_\alpha \tau) \right] ,$$
$$\frac{\langle m_\alpha\, v_\alpha(0)\, v_\alpha(2\tau)\rangle}{\langle m_\alpha\, v_\alpha^2\rangle} = e^{-b_\alpha \tau} \left[\cos(2\Omega_\alpha \tau) - \frac{b_\alpha}{2\Omega_\alpha} \sin(2\Omega_\alpha \tau) \right] . \quad (93)$$

For $\tau \gg \tau_0$, both of the above quantities can be omitted, such that $\left\langle \frac{b_\alpha}{\Omega_\alpha} \sin(\Omega_\alpha \tau) \right\rangle_\alpha$ and $\left\langle \cos(2\Omega_\alpha \tau) - \frac{b_\alpha}{2\Omega_\alpha} \sin(2\Omega_\alpha \tau) \right\rangle_\alpha$ can also be assumed to be negligible. Accordingly, we approximate (91)

$$\Delta T(3\tau/2) \approx \frac{T_0}{2} \left\langle e^{-b_\alpha \tau} \frac{\omega_\alpha^2}{\Omega_\alpha^2} \right\rangle_\alpha . \quad (94)$$

In the limit $b_\alpha \ll \omega_\alpha$ and choosing b_α the same constant b_0 for all α, follows

$$\Delta T(\tau) \approx \frac{T_0}{2}\, e^{-b_0 \tau} . \quad (95)$$

This expression for the echo depth predicts correctly the exponential dependence on τ which results from molecular dynamics simulations as shown in Fig. 7. Match of the simulation data yields a friction constant $b_0 = 1.43\ \mathrm{ps}^{-1}$ at $T_0 = 300\,\mathrm{K}$. This value characterizes the vast majority of protein modes as underdamped, justifying our earlier assumption.

6 Temperature Echo in a One-Dimensional Chain

In our models above we have assumed ensembles of harmonic oscillators with given frequencies ω_α. These frequencies result, however, from the dynamics of harmonically coupled particles. The simplest case is a one-dimensional chain of N atoms with momenta p_n and positions x_n governed by a Hamiltonian

$$H = \sum_{n=1}^{N} \frac{p_n^2}{2m} + \sum_{n=1}^{N+1} \frac{1}{2} m\tilde{\omega}^2 (x_n - x_{n-1})^2 , \quad x_0\,, x_{N+1}\ \text{fixed} . \quad (96)$$

The Hamiltonian can be expressed as a sum of quadratic forms $H = T + V + W$ where

$$T = \sum_{n=1}^{N} t_{nn} p_n^2 , \quad t_{nm} = \frac{1}{2m} \delta_{nm} \quad (97)$$

$$V = \sum_{n=1}^{N} v_{nn} x_n^2 , \quad v_{nm} = m\tilde{\omega}^2\, \delta_{nm} \quad (98)$$

$$W = \sum_{n,m=1}^{N} w_{nm} x_n x_m , \quad w_{nm} = -\frac{1}{2} m\tilde{\omega}^2 \left(\delta_{n,m-1} + \delta_{n,m+1}\right) . \quad (99)$$

Denoting by $\hat{t}, \hat{v}, \hat{w}$ the matrices with elements t_{nm}, v_{nm}, w_{nm}, one seeks a similarity transformation $\mathcal{A}$ which makes these matrices simultaneously diagonal. One can readily show that any $\mathcal{A}$ leaves $\tilde{t} = \mathcal{A}\hat{t}\mathcal{A}^{-1}$ and $\tilde{v} = \mathcal{A}\hat{v}\mathcal{A}^{-1}$ diagonal. Defining for $\alpha = 1, 2, \ldots N$

$$q_\alpha = \sum_{n=1}^{N} \mathcal{A}_{\alpha n}\, x_n \tag{100}$$

$$\tilde{p}_\alpha = \sum_{n=1}^{N} \mathcal{A}_{\alpha n}\, p_n \tag{101}$$

one can, in fact, readily show

$$T = \sum_{\alpha=1}^{N} \tilde{t}_{\alpha\alpha}\tilde{p}_\alpha^2 \ , \quad \tilde{t}_{\alpha\beta} = \frac{1}{2m}\,\delta_{\alpha\beta} \ , \tag{102}$$

$$V = \sum_{\alpha=1}^{N} \tilde{v}_{\alpha\alpha}q_\alpha^2 \ , \quad \tilde{v}_{\alpha\beta} = m\tilde{\omega}^2\,\delta_{\alpha\beta} \ . \tag{103}$$

In order to render $\tilde{w} = \mathcal{A}\hat{w}\mathcal{A}^{-1}$ diagonal one chooses the well known matrix

$$\mathcal{A}_{\alpha n} = \sqrt{\frac{2}{N}}\,\sin\left(\frac{\alpha n \pi}{N+1}\right) \ . \tag{104}$$

One can verify

$$W = \sum_{\alpha=1}^{N} \tilde{w}_{\alpha\alpha}q_\alpha^2 \ , \quad \tilde{w}_{\alpha\beta} = m\tilde{\omega}^2\,\cos\left(\frac{\alpha\pi}{N+1}\right)\,\delta_{\alpha\beta} \ . \tag{105}$$

This matrix is orthogonal, i.e., $\mathcal{A}^T$ is the inverse of $\mathcal{A}$. Using $1 + \cos(\alpha\pi/N{+}1) = 2\cos^2(\alpha\pi/2(N+1))$ one can finally write the Hamiltonian (96)

$$H = \sum_{\alpha=1}^{N} \left(\frac{\tilde{p}_\alpha^2}{2m} + \frac{1}{2}m\omega_\alpha^2 q_\alpha^2\right) \tag{106}$$

where

$$\omega_\alpha = 2\tilde{\omega}\,\cos\frac{\alpha\pi}{2(N+1)} \ . \tag{107}$$

Having derived the normal modes, we can now describe temperature echoes for the one-dimensional chain. We will consider in the following solely the case of temperature quench echoes. We adopt a description which applies both to all atom echoes as well as to local echoes, i.e., we will quench the kinetic energy of atoms $n = 1, 2, \ldots r$ and observe the response of the particles $m = s \ldots t$ where $r \leq N$, $1 \leq s \leq N$, $s \leq t \leq N$.

(0) Before the first quench. Before the first quench, i.e., for $t < 0$, the position and corresponding velocity of the α-th normal mode can be expressed [c.f. Eqs. (50) and (52)]

$$q_\alpha{}^{(0)}(t) = A_\alpha{}^{(0)} \cos(\omega_\alpha t + \theta_\alpha{}^{(0)}) \tag{108}$$

$$v_\alpha{}^{(0)}(t) = -A_\alpha{}^{(0)} \omega_\alpha \sin(\omega_\alpha t + \theta_\alpha{}^{(0)}) \tag{109}$$

where the amplitudes $A_\alpha{}^{(0)}$ are distributed according to the Rayleigh distribution (19) and where the phases $\theta_\alpha{}^{(0)}$ are homogenously distributed in the interval $[0, 2\pi]$.

(1) After the first quench but before the second. One needs to express at this point the effect of the quench, executed on the velocities $\dot{x}_n$, in terms of the velocities v_α of the individual modes α. For this purpose we define the quench matrix $\mathcal{Q}$

$$\mathcal{Q}_{nm} = \begin{cases} 1 \text{ if } n = m \text{ and } r+1 \leq n \leq N \\ 0 \text{ otherwise} \end{cases} \tag{110}$$

which, applied to the vector of atomic velocities $(\dot{x}_1, \dot{x}_2, \ldots \dot{x}_N)^T$, describes the first quench. The effect of the quench on the normal mode velocities v_α is then described by

$$\mathcal{B} = \mathcal{A}\mathcal{Q}\mathcal{A}^T \ . \tag{111}$$

We express the motion of the normal modes at $t > 0$ in the by now familiar form

$$q_\alpha{}^{(1)}(t) = A_\alpha{}^{(1)} \cos(\omega_\alpha t + \theta_\alpha{}^{(1)}) \tag{112}$$

$$v_\alpha{}^{(1)}(t) = -A_\alpha{}^{(1)} \omega_\alpha \sin(\omega_\alpha t + \theta_\alpha{}^{(1)}) \ . \tag{113}$$

The new amplitudes $A_\alpha{}^{(1)}$ and phases $\theta_\alpha{}^{(1)}$ are determined through the two conditions

$$q_\alpha{}^{(1)}(0) = q_\alpha{}^{(0)}(0) \tag{114}$$

$$v_\alpha{}^{(1)}(0) = \sum \mathcal{B}_{\alpha\beta}\, v_\beta{}^{(0)}(0) \ . \tag{115}$$

From this follows

$$A_\alpha{}^{(1)} \cos(\theta_\alpha{}^{(1)}) = A_\alpha{}^{(0)} \cos(\theta_\alpha{}^{(0)}) \tag{116}$$

$$-\omega_\alpha A_\alpha{}^{(2)} \sin(\theta_\alpha{}^{(1)}) = -\sum_{\beta=1}^{N} \mathcal{B}_{\alpha\beta}\omega_\beta A_\beta{}^{(0)} \sin(\theta_\beta{}^{(0)}) \ . \tag{117}$$

This yields for $A_\alpha{}^{(1)}$ and $\theta_\alpha{}^{(1)}$

$$\theta_\alpha{}^{(1)} = \arctan\left(\frac{\sum_{\beta=1}^{N} \mathcal{B}_{\alpha\beta}\, \omega_\beta A_\beta{}^{(0)} \sin(\theta_\beta{}^{(0)})}{\omega_\alpha A_\alpha{}^{(0)} \cos(\theta_\alpha{}^{(0)})}\right) \tag{118}$$

$$A_\alpha{}^{(1)} = \left\{ [A_\alpha{}^{(0)}]\cos^2(\theta_\alpha{}^{(0)}) + \left[\sum_{\beta=1}^{N} \mathcal{B}_{\alpha\beta}\, \omega_\beta A_\beta{}^{(0)} \sin(\theta_\beta{}^{(0)})/\omega_\alpha\right]^2 \right\}^{\frac{1}{2}} \tag{119}$$

(2) After the second quench. At $t = \tau$, the kinetic energy of the particles $1, \ldots, r$ is quenched again. Expressing

$$q_\alpha{}^{(2)}(t) = A_\alpha{}^{(2)} \cos(\omega_\alpha t + \theta_\alpha{}^{(2)}) \tag{120}$$

$$v_\alpha{}^{(2)}(t) = -A_\alpha{}^{(2)} \omega_\alpha \sin(\omega_\alpha t + \theta_\alpha{}^{(2)}) \tag{121}$$

one can again determine the new amplitudes and phases from the conditions

$$q_\alpha{}^{(2)}(\tau) = q_\alpha{}^{(1)}(\tau) \tag{122}$$

$$v_\alpha{}^{(2)}(\tau) = \sum \mathcal{B}_{\alpha\beta}\, v_\beta{}^{(1)}(\tau) \tag{123}$$

from which follows

$$\theta_\alpha{}^{(2)} = \arctan\left(\frac{\sum_{\beta=1}^{N} \mathcal{B}_{\alpha\beta}\, \omega_\beta A_\beta{}^{(1)} \sin(\omega_\beta \tau + \theta_\beta{}^{(1)})}{\omega_\alpha A_\alpha{}^{(1)} \cos(\omega_\beta \tau + \theta_\alpha{}^{(1)})}\right) \tag{124}$$

$$A_\alpha{}^{(2)} = \Bigg\{ [A_\alpha{}^{(1)}] \cos^2(\omega_\alpha \tau + \theta_\alpha{}^{(1)}) + \tag{125}$$

$$+ \left[\sum_{\beta=1}^{N} \mathcal{B}_{\alpha\beta}\, \omega_\beta A_\beta{}^{(1)} \sin(\omega_\beta \tau + \theta_\beta{}^{(0)})/\omega_\alpha\right]^2 \Bigg\}^{\frac{1}{2}} .$$

The velocity and, hence, the kinetic energy of the chain at time $t > \tau$ can be determined from Eqs. (118), (119), (124), and (125). In the present case we focus on the temperature of the particles $n, s \leq n \leq t$ which is given by

$$T(t) = \frac{2}{k_B(t - s + 1)} \sum_{i=s}^{t} \frac{1}{2} m_i \dot{x}_i^2(t) \ . \tag{126}$$

Echoes in one-dimensional chain. We have evaluated the all atom as well as local echoes for a linear harmonic chain of 200 atoms as described above. For this purpose we have carried out a numerical average, choosing initial amplitudes $A_\alpha^{(0)}$ and initial phases $\theta_\alpha^{(0)}$ employing suitable random number generators. We have employed samples of 50 such random choices. We employed Eqs. (118), (119), (124) and (125) to determine $A_\alpha^{(0)}$ and $\theta_\alpha^{(0)}$ and evaluated the temperature trace $T(t)$ using Eqs. (126) and (121). In this case atoms $1, 2, \ldots M$ were quenched and monitored through their total kinetic energy. The resulting behaviour of $T(t)$ is shown in Fig. 10 for $M = 200$ (all atom echo) as well as for $M = 100$ and $M = 50$. One can recognize that the local quenches produce only weak echoes in the present case, e.g., in case of $M = 100, 50$ the echo depth measures 37 K and 15 K. This is due to the fact that in a homogenous linear chain vibrational modes extend over the complete chain. The appearance of stronger local echoes in proteins is due to the inhomogeneity of proteins which results in localized vibrational motion. One can recognize in Fig. 10 (b) (c) clearly the diffusion of temperature into the quenched part of the chain.

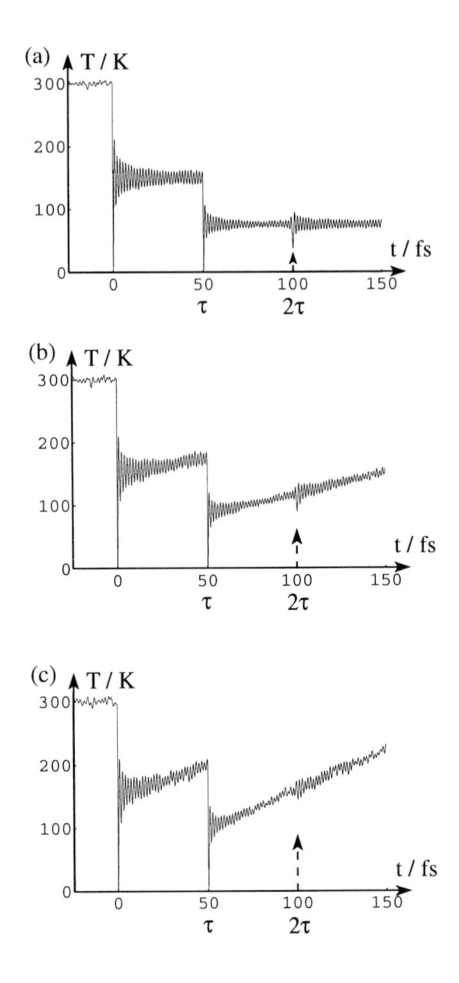

Fig. 10. Kinetic energy of a one-dimensional chain of 200 oscillators for an all atom and a local quench at $t = 0$ and $t = \tau$ with $t = 100$ (arbitrary units) and $T_0 = 300\,\mathrm{K}$, $T_1 = 0$, $T_2 = 0$; (a) all atoms quenched; (b) the first 100 oscillators quenched; shown is the kinetic energy of these oscillators; (c) the first 50 oscillators quenched; shown is the kinetic energy of these oscillators. In the calculations each mode had been sampled through 50 initial amplitudes and phases which were randomly assigned.

7 Conclusion

Proteins are materials which do not possess narrowly confined equilibria at physiological temperatures. Rather, the nature of their role in biological cells requires proteins to exhibit flexibility. As a result, proteins are restless wanderers exploring the energy landscape characterizing their conformational universe. The lack of a deep, confining minimum implies that proteins do not exhibit harmonic behaviour with vibrational modes corresponding to a local quadratic Hamiltonian characterizing the minimum. Rather, proteins live near the barriers of metastable states such that their local Hamiltonians involve significant non-harmonic terms and that jumps to new metastable states occur, leading again and again to new local Hamiltonians. It is a considerable and important challenge to characterize motions of such systems. This chapter has shown how a combination of molecular dynamics simulation and mathematical analysis allows one to characterize protein motion.

To identify approximate protein modes one can either determine a local quadratic Hamiltonian and bring it to diagonal form or sample the covariance matrix $\langle x_j \, x_k \rangle$ and determine its eigenvectors, the so-called principle components [9]. One finds, however, that over the time course of 100 ps the principle components drift, reflecting the protein's "wanderlust" [9]. The question arises, if the modes identified such exhibit, at least over short times, the behaviour of normal modes in that coherent phases are maintained. Identifying coherency through temperature echoes, this chapter provided an answer to this question: over a time scale of about 1 ps coherency is indeed maintained. This finding is relevant since reaction processes in proteins, e.g., electron transfer, spectral transitions or dipolar relaxation, are governed by normal modes [50, 11].

The mathematical analysis of temperature echoes introduced above shows that a description in terms of pure harmonic modes does not suffice, but that the relevant protein motion can be expressed well in terms of distributions of Langevin oscillators. This analysis extends beyond the realm of temperature echoes covering, for example, dipolar relaxation [34]. The Langevin oscillator expansion employed here and the combination of simulation and theory chosen here can serve researchers in future studies of protein function and dynamics and may be considered of equal importance for proteins as normal mode analysis is for small molecules and crystals.

Acknowledgements

The authors express their great indebtedness to Dong Xu who, together with K.S., had started the research on temperature echoes in our group, who had provided us with needed simulation programs, and who had introduced H.L. to molecular dynamics of temperature echoes. The authors like to thank also Jim Phillips for help with the preparation and the revision of the manuscript. This work has been supported by the National Institutes of Health grant P41RRO5969.

References

[1] C. Jungnickel and R. McCormmach. *Intellectual Mastery of Nature - Theoretical Physics from Ohm to Einstein, Vol. 1: The Torch of Mathematics 1800-1870.* Chicago University Press, Chicago, 1986.

[2] C. Jungnickel and R. McCormmach. *Intellectual Mastery of Nature - Theoretical Physics from Ohm to Einstein, Vol. 2: The Now Mighty Theoretical Physics 1870-1925.* Chicago University Press, Chicago, 1986.

[3] Thomas Bishop and Klaus Schulten. Molecular dynamics study of glucocorticoid receptor–DNA binding. *Proteins*, 24(1):115–133, 1996.
[4] Helmut Heller, Michael Schaefer, and Klaus Schulten. Molecular dynamics simulation of a bilayer of 200 lipids in the gel and in the liquid crystal-phases. *J. Phys. Chem.*, 97:8343–8360, 1993.
[5] Feng Zhou and Klaus Schulten. Molecular dynamics study of the activation of phospholipase A_2 on a membrane surface. *Proteins*. In press.
[6] Klaus Schulten, William Humphrey, Ilya Logunov, Mordechai Sheves, and Dong Xu. Molecular dynamics studies of bacteriorhodopsin's photocycles. *Israel Journal of Chemistry*, 35:447–464, 1995.
[7] R. Car and M. Parrinello. Structual, dynamics and electronic properties of amorphous silicon, an ab initio molecular-dynamics study. *Phys. Rev. Lett.*, 60:204–207, 1988.
[8] Helmut Grubmüller, Helmut Heller, Andreas Windemuth, and Klaus Schulten. Generalized Verlet algorithm for efficient molecular dynamics simulations with long-range interactions. *Molecular Simulation*, 6:121–142, 1991.
[9] Manel A. Balsera, Willy Wriggers, Yoshitsugu Oono, and Klaus Schulten. Principal component analysis and long time protein dynamics. *J. Phys. Chem.*, 100(7):2567–2572, 1996.
[10] Willy Wriggers and Klaus Schulten. A novel algorithm to investigate domain motions in proteins. To be submitted.
[11] Klaus Schulten. Curve crossing in a protein: Coupling of the elementary quantum process to motions of the protein. In D. Bicout and M. J. Field, editors, *Proceedings of the Ecole de Physique des Houches*, pages 85–118, Paris, 1995. Les Editions de Physique, Springer.
[12] Feng Zhou and Klaus Schulten. Molecular dynamics study of a membrane–water interface. *J. Phys. Chem.*, 99:2194–2208, 1995.
[13] M. P. Allen and D. J. Tildesley. *Computer Simulation of Liquids.* Oxford University Press, New York, 1987.
[14] J. P. Boon and S. Yip. *Molecular Hydrodynamics.* Dover Publications, Inc, New York, 1980.
[15] N. E. Cusack. *The Physics of Structurally Disordered Matter.* Adam Hilger, Bristol, 1987.
[16] Neil W. Ashcroft and N. David Mermin. *Solid State Physics.* W. B. Saunders, Philadelphia, 1976.
[17] Rita G. Lerner and George L. Trigg. *Encyclopedia of Physics.* VCH Publisher, Inc., New York, 2 edition, 1990.
[18] T. Noguti and N. Gō. Collective variable description of small-amplitude conformational fluctuations in a globular protein. *Nature*, 296:776, 1982.
[19] B. R. Brooks and M. Karplus. *Proc. Natl. Acad. Sci. USA*, 80:6571–6575, 1983.
[20] M. Levitt, C. Sander, and P. S. Stern. Protein normal-mode dynamics: Trypsin inhibitor, crambin, ribonuclease and lysozyme. *J. Mol. Biol.*, 181:423–447, 1985.
[21] N Gō, T. Noguti, and T. Nishikawa. Dynamics of a small globular protein, in terms of low-frequency vibrational modes. *Proc. Natl. Acad. Sci. USA*, 80:3696, 1983.
[22] R. M. Levy, A. R. Srinivasan, W. K. Olson, and J. A. McCammon. Quasi-harmonic method for studying very low frequency modes in proteins. *Biopolymers*, 23:1099, 1984.
[23] R. M. Levy, O. de la Luz Rojas, and R. A. Friesner. Quasi-harmonic method for calculating vibrational spectra from classical simulations on multidimensional anharmonic potential surfaces. *J. Phys. Chem.*, 88:4233, 1984.

[24] E.R. Henry, W.R. Eaton, and R.M. Hochstrasser. Molecular dynamics simulations of cooling in laser-excited heme proteins. *Proc. Natl. Acad. Sci. USA*, 83:8982–8986, 1986.

[25] J.F. Gibrat and N. Gō. Normal mode analysis of human lysozyme: Study of the relative motion of the two domains and characterization of harmonic motion. *Proteins: Structure, Function, and Genetics*, 8:258–279, 1990.

[26] B. R. Brooks and M. Karplus. Normal modes for specific motions of macromolecules: Application to the hinge-bending mode of lysozyme. *Proc. Natl. Acad. Sci. USA*, 82:4995, 1985.

[27] William Humphrey, Ilya Logunov, Klaus Schulten, and Mordechai Sheves. Molecular dynamics study of bacteriorhodopsin and artificial pigments. *Biochemistry*, 33:3668–3678, 1994.

[28] G. S. Grest, S. R. Nagel, and A. Rahman. Quench echoes in molecular dynamics — a new phonon spectroscopy. *Solid State Communications*, 36:875–879, 1980.

[29] G. S. Grest, S. R. Nagel, A. Rahman, and T. W. Witten, Jr. Density of states and the velocity autocorrelation function derived from quench studies. *J. Chem. Phys.*, 74(6):3532–3534, 1981.

[30] S. R. Nagel, A. Rahman, and G. S. Grest. Normal-mode analysis by quench-echo techniques: Localization in a amorphous solid. *Phys. Rev. Lett.*, 47(23):1665–1668, 1981.

[31] S. R. Nagel, G. S. Grest, and A. Rahman. Quench echoes. *Physics Today*, October:24–32, 10 1983.

[32] O. M. Becker and M. Karplus. Temperature echoes in molecular dynamics simulations of proteins. *Phys. Rev. Lett.*, 70(22):3514–3517, 1993.

[33] Dong Xu, Klaus Schulten, Oren M. Becker, and Martin Karplus. Temperature quench echoes in proteins. *J. Chem. Phys.*, 103:3112–3123, 1995.

[34] Dong Xu, James Christopher Phillips, and Klaus Schulten. Protein response to external electric fields: Relaxation, hysteresis, echo. *J. Phys. Chem.* In press.[Beckman Institute Technical Report TB-95-09].

[35] Dong Xu and Klaus Schulten. Velocity reassignment echoes in proteins. *J. Chem. Phys.*, 103:3124–3139, 1995.

[36] Axel T. Brünger. *X-PLOR*. The Howard Hughes Medical Institute and Department of Molecular Biophysics and Biochemistry, Yale University, New Haven, CT, May 1988.

[37] Bernard R. Brooks, Robert E. Bruccoleri, Barry D. Olafson, David J. States, S. Swaminathan, and Martin Karplus. CHARMm: a program for macromolecular energy, minimization, and dynamics calculations. *J. Comp. Chem.*, 4(2):187–217, 1983.

[38] S. Lifson and P. S. Stern. Born-oppenheimer energy surfaces of similar molecules – interrelations between bond lengths, bond angles, and frequencies of normal vibrations in alkanes. *J. Chem. Phys.*, 77:4542–4550, 1982.

[39] Lord Rayleigh. *Scientific Papers, Vol. I, p491 and Vol. IV, p370.* Cambridge University Press, Cambridge, England, 1899–1920.

[40] Lord Rayleigh. *The Theory of Sound, Vol. I, 2nd ed.* MacMillan and Company Ltd., London, 1894.

[41] Andreas Windemuth and Klaus Schulten. Stochastic dynamics simulation for macromolecules. Beckman Institute Technical Report TB-91-19, University of Illinois, 1991.

[42] C. W. Gardiner. *Handbook of Stochastic Methods.* Springer, New York, 1983.

[43] H. Risken. *The Fokker-Planck Equation: Methods of Solution and Applications, 2nd Ed.* Springer, New York, 1989.
[44] N. G. van Kampen. *Stochastic Processes in Physics and Chemistry.* North-Holland, Amsterdam, New York, 1992.
[45] H. Frauenfelder, F. Parak, and R.D. Young. Conformational substates in proteins. *Ann. Rev. Biophys. Biophys. Chem.*, 17:451–79, 1988.
[46] B. Cartling. From short-time molecular dynamics to long-time stochastic dynamics of proteins. *J. Chem. Phys.*, 91:427–438, 1989.
[47] Bo Cartling. Stochastic model of intermode couplings in protein dynamics. *J. Chem. Phys.*, 94:6203–6210, 1991.
[48] S. Chandrasekhar. Stochastic problems in physics and astronomy. *Rev. Mod. Phys.*, 15:1–89, 1943.
[49] M. C. Wang and G. E. Uhlenbeck. On the theory of the brownian motion II. *Rev. Mod. Phys.*, 17(2):323–342, 1945.
[50] Dong Xu and Klaus Schulten. Coupling of protein motion to electron transfer in a photosynthetic reaction center: Investigating the low temperature behaviour in the framework of the spin-boson model. *Chem. Phys.*, 182:91–117, 1994.

Part III

Motors, Membranes, Microtubules

Motor Proteins

Jonathon Howard and Frederick Gittes

Department of Physiology and Biophysics, University of Washington, Box 357290, Seattle, WA 98195-7290, U.S.A

1 Introduction

The human body is made up of perhaps 10^{12} cells, all of which have many internal features in common. These cells are each surrounded by a membrane, they all have at least one nucleus which contains DNA, and they all contain in their cytoplasm numerous smaller structures called organelles, which are involved in protein synthesis, energy production, motility, and a host of other specialized processes. But there are also different types of cells: for example; the human body contains over 200 major cell types that can have strikingly different morphology (Alberts et al. 1994). Perhaps the most important problem in biology today is how the structure of a cell, which can be 10 μm or 1000 μm in size, is determined by its constituent molecules–proteins, lipids and sugars–which are 1 nm to 100 nm in size. Somehow these small molecules implement genetic instructions to build and maintain a structure thousands to millions of times larger than themselves. To find out how this is done, we are studying (a) the structural elements of the cytoskeleton, which is a network of protein filaments that fills the cytoplasm, and (b) the motor proteins which, powered by splitting the fuel molecule ATP (adenosine triphosphate), can actively transport molecules and organelles along the cytoskeleton from one part of the cell to another.

2 The Biology of Motors

2.1 Cells are a meshwork of cytoskeletal filaments

Is cytoplasm a solid or a fluid? Early experiments to address this question were performed on the giant axon of the squid, which is an enormous membrane-bounded cylinder of cytoplasm (about 1 millimeter in diameter, large enough to allow very fast transmission of nerve impulses). Small metal spheres dropped into a vertical segment of the giant axon remained suspended in the cytoplasm, showing that cytoplasm has a gel-like nature (see Hodgkin, 1964). More definitive measurements gave results as shown in Fig. 1a: For molecules smaller than about 10 nm, diffusion is only slowed by a factor of 2 to 5 compared to free solution. For example, a burst of small molecules released by an organelle can permeate every region of a cell within a fraction of a second and, in fact, hundreds of different small molecules form complex systems of metabolism and signaling in

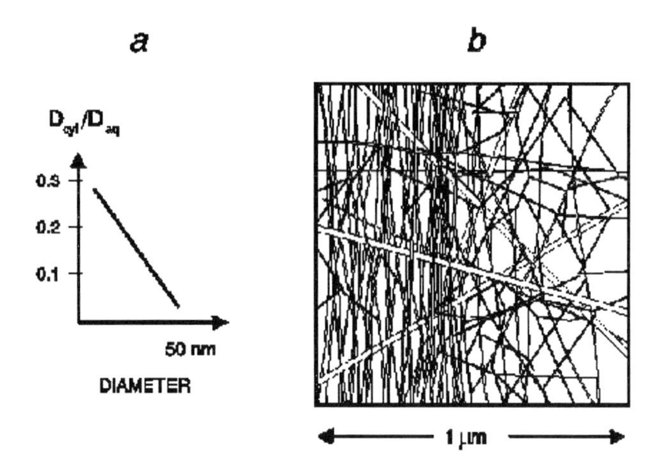

Fig. 1. (a) Relative diffusivity or mobility, compared to aqueous solution, of fluorescent, spherical polysaccharides of various diameters injected into living cells by Luby-Phelps et al. (1987), as a function of their diameter. (b) A drawing of a region of the cytoskeleton of a fibroblast, depicting a field about 1 μm in width and showing the three principal cytoskeletal filaments. A few microtubules are the thickest filaments visible; actin filaments are here shown as relatively well aligned, and the remaining filaments, forming an irregular meshwork, are largely intermediate filaments (drawn in solid black). For detailed electron micrographs of the cytoskeleton, see for example Heuser and Kirschner (1980).

cells. But for bigger objects, mobility drops precipitously. Molecules larger than about 50 nm are immobile.

We now know that the cytoplasm is filled with a network of filaments, the **cytoskeleton**, in which large objects are immobilized while small objects are relatively free to diffuse. One spectacular discovery of the electron microscope was this cytoplasmic meshwork (Fig. 1b). There are three classes of filaments in the cytoskeleton (Amos and Amos 1991) as well as a number of accessory proteins that bind to them:

Actin filaments have a diameter of roughly 6 nm and are composed of the monomers of the protein actin. Each subunit addition of an actin subunit requires the hydrolysis (splitting) of an ATP molecule. Actin subunits polymerize head-to-tail, in staggered pairs, to form a long actin filament that resembles a pair of gently twisted wires.

Microtubules are the stiffest cytoskeletal component. They are hollow cylinders with an outer diameter of about 30 nm, composed of dimers of the protein tubulin. Their stiffness derives from their hollow cylindrical structure. To form microtubules, dimers of the α and β subunits of tubulin align head-to-tail into protofilaments, and then 13 (usually) of these protofilaments associate laterally to make a hollow cylinder. Each subunit addition of a tubulin dimer requires the hydrolysis of a GTP molecule (guanosine triphosphate, similar to ATP).

Intermediate filaments have a diameter of about 10 nm and can be composed of various types of intermediate filament proteins; the subunits are aligned head-to-head so that the structure has no overall polarity. In specializations of their cytoskeletal role, the various intermediate filaments include keratin filaments, which lend substance to skin, hair and fingernails, and neurofilaments in nerve cells.

Each of these cytoskeletal filaments can grow as long as several microns; microtubules can grow up to 1 millimeter in some protozoa. Each micron of actin filament contains about 360 subunits, while each micron of microtubule contains about 1,600 tubulin dimers. In the cell, there are also tens or hundreds of cellular proteins that bind to each of these filaments, which may assist in polymerization, depolymerization, severing, and cross linking. Many cellular enzymes such as those involved in glycolysis (making ATP by splitting glucose) were once thought to act while in solution, but are now known to be bound to the cytoskeleton. And, finally, there are families of motor proteins that will be of interest to us here.

Microtubules and actin filaments are of central importance to cell organization because (unlike intermediate filaments) they are *polar* due to the head-to-tail arrangement of their constituent monomers. Polymerization rates at the two ends of these filaments are different and the faster- and slower-growing ends are called the **plus end** and the **minus end**, respectively. As a result of their polarity, oriented arrays of these filaments can be used to establish directionality within the cell. Orientation and polarity of microtubules plays a fundamental role in cell organization (Fig. 2).

2.2 Cellular motion results from motor proteins following cytoskeletal tracks

In the cytoskeletal meshwork of the cell, large macromolecular complexes such as polyribosomes will be severely restrained. Organelles such as mitochondria (which produce the fuel molecule ATP through respiration) can be 1 μm or more and will be essentially immobile. Immobility is a good thing because it means that any particular arrangement of organelles will be stable against diffusion. But it also raises the question of how organelles can bc actively moved about by cells, which is often seen to happen. Under changing conditions, organelles may move in or out along the microtubule array that is shown in Fig. 2a. For example, radial motion of pigment granules in fish scale cells allows the scales to change color dramatically within a matter of seconds. A closely related process is mitosis (Fig. 2b), the segregation of the DNA-containing chromosomes prior to splitting into two daughter cells; here microtubules make up the mitotic spindle.

The transport of organelles along the filaments of the cytoskeleton seems generally to be accomplished by means of **motor proteins**. Actin filaments and microtubules each have entire families of related motor proteins associated with them, which move in a directed manner along the filaments and use the fuel molecule ATP (adenosine triphosphate) as the energy source. Each individual motor protein moves in a specific direction with respect to its respective filament.

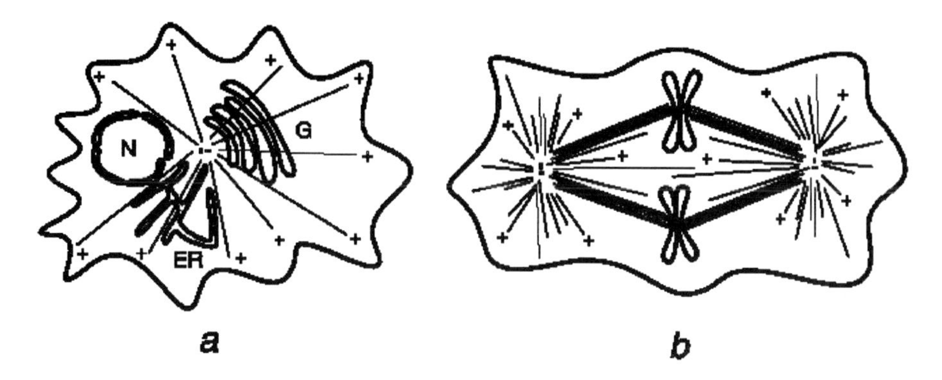

Fig. 2. (a) Schematic organization of a typical eukaryotic (nucleated) cell. Microtubules radiate with their plus ends outward from a region called the centrosome, usually located near the nucleus (N), where the minus ends of the microtubules are stabilized. This radial array is quite dynamic, and the individual microtubules within it grow and shrink with a turnover time of about 10 minutes. Intracellular structures and organelles, such as the endoplasmic reticulum (ER) and the Golgi apparatus (G), take up characteristic radial positions and orientations with respect to the radial array of microtubules. (b) Mitosis. In preparation for mitosis, the centrosome divides and the two resulting centrosomes migrate to opposite ends of the cell. The microtubular array then becomes bilaterally symmetric, and the chromosomes take up particular positions and orientations with respect to this bipolar array, called the **mitotic spindle**. Subsequently, each half of each chromosome will separate and move towards the corresponding centrosome.

Note that, in contrast, no motors can be associated with intermediate filaments because they are not polar structures, so that there can be no basis for directed motion along them.

Because the cytoskeleton can change its form dramatically through evolution, variations on this scheme of motor proteins moving along the cytoskeleton underlie a huge number of cellular mechanisms. For example, many changes in cell shape and cell crawling (Alberts et al. 1994) and the operation of the sensory cells responsible for hearing (Hudspeth 1989) are all probably accomplished by analogous motor-filament interactions. Even familiar large-scale biological motion originates from the motion of motor proteins along cytoskeletal filaments, usually in cases where the cytoskeleton has evolved into a highly specialized form. In such cases the motor proteins are not transporting organelles, but instead are acting together in great numbers to slide one filament relative to another (Fig. 3).

Most individual motor proteins identified so far have not been matched to specific tasks in intracellular transport, although these proteins can be isolated and shown to move along filaments *in vitro*. Broadly speaking, there are three families of motor proteins:

Myosin motors move along actin filaments, towards their plus ends. This is a very large family of proteins that fall into at least 10 different groups based on amino acid sequence similarity, and each group has several members.

One group of myosins includes all of the myosin isoforms found in skeletal, heart, and smooth muscle; these myosins power the contraction of muscle, which was the first-understood sliding-filament system (Fig. 3a).

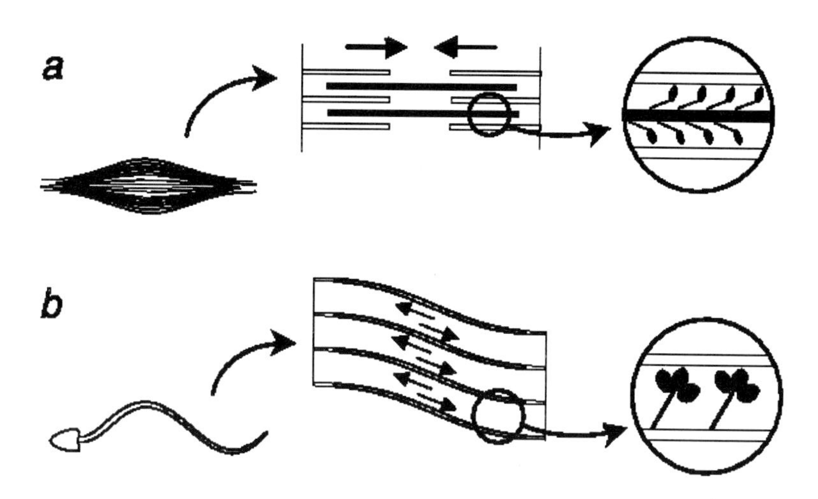

Fig. 3. (a) Muscle contraction originates with molecular motors moving along actin filaments. A muscle contains many long fibers, which are giant cells; each of these in turn contains a highly specialized cytoskeleton. On the smallest scale, within individual muscle cells, actin filaments are interdigitated with "thick filaments" composed of interwoven myosin molecules, and when these filaments slide with respect to one another the entire assembly contracts in length. (b) Flagella (here, a sperm tail) and cilia represent another sliding-filament system of microtubules, the axoneme. Dynein molecules are attached to one microtubule and move along an adjacent microtubule, creating shear forces that are non-trivially coordinated over the entire axoneme so as to cause a beating motion. In sperm, the base of the tail is packed with mitochondria to produce a huge gradient of ATP which diffuses down the tail to power the dynein motors.

Kinesin and **dynein** motors move along microtubules. These also represent large families of protein motors. For example, the fruit fly has at least 8 dyneins and 30 kinesins, and humans probably have at least this many. Dyneins all move towards the minus end of the microtubule. Each specific kinesin also moves in a particular direction but, interestingly, some kinesins move in one direction, while others move in the other. This was a very surprising finding, because the sequence homology between the kinesins is so high that they are all expected to fold into similar shapes.

Several types of dynein are involved in sperm motility and in the beating of the cilia in the airways, the reproductive tracts, and in fluid ducts in the brain. This is a second example of large-scale sliding-filament action: sperm tails and cilia both contain a specialized cytoskeletal structure, the **axoneme** (Fig. 3b), which is an array of microtubules with dynein molecules tucked in between them.

Kinesin variants have been found to be involved in mitosis (Saunders and Hoyt 1992), but their oldest-known role is in transporting vesicles down the axons of nerve cells. Axons are extensions of single neuronal cells which may be as long as a meter in nerve fibers extending from neurons in the lower spine, say, to the foot. In long axons, even the transport of small molecules cannot be left to diffusion. The diffusion coefficient D for a small molecule, such as a neurotransmitter, might be roughly $D \approx 500\ \mu m^2\ sec^{-1}$. Diffusion of a neurotransmitter from its point of synthesis (in the neuronal cell body) to the foot where it is needed would take thirty years! This problem is again solved by organelle transport along the cytoskeleton. Axons are filled with microtubules that run down it like railroad tracks. Small molecules are packaged into vesicles that are carried down axons by kinesin motors at speeds of about one μm per second.

Finally, however, we must admit that motor proteins, although extremely general, are not a universal paradigm for cell motility. As counter-examples there are certainly cases where filaments themselves do all the mechanical work through their negative free energy of polymerization (recall that actin and microtubule polymerization are coupled to ATP and GTP hydrolysis respectively). A striking illustration is the intracellular bacterium *Listeria monocytogenes*, described in Fig. 4. Similar polymerization and depolymerization effects in microtubules are probably important in mitosis as well, though this still needs to be worked out in detail. Exceptions and odd mechanisms such as *Listeria* are the rule in biological systems; in fact, there are no completely general theories in biology except the theory of evolution.

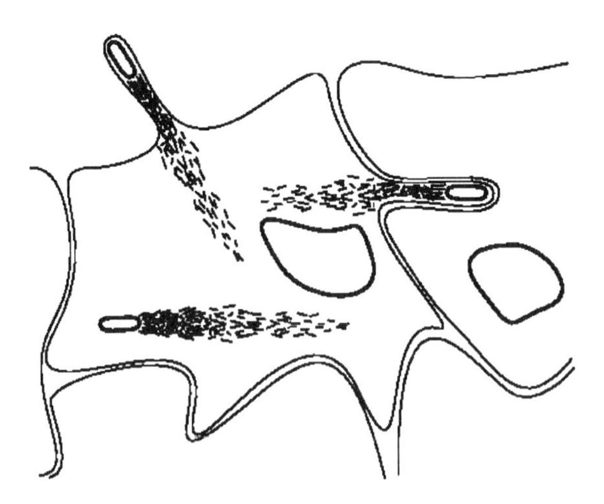

Fig. 4. A cell infected by the bacterium *Listeria monocytogenes*, which propels its way through the cytoplasm of its host cell like a jet engine, even penetrating into adjacent host cells, by inducing the polymerization of the host's super-saturated actin monomers at its rear end (Tilney and Portnoy 1989).

2.3 How can we tell which motor is doing what?

Cells are bewildering in their complexity. They are made up of thousands of copies of thousands of different proteins. As we have mentioned, motor proteins exist in large families where the members may have quite distinct roles. How can one ever figure out which protein is doing what? Although many such questions are only beginning to be answered, here is where the tools of biochemistry, genetics and molecular biology have shown their power. It is possible to separate and characterize proteins based on their molecular weight and other properties. In this way proteins can be purified from the other components, and under favorable conditions their activity can be reconstituted in the test tube. For example, tubulin can be purified from brain as a soluble dimer, then polymerized and examined under the electron microscope to reveal the hollow 30-nm-diameter cylindrical filament first identified in intact cells. Likewise kinesin can be purified, added to purified microtubules, and the microtubules can be seen to move under the microscope in the presence of ATP; in this way kinesin was shown to be a motor protein. But how do we know that kinesin is a motor in living cells, and how can we determine which organelles it carries? This is a very general problem in biology, and there are now many approaches one can take to dissect the operation of living cells. Here are a few examples.

1. Certain drugs interfere with the process under consideration. AMP-PNP is an analogue of ATP that is not well hydrolyzed by kinesin, and it is found to block organelle transplant in squid axons (Lasek and Brady 1985). This provides evidence that kinesin is in fact responsible for such transport. Even more selectively, antibodies (molecules made by the immune system that bind to specific proteins) often inhibit the protein to which they bind, and can be injected into cells to see which functions are disrupted.

2. Genetic instructions for protein sequence is copied from DNA to complementary RNA molecules, from which ribosomes produce the proteins. One can design antisense RNA which is complementary to the RNA that codes for the protein, and which will bind to this normal RNA and disable it (Melton 1988). Tissue culture cells can be transfected with DNA coding for antisense RNA, the synthesis of the specific protein can be blocked, and the effect can be observed.

3. Transgenic animals can be produced in which the gene coding for the protein of interest is removed from the genome (Alberts et al. 1994); thus no protein will be produced at all. This technique, in which a protein is first identified and then its gene is knocked out, is called reverse genetics to distinguish it from classic genetics, in which a gene is first identified and then the corresponding protein is sought.

3 Anatomy of a Motor: The Structure of Kinesin and Microtubules

We turn to the question of how motor proteins convert chemical energy to mechanical work used to power cell motility. The approach that we and many other labs are taking is to study the workings of these proteins as an engineer would try to understand the workings of any motor. In particular we wish to measure the speed, the force, and the efficiency, and we want to relate this to the motor's protein structure. In addition to being an important biological problem, these questions are also interesting from a physical point of view because motor proteins can convert chemical energy (the fuel molecule ATP; see Fig. 5) into mechanical energy *directly*, while man-made motors always use an intermediate form of energy such as heat or electrical energy.

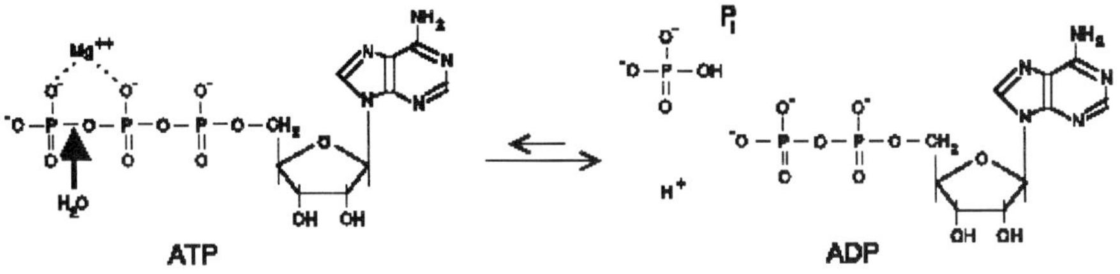

Fig. 5. Adenosine triphosphate (ATP). Many biochemical processes in the cell derive their energy from the hydrolysis of this nucleotide (or of the similar molecule guanosine triphosphate (GTP)). This reaction is: ATP $\rightarrow$ ADP + P_i, where P_i denotes inorganic phosphate, PO_4^{-2}. Hydrolysis literally means splitting by water, and the arrow indicates the bond which is, in effect, broken by the introduction of oxygen from a water molecule. Mg^{++} is also important in hydrolysis and can form a complex as indicated here. The free energy from this reaction, adjusted for typical cellular concentrations of ATP ($\sim$ 1 mM), P_i ($\sim$ 1 mM), and ADP ($\sim$ 10 μM) is $\Delta G \approx -25$ kT where G is the Gibbs free energy. Since 1 kT $\approx 4 \times 10^{-21}$ J = 4 pN nm, and proteins have dimensions of nm, we expect the forces exerted by motor proteins to be of the order of pN.

The motor protein kinesin serves as an excellent model for understanding the molecular mechanism of force generation. The geometry of the kinesin motor as it moves a load along a microtubule is illustrated in Fig. 6a. Kinesin is a dimer of two identical proteins with 975 amino acids each; the first 340 amino acids forms a globular *head domain* which has been shown to bind ATP and microtubules. Furthermore, in experiments the rest of the molecule can be removed using genetic engineering and the truncated protein still moves, albeit more slowly. Thus the first 340 amino acids by itself defines a motor domain. Why kinesin has two heads is still not fully understood; an attractive idea is that almost continuous attachment is maintained by coordination of these two domains, so that one detaches only after the other has attached.

Studying kinesin is appealing for a number of reasons:

1. The geometry of organelle transport (Fig. 6a) is relatively simple compared to the sliding-filament systems (Fig. 3) which consist of ordered arrays of thousands to billions of motors; for small organelles there is room on their surfaces for only a handful of motors. A corollary to this functional simplicity is that a single molecule of kinesin must move a large distance along a microtubule

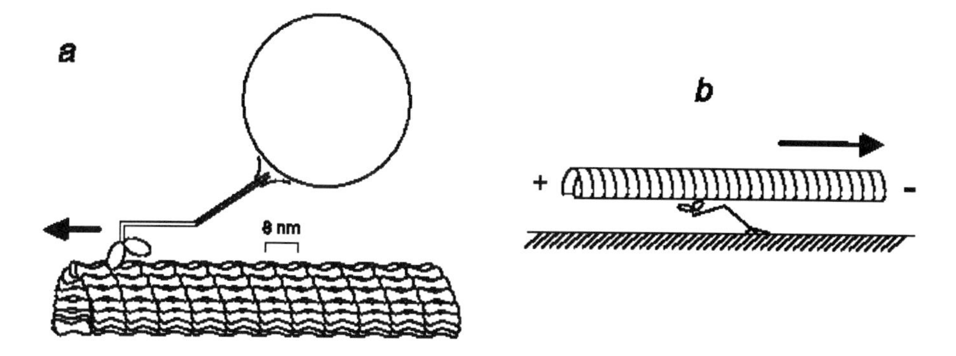

Fig. 6. (a) A kinesin dimer is shown as it transports a cellular cargo along a microtubule. Many organelles and vesicles would be larger than shown here. Kinesin has two head domains joined by intertwined alpha helices along a rodlike tail domain. The microtubule is shown to be composed of 13 protofilaments made up of oriented tubulin dimers (directionality not shown here); the protofilaments associate side-by-side with a slight offset to give a helical lattice. (b) Transport by few or even single motors was first demonstrated in *gliding assays* in which kinesin is adsorbed to a microscope slide surface and microtubules introduced into the adjacent solution. Microtubules diffuse to the surface, bind to kinesin and move at speeds of about 1 μm/s in the presence of ATP. Motion can be observed even at very low kinesin densities, in which case the motion of the microtubule clearly indicates that it is attached at only one point on the surface (Howard et al. 1989). Subsequent experiments measuring forces and displacements at high resolution establish almost beyond a doubt that a single kinesin molecule is sufficient for motility.

without falling off. We call this property *processivity*, a term first used to describe DNA replication enzymes that can move from one base to the next with only a very small chance of dissociating. On the other hand myosin and dynein require several motors for continuous movement.

2. Kinesin's motor domain is small, with about one-third the mass of a myosin head and one-tenth the mass of a dynein head. The small size of kinesin, together with the ability to observe single-motor function, has facilitated structural and functional analysis of this motor.

3. The motile machinery underlying organelle transport is *biochemically* simpler than that driving axonemal and muscle motility. Organelle transport consists of a motor, a microtubule, and possibly an additional protein that couples the motor to the organelle. Biochemical regulation of muscle contraction, in contrast, requires dozens of proteins that regulate contraction; similarly, the axoneme is an incredibly complex structure, with over 200 proteins.

3.1 Kinesin steps along a single protofilament

What path does kinesin follow along the surface of the microtubule? Does it move along the protofilaments, does it take a knights-move on the lattice, or does it perhaps move in a random fashion, switching irregularly from protofilament to protofilament? To distinguish between these possibilities (Ray et al. 1993) we synthesized special microtubules so that the protofilaments did not run exactly parallel to the axis of the microtubule, but instead twisted in shallow helical paths around the axis. These microtubules were marked so that rotation in the gliding assay could be detected. We found that these supertwisted microtubules did rotate, and that the pitch and handedness of the rotation agreed with the supertwist, as measured independently by electron microscopy. These experiments are only compatible with the hypothesis that the motor moves parallel to the filament axis; a misstep to a different protofilament is rare and occurs less than twice per hundred forward steps.

An important implication of the rotation experiment is that the distance between consecutive kinesin binding sites on the surface of the microtubule, the step size, must be a multiple of 4 nm, the inter-monomer spacing along the protofilament, or of 8 nm, the inter-dimer spacing. Binding studies show that there is only one motor-binding site per dimer, indicating that the step size is a multiple of 8 nm. Since the kinesin head is only about 10 nm long, a step of 16 nm or more is unlikely. In fact, high resolution tracking experiments by Svoboda et al. (1993) offer direct support for 8 nm steps.

3.2 Models for motor stepping

How does the motor get from one binding site to the next one 8 nm away? The concept that emerged from experiments on muscle was that myosin underwent a conformational change, referred to as the **power stroke**, and that the subsequent motion of the filament brought the next binding site to the myosin in question. The original support came from electron micrographs of insect flight muscle which showed that in the presence of ATP the myosin heads that cross-bridge to the actin filaments were aligned approximately perpendicular to the filaments, while in the absence of ATP, the heads were at a 45- degree angle (Reedy, Holmes and Tregear 1965). The idea then was that as ATP bound to myosin, was hydrolyzed, then dissociated, the head cycled through these mechanical states. This idea found good experimental support. Lymn and Taylor (1971) found that affinity between myosin and actin was greatly altered by ATP binding to myosin, which very rapidly dissociates the two proteins prior to hydrolysis occurring. Second, mechanical measurements on whole muscle indicated that there were at least two mechanical states of myosin that were separated by about 4 nm (Huxley and Simmons 1973), the first estimate of the power stroke distance. The recent atomic structure of myosin supports the notion of a conformational change of this magnitude: the myosin head has a large alpha-helical domain starts near the ATP-binding domain and projects away the actin filament. The rotation of this domain through 45 degrees due to the release of ADP would swing this domain through as much as 6 nm (Rayment et al. 1993).

The power stroke theory is the standard model in the myosin field, and it is expected by most to apply to kinesin and dynein as well. This is despite the failure of many workers on myosin over many years to directly resolve a power stroke, for example by looking for rotation of fluorescent markers in the alpha-helical domain.

Some theoretical models are well known to physiologists (e.g. Huxley 1957, Hill 1974), but contained entire functions (transition rates) as free parameters. Recently, Leibler and Huse (1993) formulated the power stroke concept using a minimum number of parameters, and within this model identified two general classes of motors: "porters" like kinesin, that are attached most of the time, and "rowers" like myosin that spend only a small fraction of time bound.

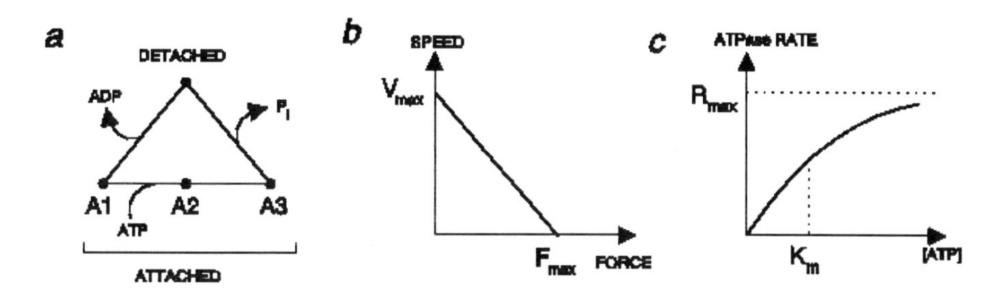

Fig. 7. (a) Power-stroke model of Leibler and Huse (1993). This cycle can be chemically interpreted either in terms of myosin or kinesin; the kinesin interpretation is shown here. The transition between the first attached state, A1 and the second attached state A2 is associated with an irreversible development of internal strain in the motor. It is the release of this strain that is the driving force of the motion. A third attached state, A3, is included to ensure that the motor spends enough time attached for a significant motion to occur before it detaches. Steady-state populations within the model are easily solved to give (b) a linear dependence of the speed of movement on the load and (c) the dependence of ATP hydrolysis R on the ATP concentration

Some recent work has discussed how motion can generally result from diffusion within asymmetric (ratchet-like) spatial potentials in the presence of irreversibility and dissipation (e.g., Prost et al. 1994). But how this concept really differs from the power stroke concept is not clear, since the power stroke models intrinsically include spatial asymmetry and ATP hydrolysis provides irreversibility. Thus it seems that the essential elements of dissipative ratchet models are contained within any protein-protein reaction.

4 Measuring Kinesin Movement and Force Generation

4.1 Recording forces exerted by single motors

Our lab, and others, have developed different techniques for placing individual motors under well defined loads and measure their output. One of the most important results of these experiment is the dependence of motor speed on loading force. We mention three of our own experiments:

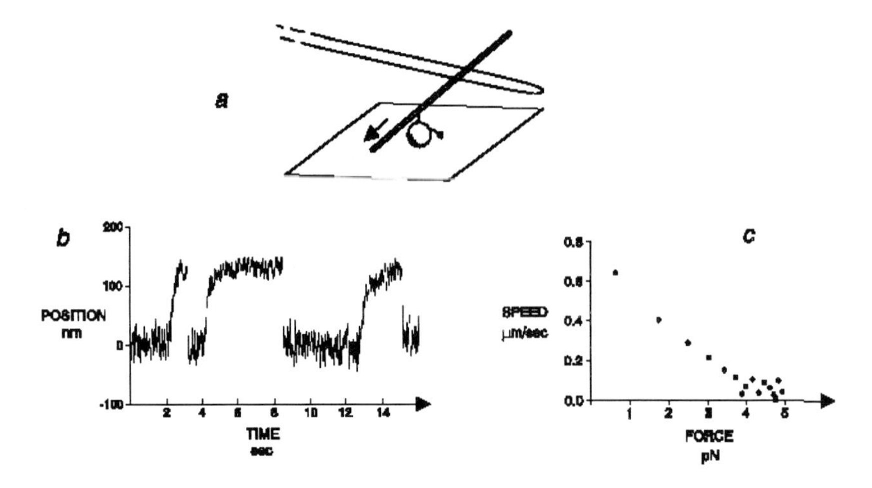

Fig. 8. (a) Bending needle assay of kinesin motor force (Meyhöfer and Howard. 1995). A microtubule is attached to a flexible glass fiber via a biotin-streptavidin connection, and then interacts with a kinesin molecule fixed to an immobile bead. The position of the glass fiber and microtubule is measured by imaging the tip of the fiber onto a split photodiode detector, with a signal resolution as small as 1 nm r.m.s. over a 1000 Hz bandwidth. The stiffness of the glass fiber is calibrated so that force is also known for a given deflection. (b) Initially the microtubule is unattached and undergoes a peak-to-peak Brownian motion of about 50 nm. When the microtubule interacts with a kinesin motor, the fluctuations decrease and the fiber starts moving, reaching a maximum displacement and load of about 120 nm and 5 pN. After sustaining this load for a second or so, the motor releases and the fiber quickly returns to its original position. The behavior is repeated several times. (c) The relation between speed and load can be deduced from displacement recordings; the speed decreases approximately linearly as the opposing force increases.

1. In one experiment (Meyhöfer and Howard 1995) the gliding assay of Fig. 6b was modified so that the motion of a microtubule could be detected with a fine glass needle. The experimental setup is described in Fig. 8a; a displacement trace from a single kinesin motor is shown in Fig. 8b and in Fig. 8c the calculated motor speed is shown as a function of the load force on the motor, and we find a linear decrease in speed, as is predicted for example by the model in Fig. 7. Similar observations were made by Svoboda and Block (1994) using an optical trap.

2. In another experiment (Hunt et al. 1994) we measured the dependence of motor speed on a drag force by performing gliding assays in a solution that was up to 100 times as viscous as water (such solutions were too thick to pour). The drag forces on different microtubules were proportional to their lengths, and plotting gliding speed versus this drag force for many microtubules again gave a linear decrease in motor speed with applied load force, with the same limiting values of unloaded speed and maximum load force as in the glass-needle experi-

ment. This agreement, interestingly, contradicts the Leibler and Huse model of Fig. 7, where stalling occurs because the motor continually slips back under high load. If a large viscous damping is present then the slippage will be less, and so a larger loading force should be needed to stall the motor. Thus slippage does not appear to be an important mechanism.

The simplest interpretation of these two experiments, taken together, is that as the load increases the rate of forward steps slows down. This is similar to the situation in whole muscle where it has been known for a long time that the hydrolysis cycle is slowed down if the muscle is prevented from contracting (Fenn 1924, Bagshaw 1993)

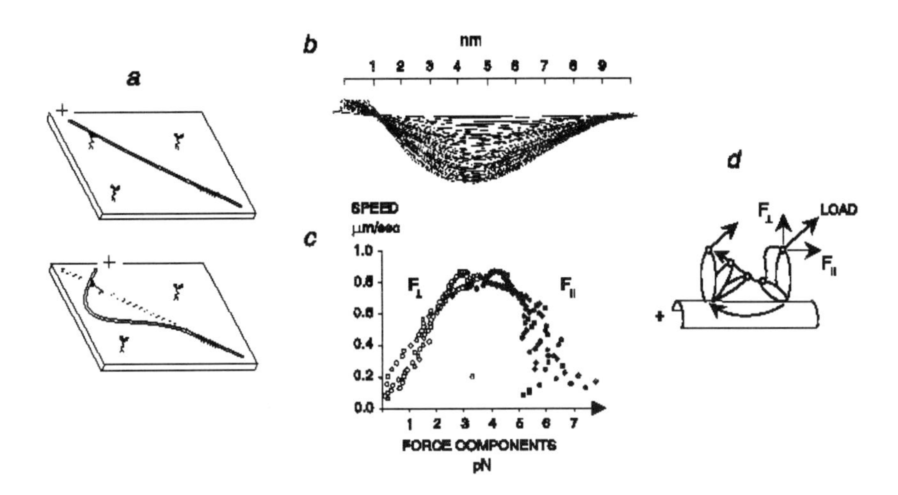

Fig. 9. (a) Buckling-microtubule assay of kinesin motor force (Gittes et al. 1996). Single kinesin motor molecules were observed to buckle the microtubules along which they moved, by means of a modified gliding assay in which one portion of each microtubule was clamped to the glass substrate via biotin-streptavidin bonds. (b) In cases where the microtubule became drastically bent without encountering other motors, the digitized shape of the buckling filament could be matched to shapes of ideal buckling filaments to deduce forces as a boundary condition on the shape. (c) Motor speed was plotted as a function of the vector load force, which developed a component perpendicular to the microtubule as the buckling event progressed (two sets of data points are superposed to show reproducibility over buckling repeats). This plot shows that in the presence of a growing perpendicular component of force (hollow circles), the motor speed reaches very high values while the parallel force (solid circles) is still large. This is different from the linear behavior found in Fig. 8c. (d) These results suggest that a force pulling the motor away from the microtubule actually enhances motor function. This *mechanical catalysis* can be understood if, microscopically, the load-bearing point follows a spatial path that is not parallel to the microtubule and motor function is rate-limited during motion directed away from the microtubule. (The "head" shown here might represent both kinesin heads working together.)

3. Another experiment (Fig. 9) arranged for single kinesin motors to buckle microtubules (Gittes et al. 1996) and from these events deduced force-velocity plots for forces not parallel to the motor motion. This work depended upon earlier estimates of the flexural rigidity of microtubules from their thermal fluctuations in shape (Gittes et al. 1993). Our results suggest that the kinesin motor appears to be *catalyzed* by a perpendicular loading force (Fig. 9d).

The lack of slipping of the kinesin motor and its sensitivity to the direction of loading illustrate the ambiguity between spatial and chemical motor function. In terms of the free energy, G, close spatial coupling between overall ATP hydrolysis (with reaction coordinate ξ) and the spatial path of the motor can be expressed by a mechanochemical coupling $(\partial G/\partial \xi)_{\mathrm{mech}} = (\partial G/\partial \xi)_{\mathrm{chem}} - \mathrm{t} \cdot \mathrm{F}$, where t is tangent to the displacement path (Gittes et al. 1996). For example, stalling of the motor occurs not at a unique force value, but whenever F is such that $(\partial G/\partial \xi)_{\mathrm{chem}} = \mathrm{t} \cdot \mathrm{F}$ is satisfied; this shows how the direction of the loading force and not just its magnitude must be considered.

5 Conclusions

We have seen how the movement of motor proteins along filaments of the cytoskeleton underlies a host of cellular processes, and we have discussed the particularly simple system of kinesin moving along microtubules. The mechanical characteristics of the kinesin motor molecule are beginning to emerge. The maximum force developed by a kinesin motor against a force antiparallel to its motion is in the neighborhood of 5 pN, and kinesin appears to step, or bind sequentially, along the tubulin dimers of a single protofilament, which have an 8 nm spacing. If such a step size occurs at the maximum loading force, the work required is about 40 pN nm $= 40 \times 10^{-21}$ J ≈ 10 kT and if this corresponds to the hydrolysis of a single ATP molecule, the efficiency of the motor is about 40 percent. We have also seen that kinesin does not appear to slip back as it moves under a load–it seems to be a well-coupled mechanochemical device. Furthermore we have even seen that a component of load pulling the kinesin molecule away from its track seems to enhance its function. Such capabilities are undoubtedly appropriate for kinesin's role in carrying cellular cargo through a cytoplasm which is full of obstructions. How these properties arise is not yet known, but kinesin quite probably relies upon the coordinated function of both of its heads, an issue that we have not touched upon here but that is becoming the next major focus of kinesin research. The problem of understanding motor function is far from solved.

Acknowledgements. We would like to acknowledge the collaboration of past and present members of this laboratory, including Sung Baek, Dave Coy, Will Hancock, Alan Hunt, Edgar Meyhöfer, Brian Mickey, Mitra Ray, and Mike Wagenbach.

References

Alberts, B., D. Bray, J. Lewis, M. Raff, K. Roberts, and J.D. Watson (1994): Molecular Biology of the Cell, 3rd edition. Garland, N.Y.

Amos, L.A., and W.B. Amos (1991): Molecules of the Cytoskeleton. The Guildford Press: New York.

Bagshaw, CR. (1993): Muscle Contraction, 2nd. edition. Chapman and Hall.

Gittes, F., E. Meyhöfer, S. Baek, and J. Howard (1996): Directional loading of the kinesin motor molecule as it buckles a microtubule. Biophys. J., in press.

Gittes, F., B. Mickey, J. Nettleton, and J. Howard. (1993): Flexural rigidity of microtubules and actin filaments measured from thermal fluctuations in shape. J. Cell. Biol. **120**, 923-934.

Fenn, W.O. (1924): The relation between the work performed and the energy liberated in muscular contraction. J. Physiol. (Lond.) **184**, 373-395.

Heuser, J.E. and M.W. Kirschner (1980): Filament organization revealed in platinum replicas of freeze-dried cytoskeletons. J. Cell Biology **86**, 212- 234.

Hill, T.L. (1974): Theoretical formulation for the sliding-filament model of contraction of striated muscle. Prog. Biophys. Mol. Biol. **28**, 267-340.

Hodgkin, A.L. (1964): The Conduction of the Nervous Impulse. Charles C. Thomas Publisher, Springfield, Illinois.

Howard, J., A.J. Hudspeth, and R.D. Vale (1989): Movement of microtubules by single kinesin molecules. Nature **342**, 154-158

Hudspeth, A.J. (1989): How the ear's works work. Nature **341**, 397-404.

Huxley, A.F. (1957): Muscle structure and theories of contraction. Prog. Biophys. Biophys. Chem. **7**, 255-318

Hunt, A. J., F. Gittes and J. Howard (1994): The force exerted by kinesin against a viscous load. Biophys. J. **67**, 766-781.

Huxley, A.F. and R.M. Simmons (1973): Mechanical transients and the origin of muscular force Cold Spring Harb. Symp. Quant. Biol. **37**, 669-680.

Lasek, R.J. and S.T. Brady (1985): Attachment of transported vesicles to microtubules in axoplasm is facilitated by AMP-PNP. Nature **316**, 645-647.

Leibler, S. and D. A. Huse (1993): Porters versus rowers: a unified stochastic model of motor proteins. J. Cell. Biol. **121**, 1357-1368.

Luby-Phelps, K., P.E. Castle, D.L. Tayler, and F. Lanni (1987): Hindered diffusion of inert tracer molecules in the cytoplasm of mouse 3T3 cells. Proc. Nat. Acad. Sci. USA **84**, 4910-4913.

Lymn, R. W. and E. W. Taylor (1971): Mechanism of adenosine triphosphate hydrolysis by actomyosin. Biochemistry **10**, A4617-A4624.

Melton, D.A. editor (1988): Antisense RNA and DNA. Curr. Comm. in Mol. Biol., Cold Spring Harbor Laboratory.

Meyhöfer, E. and J. Howard (1995): The force generated by a single molecule of kinesin against an elastic load. Proc. Nat. Acad. Sci. USA **92**, 574-578.

Prost, J., J.-F. Chauwin, L. Peliti and A. Adjari (1994): Assymetric pumping of particles. Phys. Rev. Lett. **72**, 2652.

Ray, S., E. Meyhöfer, R.A. Milligan and J. Howard 1993): Kinesin follows the microtubule's protofilament axis. J. Cell Biol. **121**, 1083-1093.

Rayment, I., H.M. Holden, M. Whittaker, C.B. Yohn, M. Lorenz, K.C. Holmes, and R.A. Milligan (1993): Structure of the actin-myosin complex and its implications for muscle contraction. Science **261**, 58-65.

Reedy, M.K., K.C. Holmes, and R.T. Tregear (1965): Induced changes in orientation of the cross bridges of glycerinated insect flight muscle. Nature (London) **207**, 1276-1280.

Saunders, W.S. and M.A. Hoyt (1992): Kinesin-related proteins required for structural integrity of the mitotic spindle. Cell **70**, 451-458.

Svoboda, K., C.F. Schmidt, B.J. Schnapp, and S.M. Block (1993): Direct observation of kinesin stepping by optical trapping interferometry. Nature **256**, 721-727.

Svoboda, K. and S.M. Block (1994): Force and velocity measured for single kinesin molecules. Cell **77**, 773-784.

Tilney, L.G. and D.A. Portnoy (1989): Actin filaments and the growth, movement and spread of the intracellular bacterial parasite, *Listeria monocytogenes*. J. Cell Biol. **109**, 1597-1608

Shapes and Fluctuations in Membranes

Luca Peliti[1,2]

[1] Dipartimento di Scienze Fisiche and Unità INFM
Università "Federico II", Mostra d'Oltremare, Pad. 19
I-80125 Napoli (Italy)
[2] Associato INFN, Sezione di Napoli. E-mail: peliti@na.infn.it

1 Membranes and Their Physical Description

The eukaryotic cell is literally packed with membranes: the cell cytoplasm is bound by a membrane called the *plasma membrane*, while the nucleus is bound by the *nuclear membrane*. Several organelles are made of folded membranes: the mithocondria, the chloroplasts (in plants), the Golgi apparatus, the endoplasmic reticulum. A great part of the transport of enzymes in the cell takes places via vesicles, small roughly spherical particles made of a "dressed" membrane, which contain the enzymes either on the interior or bound to the membrane itself. Prokaryotic cells do not show this richness, nevertheless the plasma membrane is their most prominent organ.

These observations justify the interest dedicated to membrane biophysics for the understanding of the working of the cell. The literature on this subject is immense. In these lectures I shall only consider a few problems, the ones on which I feel competent enough.

The point of view I shall adopt is the following:

1. I shall consider *model systems*, of simple and reproducible composition. Biological membranes, as emphasized in the chapter by Sackmann, have a complex composition, being made of tens (in prokaryotes) to hundreds (in eukaryotes) "principal" components, and an order of magnitude more of "minor" components. Nevertheless the understanding of the behavior of simple membranes is a first stepping stone for the description of the more complex system.
2. I shall only consider *equilibrium properties*. This is a major limitation. My only reason for it is that I do not feel competent enough to discuss nonequilibrium properties, and in particular transport along or across the membrane, and that the equilibrium phenomenology is complex enough.
3. I shall focus on one particular issue: the interplay of elasticity and geometry, dictating the spatial organization of the membrane. I shall skip the discussion of other important issues, related to the *internal* organization of the membrane, and in particular on the mechanisms by which the cell controls the fluidity of the membrane.

For an introduction to the biochemistry of biological membranes I find the chapter in Stryer's book excellent (Stryer (1988)). An introduction to the physics we are going to discuss can be found in chapters 6 and 8 of Safran's book (Safran (1994)). There are also a number of more advanced reviews on the properties of membranes considered from a point of view close to the present one: if pressed to make one title, I cannot but recommend the most recent and most comprehensive of them, contained in Lipowsky and Sackmann (1995). I also allow myself to mention my recent review (Peliti (1996)).

1.1 Biological Membranes

Our view of the basic structure of biological membranes is comparatively recent (Singer and Nicholson (1972)), and is known as the *fluid mosaic model.* The essential element is a *bilayer* of phospholipid molecules. These molecules are formed by a glycerine hinge, to which is attached on the one hand a phosphatic residue (which may itself be very complex), and on the other hand two fatty acid tails, often one saturated and one insaturated, and of different lengths. They are *amphiphilic*, i.e., they both love and hate water: the phosphatic head, polar or even charged, polarizes the small polar water molecules and is therefore strongly hydrophilic; the fatty acid tails disrupt the local order of water and are more strongly attracted by hydrocarbon tails of the same nature than by water. They are therefore considered hydrophobic.

The bilayer is an organized sheet of phospholipid molecules, of 50–100 Å thickness, made of two layers: in each layer, the phosphatic heads point towards the exterior, and the hydrocarbon tails are kept in the interior. In biological membranes, the bilayer is fluid and a typical diffusion constant for a phospholipid molecule is roughly 1 $\mu m^2 s^{-1}$. Therefore a lipid molecule can go from one end of a bacterium to the other in one second. On the other hand, the time needed for a molecule to go from one layer to the other is rather long (several hours) and is of the same order of magnitude as the characteristic time for exchanging phospholipid molecules with the solution. On a short time scale, therefore, it is safe to think that all molecules in the bilayer are bound to one layer, but rather free to wander on it.

In biological membranes, the bilayer also contains a great deal of different components which are not phospholipids: of these, the most prominent one is cholesterol, a small amphiphilic alcohol which can add up to 50% of the plasma membrane molecules (in number). Proteins can be conceived as floating on the membrane, stabilized by their hydrophobic regions: globular proteins may exhibit a preference for the inner or the outer layer, whereas some proteins can be fixed to the membrane via a comparatively short, hydrophobic, α-helix region, and extend on the exterior, often carrying oligosaccharides. These structures are called *glycoproteins* and have an important function in cellular communication. On the interior of the membrane, there is a structure made of a net of proteic filaments, anchored to the membrane via protein complexes. The most studied example is the spectrine-actine network which coates the interior of the membrane in red blood cells (erythrocytes). This structure is called the *cytoskeleton.*

One should also mention that archaebacteria are set apart from the rest of the living by the nature of their lipids: while in usual prokaryotes and in eukaryotes the fatty acid tail is ester-linked to the glycerine hinge, in archaebacteria it is ether-linked. Moreover, archaebacteria exhibit special lipids with phosphatic heads attached to both ends of the fatty acid chains (the so called *bola* lipids, by their resemblance to the "bolas" used by South-American cattle farmers). I feel that these differences make one of the deepest cleavages in the biochemistry of the living, and that understanding its origin will bring us closer to the understanding of the early organization of life.

1.2 Model Membranes and the Helfrich Hamiltonian

We now turn to the consideration of model systems, made of a single amphiphile dissolved in water. The phase behavior of this two-component system is extraordinarily rich. This is the background on which all complications related to the presence of proteins, glycoproteins and the cytoskeleton are drawn.

We shall consider one such system, made of a bilayer of phospholipid molecules belonging to a single chemical species. On the time scale of the experiment the molecules are not allowed to leave their layer to go either in the solution or in the other layer. If the bilayer is closed to form a vesicle, the enclosed volume is also a constant. The molecules are however free to rearrange within the layer and in space. The configuration they adopt corresponds to the minimum of the free energy F, taking into account the constraints of fixed enclosed volume, and of fixed number of molecules in each of the two layers. Since the membrane is fluid, we can sum over all internal degrees of freedom, such as the conformation coordinates of the fatty acid tails, and remain with a free energy which depends only on the geometrical shape of the vesicle. It will depend in particular on its area A. At equilibrium, the free energy reaches a minimum with respect to A. This implies that the surface tension γ vanishes:

$$\gamma = \left.\frac{\partial F}{\partial A}\right|_{\mathrm{eq}} = 0 \ . \tag{1}$$

We expect in fact that the free energy of a small deformation around the equilibrium value A^* of A will be given by a quadratic expression like in Hooke's law:

$$F(A) = F(A^*) + \frac{1}{2} k_{\mathrm{c}} A^* \left(\frac{A}{A^*} - 1 \right)^2 , \tag{2}$$

where we have introduced the *compression modulus* k_{c}. In phospholipid membranes, the compression modulus is quite large, of the order of a few Joule/m^2. This implies that, for a bilayer of the size of 1 μm^2 (i.e., of a bacterium), the fluctuations of the area due to thermal noise are a fraction 10^{-4} of the total area. We can thus consider fixed the area per molecule in closed vesicles.

Therefore membranes are not ordinary interfaces, whose free energy is proportional to their area. The relevant contributions to the free energy of a membrane arise from *curvature*. It is possible to give a measure of curvature by considering the deviation from the tangent plane of a small region of the membrane around a point P. Let us choose the origin of the coordinate axes (x^1, x^2, x^3) to coincide with P, and the direction in such a way that the x^1 and x^2 coordinate axes lie on the tangent plane, while the x^3 axis is normal to it and directed, say, towards the exterior of the vesicle. The shape of the membrane is locally given by an equation of the form

$$x^3 = u(x^1, x^2) . \tag{3}$$

This form is called the *Monge representation* of the surface. For small distances from P, the function $u(x^1, x^2)$ will be approximately represented by a quadratic form:

$$u(x^1, x^2) = \frac{1}{2} \sum_{i,j=1}^{2} \Omega_{ij}\, x^i x^j . \tag{4}$$

The quantity Ω_{ij} is known as the *curvature tensor*. Its eigenvalues, c_1 and c_2, are called the principal curvatures. They have the following geometrical interpretation.

Consider a plane containing the normal to the surface in P. The intersection between it and the surface forms a curve, whose radius of curvature in P is equal to R. This quantity is positive if the center of curvature lies towards the positive x^3 half-axis (i.e., towards the exterior of the vesicle) and negative otherwise. Now let the plane rotate along the normal in P. The curvature radius will vary accordingly. It is easy to convince oneself that one can have the following cases:

1. The radius R varies between two extremal values, R_1 and R_2, having the same sign. As a particular case one may have $R_1 = R_2$.
2. The radius R varies *at the exterior* of the interval bound by the values R_1 and R_2, having opposite sign. One can have in particular $R_1 = -R_2$.

The principal curvatures c_1 and c_2 are equal to the inverse of the extremal values, R_1 and R_2, of the curvature radius. The invariants of the curvature tensor, i.e., its trace $H = \mathrm{Tr}\,\Omega$ and its determinant $K = \det \Omega$, are related to them by the obvious relations $H = c_1 + c_2$ and $K = c_1 c_2$. They are known as the *mean curvature* and the *Gaussian curvature* respectively.

Let us discuss the geometrical meaning of these quantities. If $K = 0$, one of the principal curvatures vanishes, and the surface can be locally applied on a plane. In particular, for a cylinder one has $K = 0$, $H = 1/R$, where R is the radius of curvature of the cylinder, taken with its sign. To a first approximation, a whole line lies on the tangent plane in this case. If $K > 0$ the surface locally looks like a sphere, and lies on one side of the tangent plane: this corresponds to case 1. above. If $K < 0$, the surface looks like a saddle, and lies on both sides of the tangent plane. This corresponds to case 2. This is the case, in particular, when $H = 0$, which implies that the two principal curvatures are opposite to each other.

The surfaces for which $H = 0$ throughout are known as *minimal surfaces* for the reasons that follow. Let us consider a patch of surface, locally characterized by its area A and by the values H and K of its mean and Gaussian curvatures respectively. Imagine to displace each surface element by a small distance δ along the normal at each point. The area A' of the surface thus obtained will be given by

$$A' = A\left(1 + \delta\, H + \delta^2\, K + \ldots\right) \ . \tag{5}$$

We see that, if $H = 0$, the area is locally extremal under small deformations of the surface.

The Gaussian curvature K is named after Gauss, who introduced it and recognized its properties. He found them so striking that he named the corresponding theorem the "remarkable theorem" (*theorema egregium*): K can be expressed solely in terms of the internal metric structure of the surface, without considering the way it is embedded in ambient three-dimensional space. In particular, the integral of K over the whole surface is a topological invariant, being proportional to the Euler-Poincaré characteristic, i.e., to the difference between the number n_{c} of connected components and the number n_{h} of handles of the surface:

$$\oint \mathrm{d}A\, K = 4\pi\, \chi_{\mathrm{E}} = 4\pi\, (n_{\mathrm{c}} - n_{\mathrm{h}}) \ . \tag{6}$$

This result is known as the *Gauss-Bonnet theorem.*

We can now, following Helfrich (1973), attempt to write down the elastic free energy of a closed vesicle as a function of its shape. The hypotheses we make are the following:

1. The membranes are fluid, and the internal degrees of freedom are integrated over. Therefore the free energy depends only on the vesicle shape.
2. The expression is local in the internal coordinates of the surface. This implies the provisional neglect of the interaction between parts of the membrane which are close in ambient space, but far when one moves along the vesicle.
3. The membrane shape is smooth, what justifies the neglect of derivatives of order higher than the second.

The expression we obtain is determined by symmetry considerations. On the one hand, because of hypothesis 1. above, the expression should be invariant with respect to changes in the internal coordinate system of the membrane (reparametrization invariance). On the other hand, it is obvious that, in the absence of interactions, e.g., with the walls of the container, it should be invariant with respect to Euclidean transformations. It is then possible to show that the most general expression that satisfies our hypotheses is

$$F = \oint \mathrm{d}A \left[\tau + \frac{1}{2}\kappa\,(H - H_{\mathrm{s}})^2 + \bar{\kappa}K\right] \ , \tag{7}$$

where the integral runs over the surface of the vesicle, and $\mathrm{d}A$ is the area element. This expression is known as the *Helfrich Hamiltonian.* The quantity τ is called the *area coefficient.* Our discussion shows that it is strictly related to the chemical potential of the amphiphile, since the area is proportional to the number of molecules in the membrane. In a closed vesicle, it multiplies a constant (the total area of the vesicle) and does not play a role. The coefficient $\bar{\kappa}$, which multiplies the Gaussian curvature, is called the *Gaussian rigidity.* The Gauss-Bonnet theorem implies that it also multiplies a constant, as long as ne only considers continuous deformations of the vesicle.

We are left therefore with the second term of the expression. The coefficient κ is called the *bending rigidity.* It has the dimensions of an energy and its typical values are of the order of a few to a few tens of $k_{\mathrm{B}}T$. Therefore the typical curvature radii that a membrane acquires under thermal fluctuations are proportional to its linear size, with a proportionality constant $\sim (\kappa/k_{\mathrm{B}}T)^{1/2}$ of order unity. This shows that bending deformations (undulations) play the most prominent role in the physics of membranes. The parameter H_{s} is called the *spontaneous curvature.* It vanishes for a symmetric bilayer, but one has to keep in mind that biological membranes are not symmetric. In model membranes, we can set $H_{\mathrm{s}} = 0$, and consider the simplified expression of the Helfrich Hamiltonian:

$$F = \oint \mathrm{d}A \, \frac{1}{2} \kappa H^2 \ . \tag{8}$$

This expression is *scale invariant*: for example, for a sphere of radius R, one has $H = 2/R$, and therefore $F = 8\pi\kappa$, independent of R. A closer look shows that the differential element is scale invariant, and thus that the integral is invariant with respect to all transformations which locally reduce to scale transformations (conformal transformations). This observation, as we shall see, has important consequences.

We can now use this expression to discuss the phase behavior of the shapes of closed vesicles.

2 Vesicle Shapes

The theory of vesicle shapes, based on the elastic free energy for membranes, was initiated by Canham (1970) and Helfrich (1973). If one cuts a rubber balloon, and takes its inside out, and somehow heals the cut afterwards, one obtains a shape resembling a red blood cell. This suggests that considerations of minimal elastic energy alone should be sufficient to determine the shape of red blood cells. The theory was revived in recent years, following the experimental investigations of the groups led by Sackmann and by Bensimon. For a review of the experiments on vesicle shapes see Wortis et al. (1993). A brief and very readable review of membrane conformation is Lipowsky (1991).

2.1 Simple Vesicles

At a mean field level, the determination of vesicle shape reduces to the identification of the shapes for which the free energy (8) is minimal, subject to the relevant physical constraints. They are:

1. A fixed value of the enclosed volume V. This is due to the fact that, although phospholipid membranes are rather permeable to water, they are usually

impermeable to dissolved impurities, and in particular to ions. Therefore the osmotic pressure buildup blocks any volume fluctuation that attemps to change the impurity concentrations.
2. A fixed value of the vesicle area A. As we have seen, this stems from the small compressibility of the (two-dimensional) "fluid" of phospholipid molecules.
3. A fixed value of the *total curvature* $M = \oint dA\, H$ (Svetina and Žekš (1983)). This constraint arises from the fixed value of the number of phospholipid molecules present in each layer, which fixes the value of the area difference between them. One has in fact from (5)

$$\Delta A = A_+ - A_- \simeq \oint dA\,(H\delta) = M\delta \ , \tag{9}$$

where δ is the bilayer thickness.

These constraints are not absolute: in principle, one can let V, A and M fluctuate, and add a Hooke-like term to the free energy. It turns out that the effect of these modifications can be reabsorbed in a minor redefinition of the shape, except for the term involving M:

$$\Delta F = \kappa \frac{\alpha\pi}{8A\delta^2} (\Delta A - \Delta A_0)^2 \ , \tag{10}$$

where ΔA_0 is the "preferred" value of the area difference and α is a dimensionless parameter. This model is known as the *area difference elasticity* (ADE) model (Seifert et al. (1991b)). The absolute constraint on M is recovered when $\alpha \to \infty$. It is known as the *bilayer coupling* (BC) model.

At the mean field level, the rigidity κ only sets the energy scale, but does not determine shape. Since the three constrained quantities, V, A and M have the dimensions of length^3, length^2, and length respectively, one may take advantege of the scaling invariance of (8) to express the shape in terms of dimensionless quantities. One introduces the effective radius R_0 as the radius of the sphere having the same area A:

$$A = 4\pi R_0^2 \ . \tag{11}$$

The reduced volume v and reduced total curvature m are then defined by $v = 3V/4\pi R_0^3$ and $m = M/R_0$ respectively. In this way, it is possible to describe the phase diagram for vesicle shape in a two-dimensional space. The reference point is the sphere, for which $v = 1$ and $m = 4\pi$.

A scheme of the result is shown in Fig. 10 of the chapter by Sackmann. There are three major families of axisymmetric shapes: discocytes, stomatocytes and dumbbells. The *discocyte* shape is the characteristic biconcave shape of red blood cells. It is symmetric with respect to up-down reflections. *Stomatocytes* are shapes which do not exhibit this symmetry, having one concavity more marked than the other, which can even disappear. The word comes from the Greek $\sigma\tau\acute{o}\mu\alpha$, mouth. The mouth becomes deeper and deeper as m decreases, until one reaches a limit shape on the lower branch on the parabola, in which a small sphere with a "strangled neck" is contained inside a larger one. Healthy

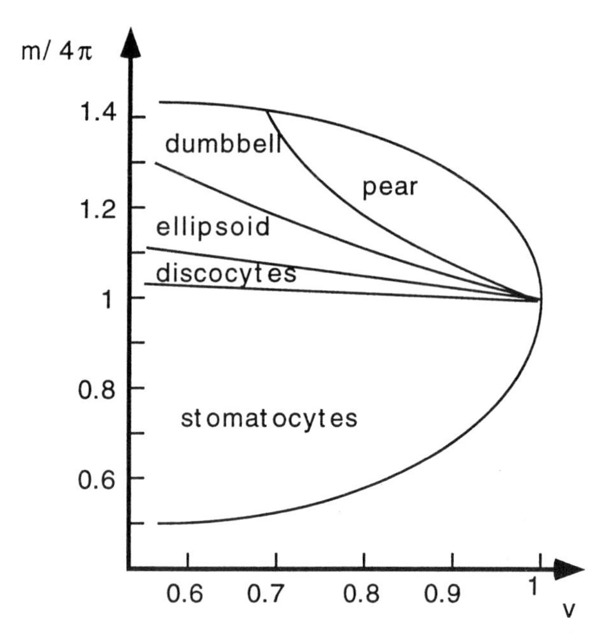

Fig. 1. Schematic phase diagram for vesicles of spherical topology, according to the bilayer coupling model. The external parabola corresponds to limit shapes ("budding"). On its exterior, equilibrium shapes are not known. In the region marked "ellipsoid", equilibrium shapes are nonaxisymmetric ellipsoids. The symmetry breaking lines separating discocytes from stomatocytes and dumbbell from pear shapes correspond to continuous transitions. Redrawn from Seifert et al. (1991b).

red blood cells sometimes acquire the shape of stomatocytes, and the form had already been identified by Deuling and Helfrich (1976).

For larger values of m, one goes from the discocyte to the ellipsoid region. Ellipsoids are first oblate, then prolate, until they exhibit a neck in the middle which becomes more and more pronounced as m increases. One thus obtains *dumbbells*, which still exhibit up-down symmetry along the axis of rotational symmetry. The up-down symmetry is broken along the line which separates them from "pear" shapes. One then reaches limit shapes, in which the neck becomes strangled. In particular, the limit shape of "pears" correspond to a small roughly spherical vesicle sitting on top of a larger one, while the limit shape for dumbbells corresponds to two equal vesicles kissing each other on a strangled neck.

It is fascinating to follow the changes of shape of individual vesicles upon heating or cooling, as reported, e.g., by Berndl, et al. (1990). Vesicles produced in the same solution evolve towards different shapes, some towards stomatocytes, some exhibiting a reentrant symmetry breaking (going from dumbbell to pear and back to dumbbell), some going all the way to budding. The principal effect of the change in temperature is the change of area. However, in order to explain the observed behavior of vesicles, the authors have been led to postulate a different thermal expansion coefficient for each of the two phospholipid layers. This is a property of the individual vesicle (which changes itself slowly with time). Its origin is at present unknown, although a few mechanisms can be proposed.

The very existence of a strangled neck is somehow paradoxical, since it appears to correspond to infinite curvature. In fact, there are *two* infinite principal curvatures, with opposite sign, so that the free energy contribution of the strangled neck is finite. However, it is likely that the simple model of curvature elasticity that we have used so far breaks down, since higher order contributions must be taken into account.

2.2 Impurities and Curvature

The formation of these limit forms has aroused much interest, since it is an equilibrium analogue of the dynamical process by which small vesicles are formed out of a mother membrane ("vesiculation", or, for the plasma membrane, "exocytosis"). These processes involve, in the cell, membranes made of several components. Some components which are preferentially adsorbed in one of the two layers, can modify the local value of the spontaneous curvature. This corresponds to an extra term $\lambda\rho H$ in the Helfrich Hamiltonian, involving the local concentration of impurities ρ. If one is close to phase separation of the two-dimensional fluid, there is a complex feedback between shape and composition, since the components which favor higher values of H tend to segregate, producing higher values of the curvature, and eventually budding. These phenomena have been theoretically studied by Lipowsky (1992), Kawakatsu et al. (1993) and Jülicher and Lipowsky (1993).

Even impurities which are symmetrically adsorbed in the two layers can have an important effect on shape, as pointed out by Leibler (1986). In this case, the fluctuations of the local concentration of impurities lead to a reduction of the effective bending rigidity:

$$\kappa_{\text{eff}} = \kappa - 3\lambda^2 \rho / k_{\text{B}} T \ . \tag{12}$$

The effective bending rigidity can even vanish. In this case the vesicle adopts a "spiky" form resembling a hedgehog (Leibler and Andelman (1987)). An analogous shape is sometimes found in red blood cells, and the corresponding process is called "echinocytosis".

2.3 Vcsiclcs of Higher Genus

Mutz and Bensimon (1991) first observed toroidal vesicles in partially polymerized phospholipids. Vesicles of the same or higher genus have been observed in fluid vesicles made of the same phospholipid. This observation has aroused great interest, because it opened the possibility of investigating experimentally a classic mathematical problem posed by Willmore (1982). He had asked in fact which shapes minimized, for each topological genus (number of handles), the Willmore functional $W = \oint \mathrm{d}A\, H^2$, which is proportional to the Helfrich Hamiltonian. For genus 0 the solution is shown to be the sphere, while for genus 1 Willmore conjectured that it is the Clifford torus, i.e., a circular torus in which the ratio of the inner circle radius to the outer circle radius is equal to $\sqrt{2}$.

However, due to the conformal invariance of W, all images of the Clifford torus upon conformal transformations correspond to the same value of W. There is a family of these images, which look like tori with a non-central hole, known as the Dupin cyclids. Therefore, the consideration of conformal invariance paves the way to the study of nonaxisymmetric shapes. Upon conformal transformations, the reduced volume v and total curvature m are not invariant. In particular, the reduced volume of the Dupin cyclids is always larger than the volume of the Clifford torus, which is equal to 0.71. Therefore, if Willmore's conjecture is true, the solution of the curvature energy minimization problem is found on the Dupin cyclids for reduced volume larger than 0.71. The rest of the phase diagram can be calculated by solving the differential equation and checking the stability of the solution with respect to axisymmetric perturbations (Jülicher et al. (1993a)).

Experimentally one finds that vesicles having the shape of a Clifford torus turn into nonaxisymmetric shapes (close to Dupin cyclids) upon cooling, which induces a decrease of bilayer area, and thus an increase in reduced volume. However, some of the observed shapes do not appear in the calculated phase diagram of the bilayer coupling model, and suggest that either area-difference elasticity must be taken into account, or that a full study of numerically determined non-axisymmetric shapes is needed (Michalet (1994)).

For vesicles of higher genus (with more handles) the situation changes drastically, since it is possible to find conformal transformations which keep the reduced volume v and the reduced mean curvature m invariant. Therefore there is a region in the phase diagram in which the minimal curvature energy is reached for a one-dimensional manifold of shapes, connected by conformal transformations. This region lies approximately in correspondence of the Dupin cyclids line for genus 1 vesicles (Jülicher et al. (1993b)). Since there is no restoring force, vesicles belonging to this region undergo wide fluctuations in shape (*conformal diffusion*). This phenomenon has been observed in phospholipid vesicles of genus 2. As the genus increases, the number of degrees of freedom for shape change also increases. It becomes eventually more convenient to treat the system as a "gas" of interacting handles (Michalet et al. (1994)).

3 Fluctuations

Physiologists have known for more than hundred years the intensity of shape fluctuations exhibited by red blood cells, but it was shown only twenty years ago that their origin is purely thermal. These fluctuations are at the origin of the long-range steric interaction between membranes, and govern the stability of the different phases of amphiphile solutions.

3.1 Undulations

Let us consider the fluctuations of an almost planar membrane in the Monge representation $x^3 = u(x^1, x^2) = u(\underline{x})$. We write the Helfrich Hamiltonian neglecting the Gaussian curvature term (which yields only a boundary contribution) but reintroducing the area coefficient τ. We obtain to lowest order in u

$$F = \int \mathrm{d}\underline{x} \, \frac{1}{2} \left[\tau \left(\nabla u \right)^2 + \kappa \left(\nabla^2 u \right)^2 \right] \ . \tag{13}$$

The equipartiton theorem allows us to calculate the height-height correlation function $\left\langle u_{-\underline{q}} u_{\underline{q}} \right\rangle$, where $u_{\underline{q}}$ is the Fourier transform of $u(\underline{x})$. One has

$$\left\langle u_{-\underline{q}} u_{\underline{q}} \right\rangle = \frac{k_{\mathrm{B}} T}{\tau q^2 + \kappa q^4} \ . \tag{14}$$

When τ is very small, the correlation function diverges like q^{-4} for small q. This strong divergence is at the origin of the violent shape fluctuations (undulations) exhibited by membranes. We can compute in the same limit the typical height fluctuation of patches of membranes a certain distance apart. We obtain

$$\left\langle \left(u(\underline{x}) - u(\underline{x}') \right)^2 \right\rangle \propto |\underline{x} - \underline{x}'|^{2\zeta} , \tag{15}$$

which defines the *wandering exponent* ζ. We have

$$\left\langle \left(u(\underline{x}) - u(\underline{x}') \right)^2 \right\rangle \propto \left(\frac{k_{\mathrm{B}} T}{\kappa} \right) |\underline{x} - \underline{x}'|^2 \ , \tag{16}$$

which yields the exponent $\zeta = 1$. This exponent is equal to 0 for interfaces, in which the dominant term is τq^2. Therefore the typical shape of a fluctuating membrane is self-similar, since the size of undulations is proportional to its lateral size.

Peterson et al. (1992) have performed a quantitative analysis of the thickness fluctuations of red blood cells at different points along its diameter. They have confirmed that these fluctuations are consistent with the Helfrich Hamiltonian, and in particular that the shear elasticity possibly due to the underlying spectrine network is negligible.

3.2 Steric Interaction

Helfrich (1978) first realized that undulations were responsible for a long-range effective repulsion between membranes. Consider two parallel membranes, a distance ℓ apart (on average). Because of undulations, they will come in close contact from time to time. The lateral distance L between contacts can be estimated by the condition $L^\zeta \sim \ell$. Therefore, the concentration of contacts per unit area will be proportional to $L^{-2} \sim (k_{\mathrm{B}}T/\kappa)\ell^{-2}$. Assuming that each contact leads to a finite decrease in entropy with respect to the configurations of a free membrane, the increment in free energy per unit area will be proportional to $(k_B T)^2/(\kappa \ell^2)$, where the constant of proportionality is a universal number over whose value there is still some debate.

Membranes are subject to the attractive van der Waals interaction, which decays like ℓ^{-2} at short distances and like ℓ^{-4} at longer distances. It appears therefore that the balance between the van der Waals attraction and the steric repulsion leads to a first-order transition between a state in which membranes are stuck to each other and one in which they wander freely. This transition is called the *unbinding transition.* It has been observed in a stack of fluid membranes by Mutz and Helfrich (1989). However, experiments by Larché et al. (1986) have exhibited structures of packed membranes (lamellar phases) with repeat distances as large as a few thousand Ångströms. This suggests that the transition is *continuous*, and not first-order, as the simple-minded argument just given appears to imply. A theoretical explanation of this phenomenon requires the understanding of how undulations modify the interaction between membranes, i.e., the renormalization of this interaction. It has been performed by Lipowsky and Leibler (1986), with the result that the transition can indeed be continuous, so that the average repeat distance ℓ diverges like $|W - W_0|^{-\psi}$, where W is a control parameter (it can be the chemical potential of the amphiphile) and ψ is an exponent which turns out to be close to 1. Similar results can be obtained within a Flory approach (Milner and Roux (1992)).

3.3 Renormalization of the Elastic Constants

Fluctuations have also subtle consequences on the behavior of a single membrane. The quantities τ, κ, $\bar{\kappa}$, H_s, appearing in the Helfrich Hamiltonian turn out to depend on the scale of observation. This effect is called the renormalization of the elastic constants (Helfrich (1985), Peliti and Leibler (1985)).

In order to describe this phenomenon, let us imagine a patch of membrane in the Monge representation. Its istantantaneous configuration is expressed by the equation $x^3 = u(x^1, x^2)$. The weight of this configuration is proportional to the Boltzmann factor $\exp\left(-F[u]/k_B T\right)$, where $F[u]$ is the Helfrich Hamiltonian, expressed in terms of u. This expression can be made explicit by using the formulas of differential geometry. One has for example the following expressions of the area element $\mathrm{d}A$ and the mean curvature H:

$$\mathrm{d}A = \mathrm{d}\underline{x}\,\sqrt{1 + (\nabla u)^2}\ ; \qquad H = \nabla \cdot \frac{\nabla u}{\sqrt{1 + (\nabla u)^2}}\ . \tag{17}$$

The *average* conformation U is naturally defined as

$$U(\underline{x}) = \frac{1}{Z} \int \mathcal{D}u(\underline{x}')\, u(\underline{x})\, \mathrm{e}^{-F[u]/k_B T}, \tag{18}$$

where Z is the partition function, and the integral is a functional integral over all possible shapes, parametrized by u. To be sure, the measure of integration (contained in $\mathcal{D}u$) should be considered with some care (Cai et al. (1994)), but the results we are going to quote are correct and can be obtained without additional complications.

It is in principle possible to add a term $-\int \mathrm{d}\underline{x}\,\lambda(\underline{x})U(\underline{x})$ to the Helfrich Hamiltonian, introducing an external field $\lambda(\underline{x})$, in such a way to force $U(\underline{x})$ to be equal to some preassigned function. We can then define the free energy as a functional of the external field λ as the logarithm of the partition function:

$$\mathcal{F}[\lambda] = -k_{\mathrm{B}}T \log \int \mathcal{D}u \, \exp\left[-\left(F - \int \lambda u\right)/k_{\mathrm{B}}T\right] . \tag{19}$$

The average configuration U is given by

$$U(\underline{x}) = \frac{\delta \mathcal{F}}{\delta \lambda(\underline{x})} . \tag{20}$$

One can then express the free energy as a function of the *average* conformation U by means of a Legendre transform:

$$\Gamma[U] = \mathcal{F}[\lambda] + \int \lambda U . \tag{21}$$

The functional form of $\Gamma[U]$ is dictated by the same considerations that fixed the form of the Helfrich Hamiltonian. We have therefore

$$\Gamma[U] = \int \mathrm{d}A \left[\tau_{\mathrm{eff}} + \frac{1}{2}\kappa_{\mathrm{eff}} \left(H - H_{\mathrm{s}}\right)^2 + \bar{\kappa}_{\mathrm{eff}} K\right] , \tag{22}$$

where now the area element $\mathrm{d}A$, the mean curvature H, and the Gaussian curvature K are expressed as a function of the *average* conformation U as in (17). This expression defines the effective value of the area coefficient τ, and of the bending and Gaussian rigidities, κ and $\bar{\kappa}$ respectively.

This expression implies that the effective area coefficient can be interpreted as the force per unit length that the membrane applies to a frame to which it is attached (David and Leibler (1991)). It is therefore conjugate to the *projected area* A_{p}, i.e., the area corresponding to the *average shape*, which is in general smaller than the true area A.

If we consider an isolated patch of membrane with free boundary conditions, the equilibrium configuration is reached when the free energy Γ reaches a minimum. The minimum can be reached for a nonzero value of the projected area per molecule, which is itself proportional to the ratio $a_{\mathrm{p}} = A_{\mathrm{p}}/A$ of the projected area A_{p} to the true area A. If the equilibrium value a_{p}^* is different from zero, the membrane is *flat*, and the effective area coefficient vanishes. If $a_{\mathrm{p}}^* = 0$, the membrane is *crumpled.* In general, the area coefficient for a crumpled membrane does not vanish, but one can also have the marginal case in which it vanishes.

If we consider a patch of membrane at a given length scale μ^{-1}, we can distinguish three different regimes, according to the respective values of τ_{eff}, κ_{eff}, and $k_{\mathrm{B}}T$:

1. *Tension-dominated region:*

$$\tau_{\mathrm{eff}} > \kappa_{\mathrm{eff}}\mu^2 ; \qquad \tau_{\mathrm{eff}} > k_{\mathrm{B}}T\mu^2 . \tag{23}$$

   In this region, fluctuations are governed by the tension τ, as in an ordinary interface.

2. *Rigidity-dominated region:*

$$\kappa_{\text{eff}} > k_{\text{B}}T \ ; \qquad \tau_{\text{eff}} < \kappa\mu^2 \ . \tag{24}$$

In this region, fluctuations are small, but are dominated by the rigidity term.

3. *Fluctuation-dominated region:*

$$\kappa_{\text{eff}} < k_{\text{B}}T \ ; \qquad \tau_{\text{eff}} < k_{\text{B}}T\mu^2 \ . \tag{25}$$

In this domain, perturbation theory breaks down. Steric interactions and topology changes dominate the membrane behavior.

The dependence of the effective elastic parameters on the observation scale can be theoretically realized by integrating over all fluctuations with a wavelength between a minimal one, Λ^{-1}, of the order of the lateral size of an amphiphilic molecule (i.e., a few Å) and the "reference" one, μ^{-1}, at which one performs the observation. This calculation can be performed in several equivalent ways (see, e.g., David and Leibler (1991), Peliti and Leibler (1985)). The results can be expressed in terms of *renormalization group equations.* If one fixes the upper wavenumber cutoff Λ, one introduces the scale factor s via the relation $\mu = s\Lambda$. One defines the "running parameters" $\kappa(s) = \kappa_{\text{eff}}$, $\bar{\kappa}(s) = \bar{\kappa}_{\text{eff}}$, and $\tau(s) = \tau_{\text{eff}}s^2$. It is then possible to write down a system of autonomous ordinary differential equations which express the way in which the running parameters change with s:

$$s\frac{\partial\kappa(s)}{\partial s} = -\frac{3}{4\pi}\frac{k_{\text{B}}T}{1+\tau(s)/\kappa\Lambda^2} \ ; \tag{26}$$

$$s\frac{\partial\bar{\kappa}(s)}{\partial s} = \frac{5}{6\pi}\frac{k_{\text{B}}T}{1+\tau(s)/\kappa\Lambda^2} \ ; \tag{27}$$

$$s\frac{\partial\tau(s)}{\partial s} = 2\tau(s) + \frac{\Lambda^2}{4\pi}k_{\text{B}}T\log\left(\frac{\kappa+\tau(s)/\Lambda^2}{k_{\text{B}}T}\right) \ . \tag{28}$$

These results are perturbative, to first order in an expansion in powers of $k_{\text{B}}T/\kappa$.

Let us consider a membrane, characterized, at the smallest scale Λ^{-1}, by the "bare" values of the elastic coefficients τ_0, κ_0, and $\bar{\kappa}_0$. The effective values of the coefficients can be obtained by integrating the equations above with respect to s, and by converting to τ_{eff} via the relation $\tau_{\text{eff}} = s^2\tau(s)$. One obtaines three different kinds of trajectories:

1. If the bare tension τ_0 is large enough, the whole trajectory remains in the tension-dominated region, and the membrane can be described like an ordinary interface.
2. For smaller values of the bare tension, one starts in the rigidity-dominated region with large undulations. One may then cross over to the tension-dominated region, at a crossover length ℓ_{c} such that

$$\ell_{\text{c}} = \sqrt{\frac{\kappa_{\text{eff}}}{\tau_{\text{eff}}}} \ . \tag{29}$$

For $\ell < \ell_c$ one is in the rigidity-dominated regime: tension effects can be neglected, and the effective rigidity depends on the scale ℓ:

$$\kappa(\ell) \simeq \kappa_0 - \frac{3k_B T}{4\pi} \log \Lambda\ell \ . \tag{30}$$

3. If the tension is too small, there is a length scale ξ in which one crosses over to the fluctuation-dominated regime and the model breaks down. This length can be defined by the condition that $\kappa(\ell_c)$ is of order $k_B T$. It is known as the *persistence length* and depends exponentially on the rigidity κ_0:

$$\xi \simeq \Lambda^{-1} \exp\left(\frac{4\pi\kappa_0}{3k_B T}\right) \ . \tag{31}$$

At scales larger than ξ the membrane will be crumpled. The correlation length for the orientation of the normals to the membrane is of order ξ.

3.4 Size Distribution of Vesicles

The vesicles that we have described in the previous section were out of equilibrium. They were most often obtained by shaking a solution of amphiphile in water by means of ultrasound (*sonication*). Equilibrium phases formed by vesicles in water have recently been observed, and denoted as the L_4 phase (Cantù et al. (1991), Cantù et al. (1994)). The equilibrium distribution of vesicle sizes determines the phase behavior of this system and is not yet sufficiently known experimentally.

We would like to estimate the concentration c_ν of vesicles containing ν amphiphilic molecules per unit volume. Let us denote by ϵ_ν the free energy of such a vesicle. If we consider the coexistence of vesicles with different values of ν, the free energy of such a mixture can be expressed as a function of the fraction, P_ν, of molecules that are incorporated in vesicles of aggregation number ν as follows:

$$\mathcal{F} = \sum_\nu \frac{P_\nu}{\nu} \left[k_B T \left(\log \frac{P_\nu}{\nu} - 1\right) + \epsilon_\nu\right] \ . \tag{32}$$

Equilibrium corresponds to the minimum of $\mathcal{F}$, which is reached for $P_\nu \propto \exp[-(\epsilon_\nu - \mu\nu)/k_B T]$, where μ is the chemical potential of the amphiphile in solution.

When one computes ϵ_ν, one obtains, as usual with finite-size thermodynamical systems, an expression of the form

$$\epsilon_\nu = \nu\mu^* + \alpha k_B T \log \nu \ . \tag{33}$$

The logarithmic term is the usual finite-size correction one expects in the evaluation of the free energy of a finite system. One obtains therefore a size distribution of the form $\nu^{-\alpha} \exp(-\nu/\nu^*)$, where $1/\nu^* = (\mu^* - \mu)/k_B T$. Now, ν^* is essentially determined by the constraint on the total quantity of amphiphile, so that all the interesting physics is hidden in the power-law correction $\nu^{-\alpha}$. There is some debate on the correct value of α (Simons and Cates (1992), Morse and Milner (1994)), which also has some important implications for the stability of the vesicle phase. This is one point in which the statistical physics of amphiphilic membranes helps in revealing the subtleties of calculations which are usually taken for granted.

3.5 Topological Instabilities

The phase behavior of a solution of amphiphile in water is also determined by the Gaussian rigidity $\bar{\kappa}$. This quantity may be considered as a chemical potential for "handles": when it is negative, it favors the formation of disconnected components of the membrane, each of spherical topology (vesicles). When it is positive, it favors the formation of a single connected component with a high topological genus. In the first case, steric interaction may favor a cubic structure of stuck vesicles, or, if the amphiphile concentration is large enough, of a lamellar phase. In the other case, one may observe a regular structure, again of cubic symmetry (the so called *plumber's nightmare* structures), or, at lower amphiphile concentrations, an irregular structure which separates the space in two infinite connected components (*sponge phases*). The situation is made more interesting by the fact that both the Gaussian and the bending rigidity get renormalized, and that therefore while, e.g., lamellar structures can be favored at short wavelength, the preferred structures at longer wavelength can belong to a different topological class. There has been a number of recent studies of these *topological instabilities*, but, since the space allotted to me has already been used up, I can only suggest the interested reader to consult my review (Peliti (1996)), and the references quoted in it.

3.6 Conclusions

The ambition of this text is to raise some interest in the fascinating physics of fluctuating membranes. There are a number of interesting aspects that I have not been able to cover—the most interesting one is probably the strange kind of symmetry breaking which appears in the sponge phases (Roux et al. (1992)). I urge my readers with an interest in the bases of thermodynamics to read this review which will give them food for thought. I have no doubts that more surprises are hidden in these systems, which to me are the paradigm of Biologically Inspired Physics.

Acknowledgments

These notes were written with the collaboration of Roberta Donato. I thank X. Michalet and U. Seifert for sending me their theses, which have been of great help. I am deeply grateful to the organizers of the School, and in particular to H. Flyvbjerg and O. Mouritsen, for inviting me to Humlebæk and for having made the School such a pleasant and exciting event.

References

Berndl, K., Käs, J., Lipowsky, R., Sackmann, E., Seifert, U. (1990): *Europhys. Lett.* **13** 690
Cai, W., Lubensky, T.C., Nelson, P., Powers, T. (1994): *J. Phys. II France* **4** 931
Canham, P.B. (1970): *J. Theor. Biol.* **26** 61
Cantù, L., Corti, M., Musolino, M., Salina, P. (1991): *Europhys. Lett.* **13** 561
Cantù, L., Corti, M., Del Favero, E., Raudino, P. (1994): *J. Phys. II France* **4** 1585
David, F., Leibler, S. (1991): *J. Phys. I France* **1** 959
Deuling, H.J., Helfrich, W. (1976): *J. Phys. France* **37** 1335
Helfrich, W., (1973): *Z. Naturforsch.* **28c** 693
Helfrich, W. (1978): *Z. Naturforsch.* **33a** 305
Helfrich, W. (1985): *J. Phys. France* **46** 1263
Jülicher, F., Lipowsky, R. (1993): *Phys. Rev. Lett.* **70** 2964
Jülicher, F., Seifert, U., Lipowsky, R. (1993a): *J. Phys. II France* **3** 1681
Jülicher, F., Seifert, U., Lipowsky, R. (1993b): *Phys. Rev. Lett.* **71** 452
Kawakatsu, T., Andelman, D., Kawasaki, K., Taniguchi, T. (1993): *J. Phys. II France* **3** 971
Larché, F., Appell, J., Porte, G., Bassereau, P., Marignan, J. (1986): *Phys. Rev. Lett.* **56** 1700
Leibler, S. (1986): *J. Phys. France* **41** 109
Leibler, S., Andelman, D. (1987): *J. Phys. France* **48** 2013
Lipowsky, R. (1991): *Nature* **349** 475
Lipowsky, R. (1992) *J. Phys. II France* **2** 1825
Lipowsky, R., Leibler, S. (1986): *Phys. Rev. Lett.* **56** 2541
Lipowsky, R., Sackmann, E. (1995): *Structure and Dynamics of Membranes*, in: Hoff, A.J. (ed.): *Handbook of Biological Physics*, Vol. 1, (Elsevier: Amsterdam)
Michalet, X. (1994) Etude Expérimentale de Vésicules Phospholipidiques de Genre Topologique Non-Sphérique, Thesis, University Paris VII
Michalet, X., Bensimon, D., Fourcade, B. (1994): *Phys. Rev. Lett.* **72** 168
Milner, S.T., Roux, D. (1992): *J. Phys. I France* **2** 1741
Morse, D.C., Milner, S.T. (1994): *Europhys. Lett.* **26** 565
Mutz, M., Bensimon, D. (1991): *Phys. Rev.* **A43** 4525
Mutz, M., Helfrich, H. (1989): *Phys. Rev. Lett.* **62** 2881
Peliti, L. (1996): Amphiphilic Membranes, in: David, F. and Ginsparg, P. (eds.): *Fluctuating Geometries in Field Theory and Statistical Mechanics*, Les Houches Series (North Holland: Amsterdam)
Peliti, L., Leibler, S. (1985): *Phys. Rev. Lett.* **54** 1690
Peterson, M.A., Strey, H., Sackmann, E. (1992): *J. Phys. II France* **2** 1273
Roux, D., Coulon, C., Cates, M.E. (1992): *J. Phys. Chem.* **96** 4174
Safran, S.A. (1994): *Statistical Thermodynamics of Surfaces, Interfaces, and Membranes*, Frontiers in Physics, Vol. 90 (Addison-Wesley: Reading, Mass.)
Seifert, U., Berndl, K., Lipowsky, R. (1991a): *Phys. Rev.* **A44** 1182
Seifert, U., Miao, L., Döbereiner, H.-G., Wortis, M. (1991b): Budding Transition for Bilayer Fluid vesicles with Area-Difference Elasticity, in: Lipowsky, R., Richter, D. and Kremer, K. (eds.): *The Structure and Conformation of Amphiphilic Membranes*, Springer Proceedings in Physics, Vol. 66 (Springer: Berlin)
Simons, B.D., Cates, M.E. (1992): *J. Phys. II France* **2** 1439
Singer, S. J., Nicholson, G.L. (1972): *Science* **175**, 720
Stryer, L. (1988): *Biochemistry* (Freeman: New York) Chap. 12

Svetina, S., Žekš, B. (1983): *Biochim. Biophys. Acta* **42** 86
Willmore T. J. (1982): *Total Curvature in Riemannian Geometry* (Horwood: New York)
Wortis, M., Seifert, U., Berndl, K., Fourcade, B., Miao, L., Rao, M., Zia, R.K.P. (1993): Curvature-Controlled Shapes of Lipid-Bilayer Vesicles: Budding, Vesiculation, and Other Phase Transitions, in: Beysens, D., Boccara, N. and Forgács, G. (eds.): *Dynamical Phenomena at Interfaces, Surfaces and Membranes* (Nova: New York) p. 221

Bending Energy Concept of Vesicle and Cell Shapes and Shape Transitions

Erich Sackmann

Physik-Department (Biophysics Group E22)
Technische Universität München
D-85747 Garching, Germany

1 Introductory Remarks

This chapter deals with attempts to understand complex membrane processes such as shape transitions of plasma membranes or the vesicle budding and fission from intracellular compartments in terms of universal physical laws which do not depend on the detailed structure of biomembranes.

Fig. 1 shows a cartoon-like image of a plasma membrane. It is a composite type of material made-up of three types of materials: (1) the central lipid/protein bilayer (a two-dimensional smectic liquid crystal), (2) the associated cytoskeleton (a two- dimensional macromolecular network) and (3) the glycocalix (a grafted macromolecular layer).

By coupling of the membrane associated cytoskeleton to the intracellular cytoskeleton and the glycocalix to the extracellular matrix, the elastic properties of tissue may be controlled over a large range. The plasma membrane is extremely soft with respect to bending and shearing but practically incompressible with respect to lateral stretching. This unique combination of elastic properties allows cells such as erythrocytes to travel for several hundred kilometers through our body without loss of material. The most intriguing aspect is, however, that the lateral incompressibility of cell membranes (which prevents the loss of ions during this transport) as well as the high flexibility of the composite cell plasma membrane are largely determined by the lipid/protein bilayer of the plasma membrane.

From the point of view of membrane physics most fascinating, however, is the vesicle– mediated material transport between the various intracellular compartments and between these and the plasma membrane illustrated in Fig. 2. It involves a manifold of shape changes and shape instabilities, including the vesicle budding-fission- fusion chain of events.

One purpose of this chapter is to show that many of these complex processes may be understood in terms of the bending elasticity concept of bilayer shells. In the first part I will briefly summarize the essential role of lipid bilayer elasticity for the self organization of biomembranes and the formation of specialized domains such as coated pits and buds. The second part deals with the minimum–bending–energy concept of shape transitions of pure one–component, lipid vesicles and the essential role of the constraints imposed by the closure of the shells and their stratified composition. In the third part I will introduce the enormous richness of shape transitions of membranes composed of lipid/protein–alloys which is determined by the coupling between curvature and phase separation and chemically induced bending moments. The last part deals with thermally excited undulations of membranes and their possible biological functions.

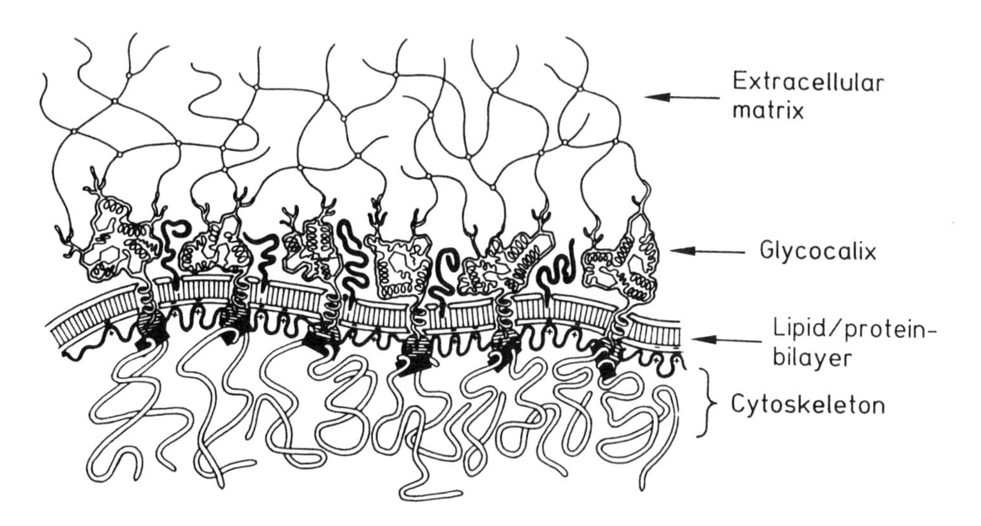

Fig. 1. Cartoon of coarse–structure of cell plasma membrane and associated networks of the intracellular cytoskeleton and the extracellular matrix. The caricature emphasizes the three–layered structure of the membrane consisting of three types of materials: The lipid/protein-bilayer which may be considered as a two-dimensional smectic liquid crystal, the membrane associated cytoskeleton which forms a two-dimensional macromolecular network, and the glycocalix which is essentially a two-dimensional macromolecular layer. The membrane is a composite material since a structural change in one layer is supposed to induce a corresponding change in the other two.

2 Principles of Vesicle Self Assembly and Stabilization

Nature was extremely clever by choosing two-chain lipids as basic building units of membranes for the following reasons:

1. They associate in water at extremely low concentrations $\left(< 10^{-12}\mathrm{M}\right)$. This is a consequence of the strong hydrophobic effect and the well known exponential dependence of the critical concentration of association c^* on the work (= chemical potential $\Delta\mu$) required to transfer a lipid molecule from the bilayer (or a micelle) into the aqueous phase (Cevc and Marsh (1978))

$$c^* = c_0 \;\; \exp\left\{\frac{-\Delta\mu}{\mathrm{k_B}T}\right\}$$

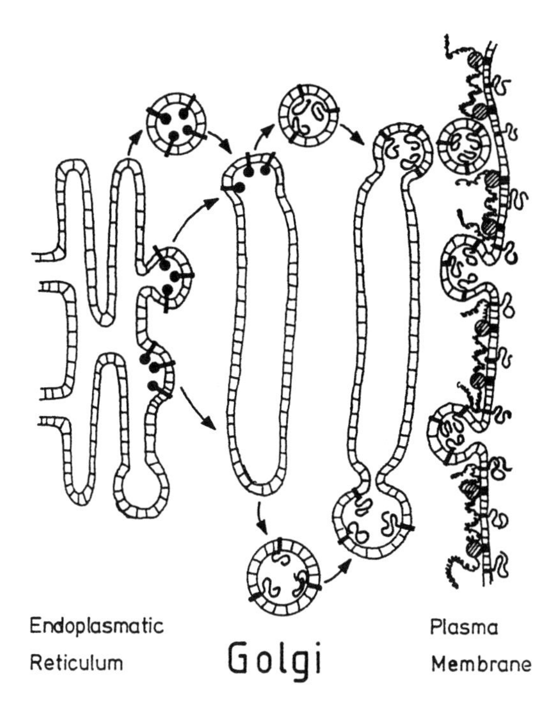

Fig. 2. Illustration of (one way of) intracellular transport from Endoplasmatic Reticulum to the plasma membrane mediated by vesicles and demonstrating the necessity of vesicle budding-fission-fusion chain of events. Note that fusion with the plasma membrane requires the transient decoupling of the cytoskeleton from the bilayer.

$\Delta\mu$ is roughly proportional to the total area of the surface of the hydrocarbon chain and thus to the chain length $n_{\mathrm{CH_2}}$. It can be estimated from the rule of thumb $\Delta\mu = 2\,(11 - 3n_{\mathrm{CH_2}})$ (Tanford (1980); Cevc and Marsh (1978)). The chemical potential differences for two-chain lipids are very high: typically $\Delta\mu = -75\frac{\mathrm{kJ}}{\mathrm{M}}$ (or $\approx 30\mathrm{k_B}T$) for a chain of $16\mathrm{CH_2}$-Groups. It is due to the exponential law, that the critical miccllc concentration is so much higher for lyso-lipids ($c^* \approx 2{\cdot}10^{-5}$M for Lyso-dimyristoylphosphatidylcholine (Lyso–DMPC)) than for normal lipids ($c^* \approx 2\cdot 10^{-12}$M for dipalmitoylphosphatidylcholine (DPPC)).

2. A consequence of the strong hydrophobic effect is a very high (and negative) internal lateral tension within bilayers of the order of $30\frac{\mathrm{mN}}{\mathrm{m}}$. It is for that reason that bilayers exhibit strong self-healing effects.
3. Bending energy concept of vesicle closure: In excess of water, most lipids spontaneously form closed shells which exhibit an astonishing high stability as illustrated in Fig. 3. Vesicle formation may be understood as a pay-off between the energy required to bend the bilayer and the energy gained by avoiding the exposure of the hydrophobic interior of the bilayer to water (Helfrich (1974)). Let us have a closer look at the process which allows si-

multaneously to introduce the bending energy concept. In order to bend the bilayer a positive tension (σ_+) has to be applied at one monolayer and a negative (σ_-) at the other which results in a bending moment $M = d_m (\sigma_+ - \sigma_-)$ (where d_m is the membrane thickness). Any curved surface may be characterized by the average curvature $\left(\frac{1}{R_1} + \frac{1}{R_2}\right)$ where R_1 and R_2 are the principal radii of curvature (measured along two perpendicular directions). Hooke's law states that

$$M = K_\mathrm{c} \left(\frac{1}{R_1} + \frac{1}{R_2}\right)$$

where K_c is the bending elastic modulus (which is measured in units of energy since the dimension of σ is $\frac{\mathrm{mN}}{\mathrm{m}}$). The total energy associated with bending is given by

$$G_\mathrm{bend} = \frac{1}{2} K_\mathrm{c} \int \left(\frac{1}{R_1} + \frac{1}{R_2}\right)^2 \mathrm{d}O \tag{1}$$

where dO is a surface element and the integral has to be performed over the whole bilayer surface. The most remarkable consequence of Eq(1) is that the bending energy of a sphere does not depend on its size (that is, it is scale invariant).

In order to assess the interfacial energy at the edge of an open bilayer, consider a partially closed shell with an opening pore of radius ρ. The total energy may be expressed as:

$$G_\mathrm{pore} = 2\pi\gamma d_\mathrm{m}\rho$$

where γ is the line tension along the rim of the pore (measured in $\frac{\mathrm{N}}{\mathrm{m}}$). Minimizing the total energy yields for the vesicle radius

$$R_0 = \frac{8K_\mathrm{c}}{\gamma} d_\mathrm{m} \tag{2}$$

The line energy per unit length γd_m of the opening may be estimated from the chemical potential of the hydrophobic effect by assuming that half of the surface of the lipid chains at the edge of the bilayer is exposed to water. For DMPC this would correspond to $\Delta\mu = 5 \cdot 10^{-20}$J per molecule or $\gamma d_\mathrm{m} = 5 \cdot 10^{-11}$N. The edge energy of a flat piece of membrane of 10μm diameter would thus be of the order of $3 \cdot 10^5$ $k_\mathrm{B}T$. The bending energy of the vesicles (of all sizes) is about 100 $k_\mathrm{B}T$. Hence the driving force for vesicle closure is very high.

The key point is that both the pore energy and the bending energy can be drastically reduced by suitable solutes.

1. The pore energy may be reduced by surfactants which can form a hydrophilic cap at the rim of the pore. Large pores can for instance be stabilized by digitonin or filipin in the presence of cholesterol. Even discs of bilayers may be formed at high concentrations of cholate (Fromherz et al. (1986)).

2. The bending energy may be drastically reduced by introduction of a spontaneous curvature C_0 which reduces the bending energy for the sphere to

$$G_{\mathrm{el}} = \frac{1}{2}\left(\frac{2}{R} - 2C_0\right)^2 \tag{3}$$

A spontaneous curvature can be generated by any gradient of the lateral pressure $\mathrm{d}\pi/\mathrm{d}z$ in the direction (z) of the bilayer normal which creates a bending moment (cf. Fig. 4)

$$M_0 = \int_{-d_{\mathrm{m}}/2}^{+d_{\mathrm{m}}/2} z\pi\,(z)\,\mathrm{d}z \tag{4}$$

resulting in a spontaneous curvature $C_0 = \frac{M_0}{K_c}$.

There are many ways to create intrinsic bending moments, in particular by the adsorption of proteins (as in coated pits) or by changing the surface charge density (eg. by the binding of Ca^{++}). The spontaneous curvature concept is also extremely helpful for the understanding of the micelle-bilayer transition (Gruner et al. (1980)).

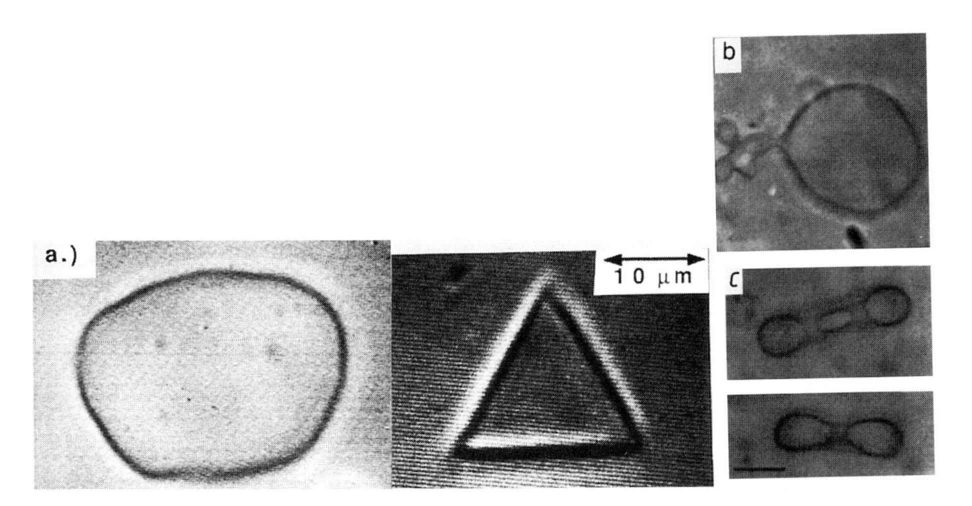

Fig. 3. Demonstration of stability of single bilayer vesicles: (a) Shape change of giant DMPC vesicle in 500mM Na Cl buffer at fluid to gel transition. Note the astonishing stability of the vesicle composed of a crystalline shell. (b) Spontaneous budding of DMPC vesicle leading to tube-like (sometimes branched) protrusions. (c) Torus shaped vesicle of diacetylene-PC according to Mutz et al. (1991) Bar corresponds to 10mm.

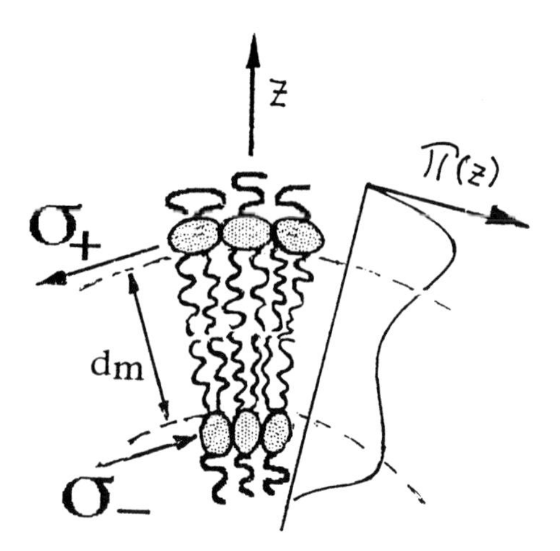

Fig. 4. Illustration of the generation of (one-dimensional) intrinsic bending moment, M, by gradient of lateral pressure $\pi(z)$ across the bilayer. The bending moment may also be represented by a couple of opposite tensions (measured in mN/m)

3 Elasticity of Bilayer-Cytoskeleton Composite Membranes

3.1 Modes of Deformation

Any deformation of cell membranes can be described in terms of a superposition of the three modes of deformation well-known from the classical theory of shells

1. pure bending (introduced above)
2. pure shearing and
3. isotropic compression.

The resistance of bilayers towards isotropic tension (compression or dilatation) is characterized by the elastic energy (per unit area)

$$g_{\mathrm{comp}} = \frac{1}{2}\kappa \left(\frac{\Delta A}{A}\right)^2 \tag{5}$$

where $\frac{\Delta A}{A}$ is the relative change in bilayer area and κ is the lateral compressibility modulus measured in $\mathrm{J/m^2}$. The tension required to maintain the area change is $\sigma = \kappa \frac{\Delta A}{A}$.

The compressibility modulus and the bending modulus are proportional to each other

$$\kappa = \frac{K_{\mathrm{c}}}{d_{\mathrm{m}}^2}$$

Pure shearing is best visualized by considering a square piece of bilayer which is stretched in one direction by a tension σ_+ to a length $L = L_0 + \delta L$ and compressed to $L = L_0 - \delta L$ in a perpendicular direction by a tension σ_- (of the same size). Clearly the area remains constant for this type of deformation. It is customary to express the strain in terms of the elongation ratio $\lambda = (L_0 - \delta L)/L_0$ and the energy is then given by (Kwok and Evans (1981))

$$g_{\text{shear}} = \frac{1}{2}\left(\lambda^2 + \lambda^{-2} - 2\right) \tag{6}$$

where μ is the shear elastic constant.

Some data: Many sophisticated techniques have been developed for high precision measurements of the elastic constants of vesicles and cell membranes such as the micropipette technique (Evans (1974)), the electric field deformation technique (Kwok and Evans (1981)) and the flicker spectroscopy (Wwang et al. (1994)). Some data are shown in Table I.

These studies have revealed outstanding elastic properties of biological membranes. This is best demonstrated by comparing the elastic constants of erythrocytes with those of a shell made of a synthetic polymer exhibiting the same size and topology (cf. data in Table I). The most remarkable finding is that the cell membrane is extremely soft with respect to bending and shearing but is laterally incompressible. This combination of elastic constant allows the cell to squeeze through very narrow channels without loss of ions (Sackmann (1990); Kwok and Evans (1981)).

Two other remarkable findings are:

1. Despite of the high cholesterol content and the bilayer-cytoskeleton coupling, the bending stiffness of the erythrocyte membrane is equal to about that of DMPC-bilayers.
2. The cell exhibits a shear free deformation regime and stiffens with increasing deformation.

Table 1. Comparison of two-dimensional elastic moduli of biological and artificial materials

	Shear modulus $\mu\,(\text{erg/cm}^2)$	Compressibility modulus $\chi\,(\text{erg/cm}^2)$	Bending modulus $K_c\,(\text{erg})$
Brass	$1.1 \cdot 10^5$	10^5	10^{-8}
Polyethylene	$5 \cdot 10^3$	$5 \cdot 10^3$	$5 \cdot 10^{-10}$
Erythrocyte	$6 \cdot 10^{-3}$	–	$7 \cdot 10^{-13}$
DMPC bilayer L_α-phase	0	145	$1.2 \cdot 10^{-12}$
DMPC with 30% cholesterol	0	647	$4 \cdot 10^{-12}$
Erythrocyte	0	–	$1 \cdot 10^{-13}$

3.2 Modulation of Bilayer Elasticity by Solutes

The membrane elasticity may be manipulated in various ways (cf. Table I):

1. Incorporation of small amounts of amphiphiles may reduce the bending stiffness dramatically. Thus, addition of 1% of cholate reduces the K_c-value of DMPC by a factor of 10. Even more dramatic effects arise if small bipolar lipids are incorporated reducing K_c to the order of thermal energy which results in extremely drastic fluctuations (Duwe and Sackmann (1990)). Thus the low bending stiffness of the erythrocyte membrane could be due to solutes.
2. The bilayer elasticity is determined by the lateral packing density. Therefore the well-known condensing effect of cholesterol leads to a strong increase of the bending and compression modulus.

3.3 Modulation of Erythrocyte Bending Stiffness

The softness of the erythrocyte membrane is still an enigma. An important observation is that the effective stiffness is remarkably increased

1. by cross-linking of the spectrin/actin network by diamid (Engelhardt and Sackmann (1988)),
2. by coupling of the network to glycophorin via band IV.1 protein (which is mediated by the binding of antibodies or wheat germ agglutinin to glycophorin),
3. by ATP-depletion, which abolishes also the transmembrane lipid asymmetry (Devaux (1988)).

These observations suggest that the membrane softness is maintained by continuous phosphorylation of the cytoskeletal proteins. Thus, phosphorylation reduces the bending of ankyrin to band III protein and of band IV.1 protein to glycophorin by a factor of about 5. A partial decoupling of the network from the bilayer (mediated by phosphorylation) appears to be essential for the maintainance of the high degree of softness (cf. Fig. 5).

4 Lateral Stucture Formation in Membranes

Over the last twenty years it has become more and more evident that biological membranes, in particular the plasma membranes, exhibit non-random lateral organization despite of the fact that the lipid bilayer moiety is in the fluid state. Indeed, direct evidence for a non-random bilayer organization has been provided by local lateral diffusion measurements and more recently by optical near field microscopy (M. Edidin, privat communication).

Several mechanisms of lateral structure formation are conceivable:

1. Coupling of membrane proteins and possibly lipids (Fig. 5) to the cytoskeleton.

2. Formation of local buds which may be induced by a protein adsorption (as in the case of coated pits) or other local bending moments (cf. Fig. 4).
3. Local phase separation within the two-dimensional lipid/protein multicomponent system.

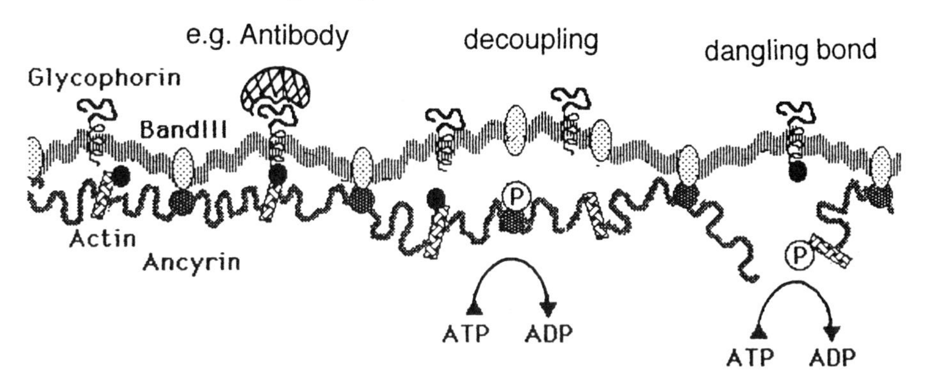

Fig. 5. Illustration of fine-tuning of coupling of lipid/protein bilayer to spectrin-actin network of erythrocytes by phosphorylation of coupling proteins ankyrin and band 4.1 protein. The strong weakening of the coupling strength after phosphorylation may result in the decoupling of the cytoskeleton from the bilayer or the formation of dangling bonds (right) within the cytoskeleton which increases the membrane flexibility. Binding of ligands (eg. antibodies) to glycophorin can result in a strengthening of the coupling and thus in an increase in stiffness.

In the following I will briefly discuss the last two mechanisms and start with the last one. A large amount of work on lipid mixtures has been accumulated (Sackmann (1990); Bloom et al. (1991)). Concerning the question of lateral structure formation in biomembranes the following findings are essential.

1. Phospholipid mixtures exhibit in general phase diagrams of the form shown in Fig. 6. Phase separation arises in general only if one of the components is in the gel state. Indeed some of the natural lipids of mammalian cells exhibit very high transition temperatures. For instance, brain sphingomyelin (SPM) exhibits a very broad regime of fluid-solid coexistence from 30 to 60°C (Döbereiner et al. (1993)) making lateral phase separation in nerve membranes at physiological temperatures quite probable.
2. In the presence of Ca^{++} the phase transition temperature of charged lipids is strongly shifted to higher temperatures (by about 50°C). Ca^{++} influx in cells can thus easily generate local domain formation in the inner leaflet of plasma membranes. Since the Ca^{++} induced domain formation exhibits a very strong hysteresis, even a transient increase of the intracellular Ca^{++}

level (eg. in synapses) can result in metastable domain formation (Sackmann (1990); Knoll et al. (1986)).

3. There is only one well-documented example of fluid-fluid immiscibility, namely the PC-cholesterol mixture (Bloom et al. (1991)). This mixture exhibits a miscibility gap between about 10 and 30 mole% cholesterol and the critical point lies about 30°C above the gel-fluid transition of the PC. Physiologically even more important is, however, that cholesterol-rich domains occur at concentrations of about 40mole% as has been conclusively shown by small angle neutron scattering studies (Sackmann (1990)). Some evidence for the segregation of cholesterol in erythrocytes is provided by the observation that erythrocytes membranes exhibit a fraction of rapidly exchangeable cholesterol while a large fraction is only slowly exchangeable. The latter fraction could be organized in clusters. We will show below that the local clustering of cholesterol may play an important role for the vesicle fission after osmotic deflation of intracellular compartments.
4. A highly interesting situation arises for lipid mixtures exhibiting a miscibility gap in the solid state with a critical point lying within or slightly below the fluid solid coexistence. For the particular case of the DMPC/DSPC mixtures it has been shown by small angle neutron scattering that the mixture exhibits lateral heterogeneity even well above the liquidus line which is due to critical demixing (Knoll et al. (1985)). Segregated domains of the high melting component exhibiting average diameters of about 300Å form roughly 10°C above the liquidus line. Very detailed Monte Carlo simulations have shown that this is in general expected for mixtures of lipids differing in chain length by more than 4C-atoms (Bloom et al. (1991)). It is therefore highly probable that critical demixing is quite common in biomembranes. The most interesting aspect is, however, that the transient domains would be stabilized by proteins, provided selective lipid/protein interaction mechanisms exist (Sackmann (1990)). This point is discussed in the following.

5 Mechanisms of Selective Lipid Protein Interaction

The local structure formation by clustering is closely related to the existence of mechanisms of selective lipid/protein interaction. Two mechanisms have been established hitherto: an electrostatic and a steric elastic mechanism (which are illustrated in Fig. 7). The former mechanism arises if the hydrophobic thicknesses of the lipid bilayer d_m and of the protein d_p are not matched. Thus, if $d_m < d_p$ the hydrocarbon chains of the lipids adjacent to the protein are stretched in order to avoid exposure of the hydrophobic parts of the protein to water. An interesting question is how far the perturbation of the lipid matrix extends. Many theoretical studies (cf. Bloom et al. (1991) for references) postulate that the perturbation decays exponentially with the distance from the surface of the protein (of radius r_p)

$$(d_m - d_p) = (d_m - d_p)_0 \, e^{-(r-R_p)/\zeta} \qquad (7)$$

where ζ is the correlation length which is about $\zeta = 30\text{Å}$ (corresponding to only 2 to 3 lipid distances). Nevertheless, the energy associated with the deformed lipid halo is substantial and depends on the bilayer compressibility modulus. For a mismatch of $(d_{\rm m} - d_{\rm P}) \approx 5\text{Å}$, the total elastic energy is about 100 $k_{\rm B}T$. This has far-reaching consequences in mixtures of lipids of different chain length since the protein will surround itself by a halo of length-adapted lipid as illustrated in Fig. 7.

The electrostatic mechanism of lipid/protein selectivity has been established in many model membrane studies. In particular, many of the actin binding proteins mediating the interaction of actin with the membranes such as talin and hisactophilin or a-actinin bind to membranes in the presence of charged lipids (cf. Kaufmann et al. (1992)). A direct interaction of talin with the lipid bilayer is also suggested by the finding that a large fraction of this protein is bound to the lipid/protein bilayer and can only be removed by surfactants (eg. Triton).

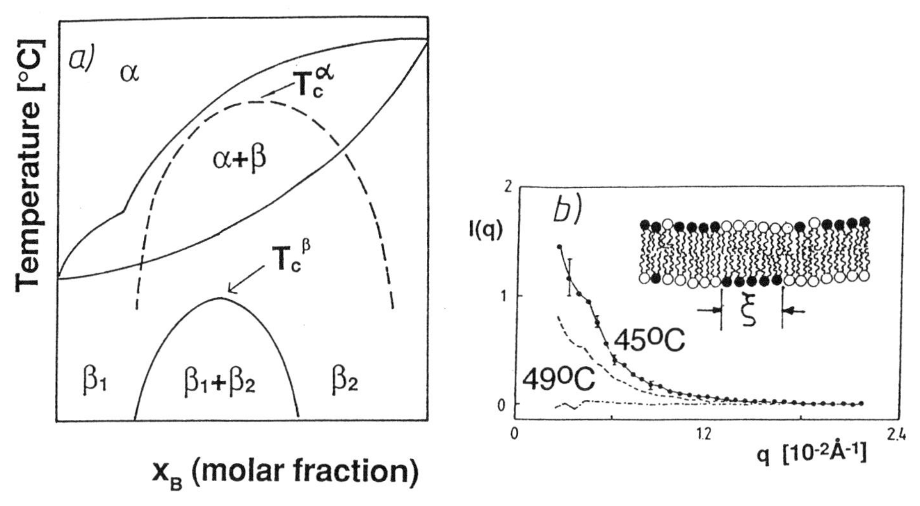

Fig. 6. (a) Hypothetical phase diagram of binary mixture of lipids. The diagram exhibits in general peritectic behaviour of the fluid-solid coexistence $(\alpha + \beta)$ if a solid-solid miscibility gap $(\beta_1 + \beta_2)$ penetrates into the coexistence region. For the same type of lipid, the critical point is shifted to higher temperatures with increasing differences in chain lengths. Most remarkably, critical demixing of the lipids is observed above the liquidus line. (b) Demonstration of critical demixing in mixture of DMPC and DSPC above the liquidus line. The plots of the scattering angle show that the average cluster size increases dramatically if the critical point is approached, The cluster size is $\xi < 500$ Å at 45°C and 300 Å at 49°C. The critical temperature is $T_c < 40$°C.

Membrane bound receptors may simultaneously exhibit the steric elastic and the electrostatic mechanism of selective lipid protein interaction. A well studied example is the transferrin receptor (Kurrle et al. (1990)). First it was shown that a substantial amount of this receptor can only be incorporated in bilayers if the hydrocarbon chains are long enough (> 16 C-atoms) and in the presence of acid lipids. As demonstraded by calorimetry and Fourier-Transform-Infrared-Spectroscopy (cf. Fig. 8b), the protein interacts preferentially with charged lipids. This is attributed to the binding of a sequence of acidic amino acids of the cytoplasmatic domain to the charged lipids (cf. Fig. 8c).

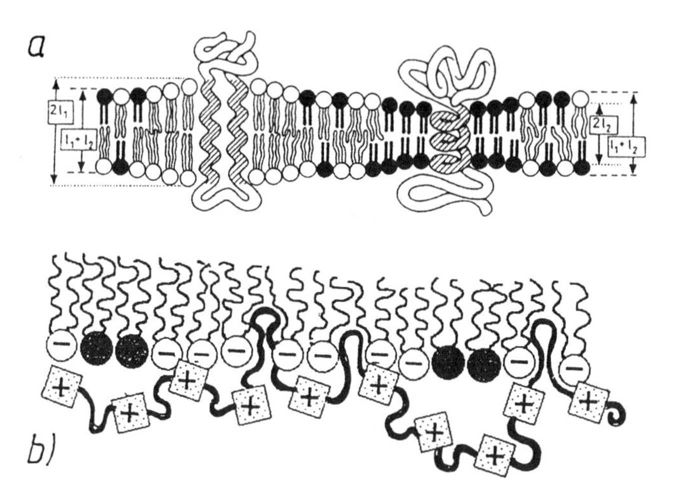

Fig. 7. On mechanisms of selective lipid-protein ineraction. (a) Selective interaction due to the mismatch of the thicknesses of the hydrophobic domain of the proteins and the lipid bilayer. (b) Electrostatic selectivity illustrating binding of filamentous protein (spectrin) to bilayers. Neutron surface scattering studies suggest that the flexible hinges between the condensed domains penetrate into the bilayer.

6 Bending Energy Concept of Vesicles and Cell Shape Changes

Giant vesicles of a single lipid component (such as DMPC) show a very rich scenario of shapes and shape transitions many of which are familiar from biology. Fig. 9 exhibits a series of characteristic and typical examples of transitions generated by variation of the excess area of the bilayer by heating: Fig. 9a shows an initially quasispherical vesicle which first changes into a discocyte which is stable over a rather large range of area-to-volume ratios before it undergoes a transition into a stomatocyte. The transition is completely reversible. As shown in Fig. 9b, a completely different behaviour is observed when the vesicle was kept under high osmotic stress before heating. Within a very narrow temperature interval (corresponding to an area increase of 7%) buds and tethers shoot out which eventually merge. Fig. 9c shows the case of a transition from a quasispherical shape to a stomatocyte caused by osmotic deflation.

In the following it will be shown that shape changes of vesicles (but also of cell membranes) can be explained in terms of the minimum bending energy hypothesis which predicts that the global shape of a soft shell is determined by the minimum of the total bending energy subjected to the constraints

1. of constant average area $\langle A \rangle$ which holds if the exchange of lipids with the environment is blocked
2. of constant area difference

$$\Delta A = A_{\rm ou} - A_{\rm i} \tag{8}$$

between the inner (i) and outer (ou) leaflet (which holds for blocked lipid exchange between the monolayers).

Two modes have been proposed: the spontaneous curvature model presented by Eq(2) and the bilayer coupling model.

The bilayer coupling model enables a simple physical interpretation of the shape changes by considering the area difference ΔA. For a sphere of radius R_0 it is $\Delta A = 8\pi R_0 d_{\rm m}$. In order to generate a stomatocyte, the outer monolayer has to be expanded with respect to the inner one. To form, for instance, two interconnected spheres each of radius $R_0/\sqrt{2}$ the area difference (at constant volume) has to be $\Delta A_0 = 8\sqrt{2}\pi R_0 d_{\rm m}$. Budding to the outside thus requires expansion of the outer monolayer with respect to the inner one.

In the framework of this model, the possible shapes of two-layered shells may be expressed in terms of only two parameters

1. the degree of deflation, expressed in terms of the ratio v of the actual volume and the volume of a spherical vesicle of the same membrane area $\langle A \rangle$.
2. the area difference ΔA normalized with respect to the value $\Delta A_0 = 8\pi R_0 d_{\rm m}$D for the sphere.

The situation may be presented in terms of the phase diagram shown in Fig. 10 in which each shape is represented by a range of values $(v, \Delta A)$. The major merit of such a phase diagram is that the physical origin of a shape change may be reconstructed from the observed pathway through the phase diagram. Thus it has been shown that the shape changes of Figs. 9a and 9b are the consequences of different coefficients of thermal expansivity at the outer and inner monolayer. In particular, it is clear that a given shape may be generated either by changing the degree of deflation v or by changing ΔA. Another important prediction is that the shape changes depend critically on the pre-history of the vesicle, that is whether one starts from a flaccid or from an osmotically stressed vesicle (Käs and Sackmann (1991)). This difference is clearly demonstrated in Fig. 9a and 9b.

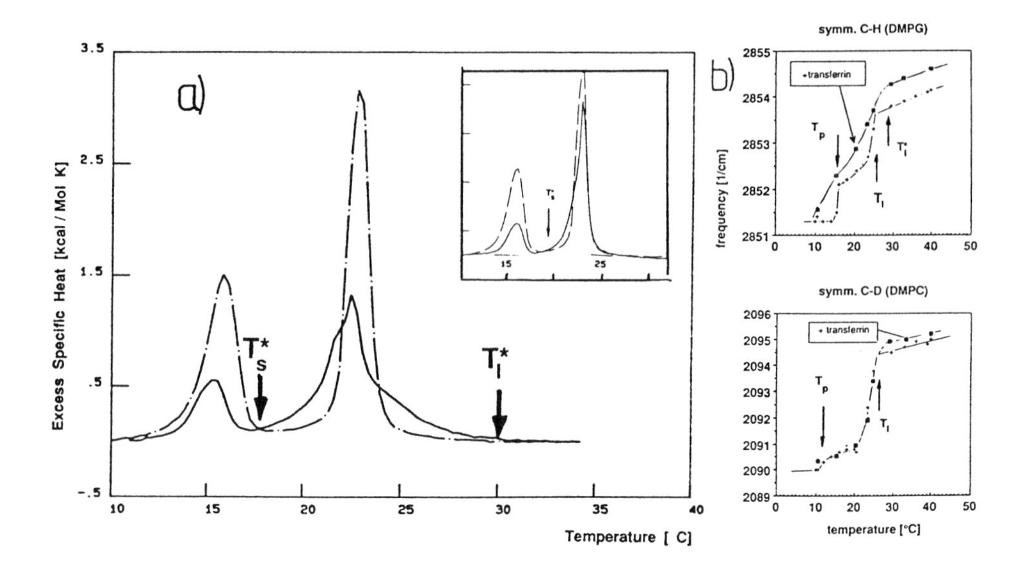

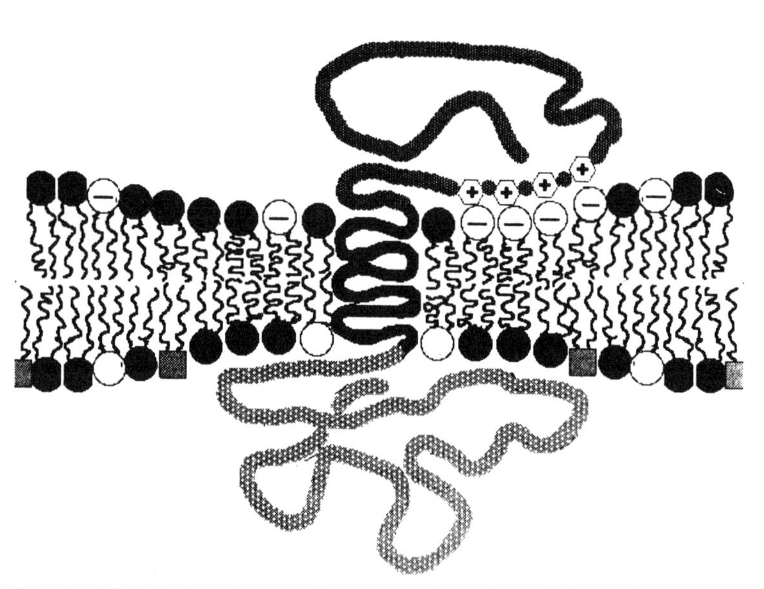

Fig. 8. (a) Calorimetic scanning curve of 1:1 DMPC/DMPG mixture in absence (—) and presence (- · -) of $x_p = 4 \cdot 10^{-4}$ transferrin receptor. The simultaneous shift of the upper transition (the liquidus line T_l) to higher and of the lower transition (the solidus line T_s) to lower temperatures is attributed to electrostatic interaction and to the elastic distortion of the hydrophobic domain, respectively. The inset shows the region of the main transition after removal of the head groups of the receptor by protease resulting in a suppression of the shift of T_l. (b) Demonstration of selective interaction of receptor with charged DMPG by Fourier Transform Infrared Spectroscopy. Note that only the phase transition of DMPG is affected by the protein (Kurrle et al. (1990)). (c) Model of combined elastic and electrostatic interaction.

7 Shape Changes of Mixed Membranes, or Coupling Between Curvature and Phase Separation

The scenario of shape changes becomes even richer for membranes composed of mixtures. The situation becomes very complex since the shapes of minimum bending energy are determined by three additional factors such as

1. the transmembrane asymmetry;
2. the different spontaneous curvatures of regions of different composition;
3. the interfacial energy between regions of different composition.

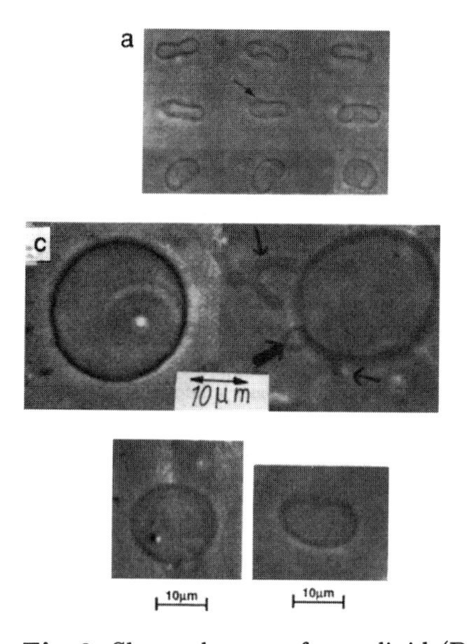

Fig. 9. Shape changes of pure lipid (DMPC) bilayer caused by increasing the excess area of the membrane by heating. (a) Discocyte-to-somatocyte transition of an initially quasispherical vesicle. (b) Blebbing transition of DMPC vesicle in pure water by increase of the area by only about 7% (according to Käs and Sackmann (1991)). (c) Example of osmotically-induced shape change of DMPC originally prepared in 200 mosm inosital. Top: outside medium replaced by 100mM NaCl. Bottom: shape observed after replacing outside medium by 178mM inosital and 11mM NaCl.

Let us consider first some experimental findings: In Fig. 11 we show that lateral phase separation may lead to a stable domain structure instead of a complete separation of the components as one would expect for ordinary mixtures. In the case of the DMPC-DMPE mixture in the fluid-solid coexistence (Fig. 11a), the DMPE-rich phase is solid and therefore exhibits completely flat domains while the DMPC-rich domains are fluid and exhibit the same average curvature as the vesicle. Fig. 11b shows the situation of lateral lipid segregation caused by the adsorption of a charged polypeptide (polylysine) to a vesicle containing charged lipids (phosphatidic acid). The polypeptide causes segregation of the charged lipid which forms, however, only small domains. Again it is important to realize that the two phases exhibit different local curvatures. Below we will argue that the stabilization of the domain structure is a consequence of an elastic interfacial energy which prevents the growth of domains.

Domain formation is essential for the stabilization of heterogeneously organized membranes since complete segregation would result in decay of the vesicles into frations of different composition (Sackmann (1990))

Fig. 12 shows an example of the coupling between curvature and phase separation in a fluid vesicle. An amphiphilic macrolipid is grafted to the outer monolayer of DMPC-vesicles by hydrophobic chains. The hydrophilic polymer chain undergoes a transition from an expanded to a collapsed state above a lower critical point at $T_c \approx 32°C$. If the polymer is in the expanded state, the vesicles tend to exhibit non-axially symmetric shapes (such as the triangular shape of Fig. 12a). Heating through the expanded-collapsed transition leads to budding of the single-walled vesicles while the vesicles with multilamellar shells form only soft protrusions.

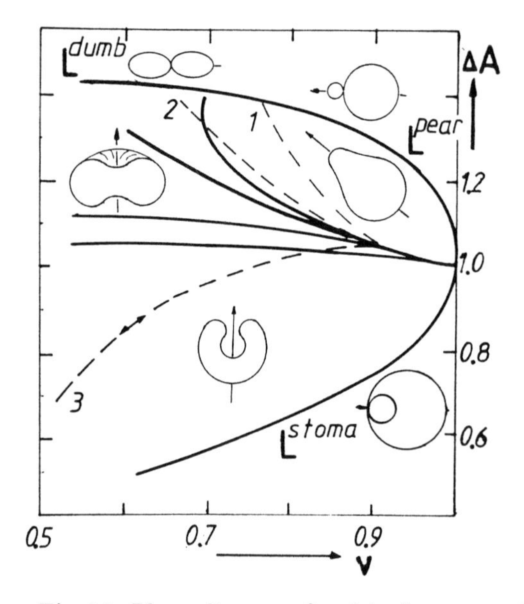

Fig. 10. Phase diagram of vesicle shapes as calculated by the bilayer coupling model. ΔA and v are reduced norm values of the area difference and the volume (cf. Berndl et al. [34]). Note that $\Delta A > 0$ corresponds to a positive and $\Delta a < 0$ to a negative spontaneous curvature. For a sphere (which is not under high osmotic tension) $\Delta A = 1$ and $v = 1$. The dashed curves 1, 2, 3 denote observed paths of shape changes (Käs and Sackmann (1991)).

The formation of vesicles of non-axially symmetric shape (which is rarely observed for one-component vesicles) is a consequence of the coupling between phase separation and curvature. The expanded macrolipids prefer areas of high convex curvature and thus tend to accumulate in the protrusions. The entropy of mixing and the repulsion between the head groups, however, counteract this lateral phase separation. Therefore only a slight distortion of the vesicle with a small non-random lateral distribution of the macrolipid is expected.

This type of shape transition can be described in a very general way by combining the Cahn-Hilliard-Langer theory of spinodal decomposition with the minimum bending energy concept. This mathematically very complex problem is treated elsewhere (Sackmann (1990); Andelmann, et al. (1992)). The complete budding caused by the collapse of the polymer is a consequence of two effects:

1. the strong positive bending moment caused by contraction of the hydrophilic head groups and
2. the strong decrease of the entropy of mixing.

Fig.12b shows that the curvature of the buds depends critically on the bending stiffness K_c of the vesicle shell. According to Eq.(3), the spontaneous curvature C_0 (or ΔA_0) is inversely proportional to K_c. This explains why the complete budding is found for unilamellar vesicles while the polymer collapse generates only softly curved protrusions in multilamellar vesicles. The bending stiffness of a shell of n bilayer is nK_c. The most intriguing aspect of the experiment shown in Fig. 12 is that the polymer induced shape changes are reminiscent of coated pit or vesicle formation (Darnell et al. (1990)).

8 Domain Stabilization Versus Vesicle Fission

The fission of budded vesicles which is seldomly observed for one-component fluid vesicles is quite common for mixed vesicles. One example is shown in Fig. 13a; another, of more direct biological relevance is shown in Fig. 13b. These findings appear to contradict the formation of the stable domain structure in vesicles undergoing phase separation. In order to understand this discrepancy, we have to consider the interfacial energy at the boundary between different phases more closely. As shown in Fig. 14, an elastic and a chemical contribution have to be considered which exert opposite effects on the vesicle stability. The former arises if the two phases exhibit different spontaneous curvatures. The local variation of the lipid orientation at the interfaces (called splay elastic deformation in liquid crystal physics) is associated with an elastic bending energy. It is easily seen that the interfacial energy increases with the radius ρ of the domains. A simple consideration yields $\gamma_{\text{if}}^{\text{el}} \approx \rho^2$. Thus, the total interfacial energy of a domain increases with its radius as $G_{\text{if}}^{\text{el}} \approx 2\pi\rho^3$. It is thus clear that the elastic interfacial energy impedes the growth of the domains.

The chemical contribution to the interfacial energy per unit length $\gamma_{\text{if}}^{\text{ch}}$ is constant and is independent of the domain size. As shown by Lipowsky (1992), the total interfacial energy $G_{\text{if}}^{\text{chem}}$ could be drastically reduced by budding in such a way that the interface is located in the narrow neck interconnecting mother and daughter vesicles. This process is only possible in vesicles exhibiting excess

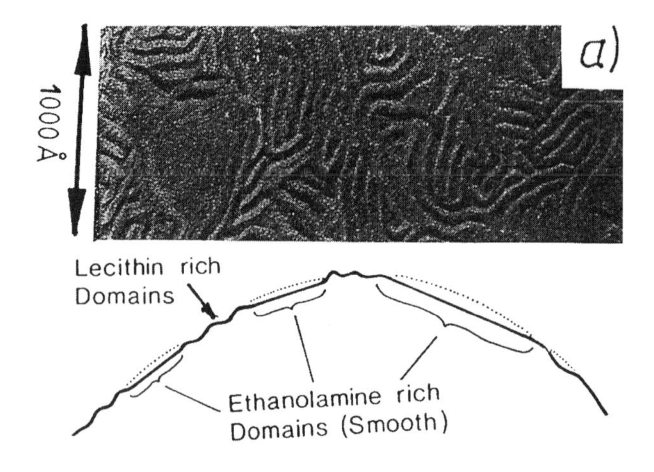

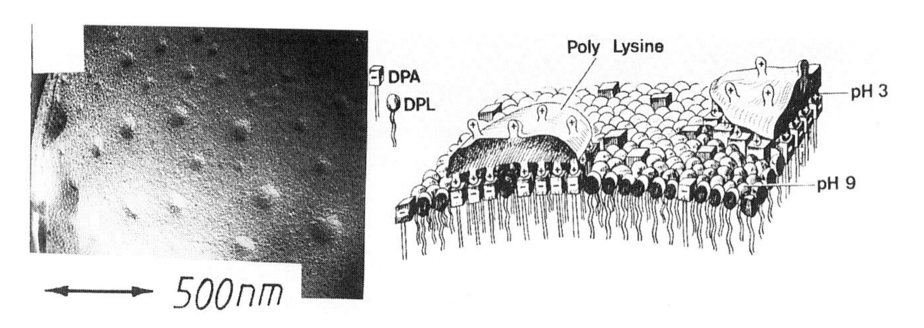

Fig. 11. (a) Stabilization of domain structure of vesicle composed of DMPC (50DMPE-rich domains are flat while the fluid DMPC-rich phase (wrinkled) exhibits the same curvature as the vesicle. (b) Domain formation in mixed vesicle containing a charged lipid component (phosphatidic acid) caused by the adsorption of randomly coiled polylysine.

membrane area. The size of the bud is determined by the condition that the bending energy associated with the bud formation is compensated by the gain in interfacial energy. A simple consideration (similar to that leading to Eq(2)) yields for the radius of the bud

$$R_{\mathrm{B}} = \frac{8K_{\mathrm{c}}}{\gamma_{\mathrm{if}}^{\mathrm{ch}}} \tag{9}$$

An example of this budding mechanism is shown in Fig. 13a. The natural brain sphingomyelin exhibits a very broad phase separation regime between 30 and 50°C (that is under physiological conditions). Heating of the vesicle generates excess membrane area and causes continuous melting of some of the lipids. The vesicle buds detach rapidly from the mother vesicle which remains essentially sherical. A closer inspection of the freeze fracture electron microscopy images shows that the composition of the buds is different from that of the mother vesicle.

The process shown in Fig. 13a could well play a role for lipid sorting in cells. As is well-known, sphingomyelin resides mainly in the plasma membrane and the lysosomes (or CURL-vesicles, Darnell et al. (1990)). Thus the question arises how this lipid is preferentially recycled to the plasma membrane after endocytosis and fusion with the cytoplasmic vesicles (endosomes). Provided that sphingomyelin has a higher tendency for budding and fission from the CURL-vesicles than the other lipids it would preferentially be transported back to the plasma membrane.

The fission of the vesicle can also be driven by local phase separation within the neck. This is strongly suggested by the finding that lecithin vesicles containing a high cholesterol content ($\geq$ 40mole%) exhibit fission after budding caused by osmotic increase in membrane excess area. This biologically relevant fission process is shown in Fig. 13b. Vesicles containing smaller cholesterol concentrations do not show fission. This is attributed to the fact that above about 40mole%, cholesterol exhibits local phase separation as was shown previously by small angle neutron scattering studies (Knoll et al. (1985)). The vesicle fission is thus attributed to the accumulation of cholesterol within the neck interconnecting the vesicles.

9 Membrane Undulations and Their Possible Role for Membrane Processes

An important consequence of the extreme softness of fluid lipid bilayers is the excitation of pronounced bending undulations (so-called flickering). These ondulations are excited by thermal fluctuations (and are thus equivalent to Brownian motion). They are strongly overdamped due to the friction caused by the coupling of the undulations to hydrodynamic flows in the aqueous phase.

In order to get a feeling for the undulation amplitudes we consider a piece of membrane of dimension $L \cdot L$. Any deflection of the membranes can then be described as a superposition of overdamped plane waves of wavelength $\Lambda = 2\pi/q$ where q is the wave vector.

$$U(\mathbf{r}, t) = \sum_q U_\mathrm{q} \cos(\mathbf{qr}) \exp\{-\omega_\mathrm{q} t\} \tag{10}$$

The bending energy is then simply given as a sum over all squared amplitudes of the individual modes of excitation

$$G_\mathrm{bend} = \frac{1}{2} K_\mathrm{c} \sum_q U_\mathrm{q}^2 q^4 L^2 \tag{11}$$

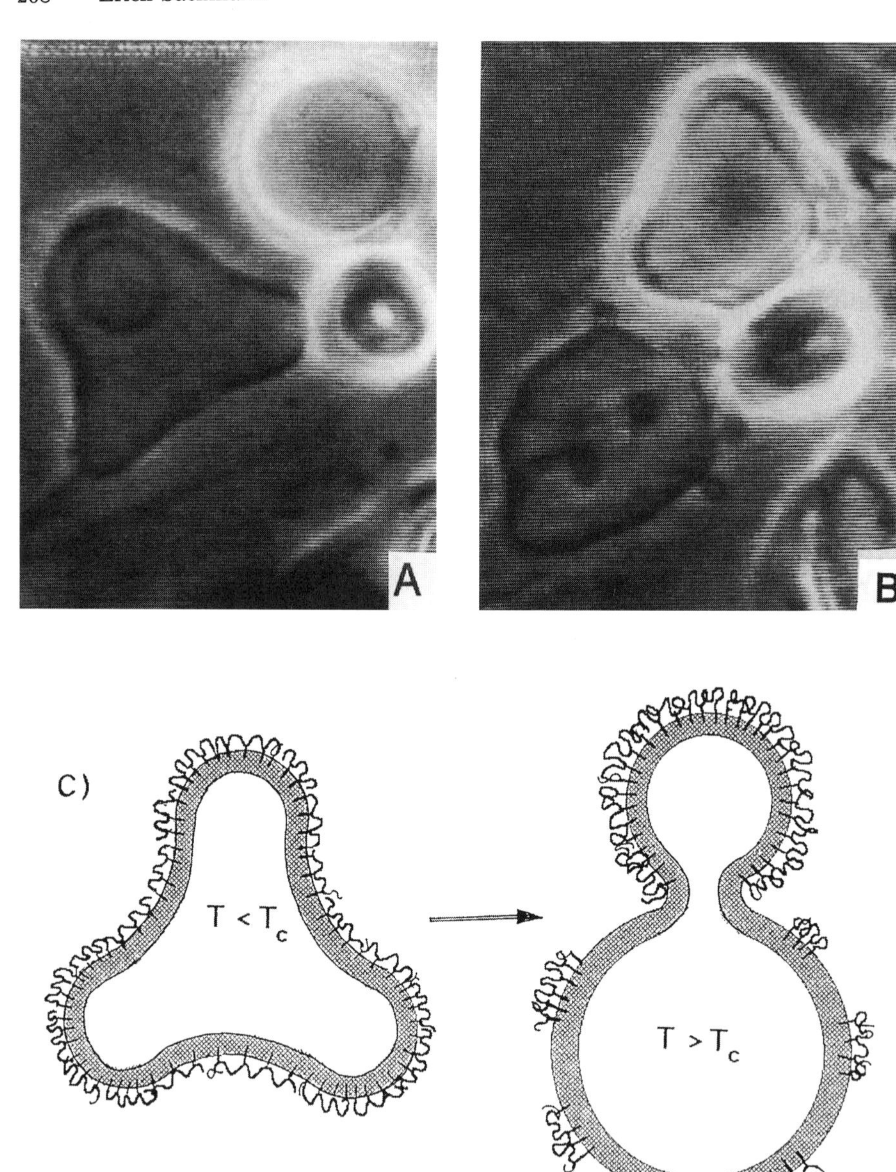

Fig. 12. Coupling between curvature and lateral segregation in vesicles of DMPC with macromolecular amphiphile anchored in the outer monolayer via hydrophobic chains and which exhibits an expanded to collapsed transition at $T > 32°C$. (a) (left side): Shape of vesicle with polymer in expanded state. Most remarkable is the non-axially symmetric (triangular) shape of the thin-walled vesicle (lower left corner). (b) (right side): Budding of vesicles after heating above the collapse point of polymer. Note that complete budding is only possible for thin-walled vesicle while for multilamellar shells only softly curved protrusions are possible.

Since each mode corresponds to a degree of freedom of the membrane, the equipartition theorem of classical statistical mechanics may be applied which predicts that the average energy per mode is $1/2k_BT$. Therefore

$$\langle U_q^2 \rangle = \frac{k_B T}{K_c q^4 L^2} \tag{12}$$

Since K_c is of the order of $10k_BT$, the amplitude of the longest wavelength mode of a bilayer of $L \approx 10\mu m$ is $\langle | U_q^2 | \rangle \approx 3\mu$m, that is the undulation amplitudes are astonishingly large.

For closed vesicles the situation is more complex. The longest wavelength exitations are suppressed since the excess area of the bilayer is in general not sufficient for their full excitation. This constraint gives rise to a lateral tension σ (Millner and Safran (1987)) and the mean square amplitude is then changed into

$$\langle U_q^2 \rangle = \frac{k_B T/L^2}{K_c q^4 + \sigma q^2} \tag{13}$$

Clearly, the amplitudes are smaller for small q (or long wavelength).

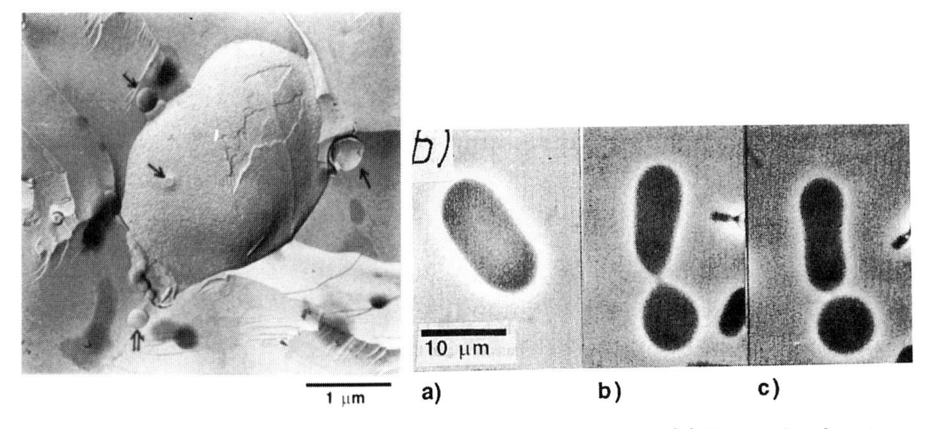

Fig. 13. Fission of budded vesicles by local phase separation (a) Example of a giant vesicle composed of brain sphingomyelin. This mixture exhibits a broad fluid-gel coexistence between 30 and 50°C. Increasing the excess area by heating leads to budding but the buds detach immediately from the mother vesicle and swim away. (b) Osmotically driven fission of a (60:40)-DMPC: cholesterol mixture. The vesicle was deflated continuously by water outflux from the vesicle.

The dynamic surface roughness associated with the undulations has many important consequences some of which will be discussed below.

1. Molecular exchange between membrane and cytoplasm: the coupling of the undulations to the aqueous environment leads to local hydrodynamic flows. These could help to facilitate the material exchange between the inner surface of plasma membranes and the cytoplasm, in particular in the erythrocytes.
2. Dynamic lateral tension: It is intuitively clear that a force is required to pull out the dynamic wrinkles. This amounts to a dynamic lateral tension as has been first shown by Helfrich (1978). This has an interesting consequence for the erythrocyte plasma membrane. As indicated in Fig. 5 the bilayer exhibits a small excess area with respect to the spectrin–actin network, and its undulations create a negative tension on the spectrin filaments (which can be considered as entropic springs Sackmann (1990)). One may have the situation of two coupled springs, one of which is stretched while the other is compressed (with respect to their resting states). Such a couple would indeed exhibit a nearly tension-free deformation regime. It is interesting to note that the tension associated with the undulations is equivalent to the entropy elasticity of macromolecules. Therefore lipid bilayers are two–dimensional analogues of semi– flexible macromolecules such as actin (Sackmann (1994)).
3. Undulation force: The most spectacular consequence of the dynamic surface roughness is a dynamic repulsion force between flickering vesicles (or cells) and a solid surface (or between two vesicles or two cells). The undulation force predicted by Helfrich (1978) can be directly observed and analysed by observation of the interaction of a vesicle with a glass substrate using reflection interference contrast microscopy (RICM).

The origin of the undulation force becomes intuitively clear if one considers a flickering vesicle which approaches the surface to a distance h of the order of the undulation amplitude. Any further decreasing requires the gradual freezing-in of long wavelength modes or of more and more degrees of freedom. This corresponds to a reduction of the entropy associated with the thermal excitations and hence to an increase of the free energy. The situation is analoguous to the pressure excerted by an ideal gas if the available volume is reduced.

The situation depends upon whether the membrane is tension free or not. In the absence of tension, Helfrich derived a dynamic repulsive (disjoining) pressure of

$$P_{\text{und}} = \frac{3}{4} \frac{(k_{\text{B}}T)^2}{K_{\text{c}} h^3} \tag{14}$$

The most remarkable aspect is, that the distance dependence is the same as for the Van der Waals force. For a membrane under a tension an exponential law holds

$$P_{\text{und}} = \left(\frac{k_{\text{B}}T\sigma}{2hK_{\text{c}}} \right) \exp\left\{ -\frac{2\pi\sigma}{3k_{\text{B}}T} h^2 \right\} \tag{15}$$

In fact, the latter situation is always expected for closed and adhering shells.

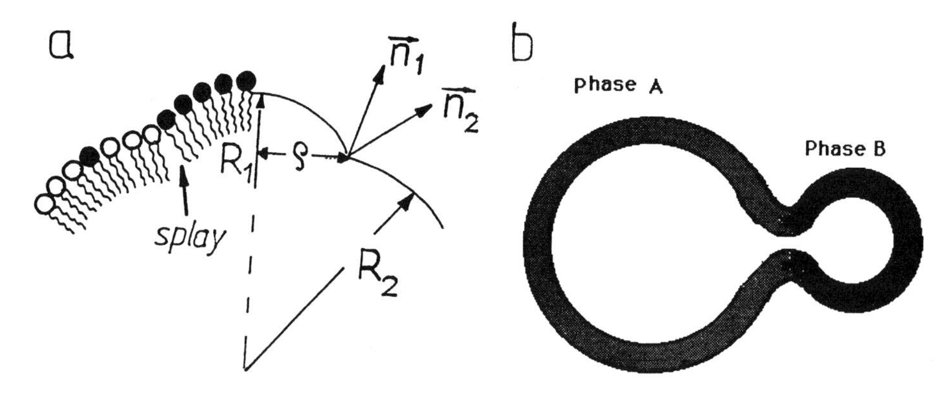

Fig. 14. Effect of elastic and chemical contribution to interfacial energy γ_{if} per unit length. (a) Splay deformation at interface between regions of different spontaneous curvature resulting in domain stabilization. Note that interfacial energy per unit length gif grows quadratically with the size, r, of the domain. (b) Reduction of chemical contribution to interfacial energy by budding resulting in a decrease of the length of the interface.

References

Andelmann, D., Kavakatsu, T., Kavasaki, K. (1992): Equilibrium shape of two- component unilamellar membranes and vesicles. Europphys. Lett. **19**, 57–62

Berndl, K., Kaes, J., Lipowsky, R., Sackmann, E., Seifert, U. (1990): Shape transformation of giant vesicles: extreme sensitivity to bilayer asymmetry. Europhys. Lett. **13**, 659–664

Bloom, M., Evans, E.A., Mouritsen, O. (1991): Physical properties of the fluid lipid-bilayer component of cell membranes: a perspective. Quart. Rev. Biophys. **24**, 293–397

Cevc, G., Marsh, D. (1978): Phospholipid bilayers: physical principles and models. Wiley, New York.

Darnell, J., Lodish, H., Baltimore, D. (1990): Molecular Cell Biology, W.H. Freemann, San Francisco

Devaux, P.F. (1988): Phosopholipid flippases. FEBS Lett. **234**, 8–12

Döbereiner, H. G., Käs, J., Noppl, D., Sprenger, I., Sackmann, E. (1993): Budding and fission of vesicles. Biophys. J. **65**, 1396

Duwe, H., Sackmann, E. (1990): Bending elasticity and thermal excitations of lipid bilayer vesicles: modulation by solutes. Physica A. **163**, 410–428

Engelhardt, H., Sackmann, E. (1988): On the measurement of shear elastic moduli and viscosities of erythrocyte plasma membranes by transient deformation in high frequency electric fields. Biophys. J. **54**,495–508

Evans, E.A., (1974): Bending resistance and chemically induced moments in membrane bilayers. Biophys. J. **14**,923–931

Fromherz, P., Röcker, C., Rüppel, D. (1986): From discoid micelles to spherical vesicles: the concept of edge energy. Farad. Disc. Chem. Soc. **81**, 39–48

Gebhardt, C., Gruler, H., Sackmann, E. (1977): On domain structure and local curvature in lipid bilayer and biological membranes. Z. Naturf. **32C**, 581–596

Gruner, A., Parsegian, A., Rand, P. (1980): Directly measured deformation energy of phosholipid H2 hexagonal phases. Farad. Disc. Chem. Soc. **81**, 267–280
Helfrich, W. (1974): The size of bilayer vesicles generated by sonication. Phys.Lett. **50A**, 115–116
Helfrich, W. (1978): Steric interaction of fluid membranes in multilayer systems. Z. Naturf. **33a**, 305–315
Hwang, J., Edidin, M., Betzig, E., Chichestin, R.J. (1994): . Biophys. J. **66**, A277
Käs, J., Sackmann, E. (1991): Shape transition and shape stability of giant phospholipid vesicles in pure water induced by area to volume changes. Biophys. J. **60**, 825–844
Kaufmann, S., Käs, J., Goldmann, W.H., Sackmann, E. Isenberg, G. (1992): Talin anchors and nucleates actin filaments at lipid membranes: a direct demonstration. FEBS Letters **314**, 203–205
Knoll, W., Schmidt, G., Ibel, K. Sackmann, E. (1985): SANS-study of lateral phase separation in DMPC-cholesterol mixed membranes. Biochemistry **24**, 5240–5246
Knoll, W., Apell, H.J., Eibl, H. Sackmann, E. (1986): Direct evidence for Ca^{++} induced lateral phase separation in BLM of lipid mixtures by analysis of gramicidin A single-channels. Eur. Biophys. J. **13**, 187–193
Kurrle, A., Rieber, P., Sackmann, E. (1990): Reconstitution of Transferrin receptor in mixed lipid vesicles: An example of the role of elastic and electrostatic forces for protein-lipid assembly. Biochemistry **29**, 8274–8282
Kwok, R., Evans, E.A. (1981): Thermoelasticity of large lecithin bilayer vesicles. Biophys. J. **35**,637–652
Lipowsky, R. (1991): The conformation of membranes. Nature **349**, 475–481
Lipowsky, R. (1992): Budding of membranes induced by intramembrane domains. J. de Phys. France **2**, 1825–1828
Millner, S., Safran, S. (1987): Dynamic fluctuations of droplet microemulsions and vesicles. Phys. Rev. A. **36**, 4371–4379
Mutz, M., Bensimon, D. (1991): Observation of toroidal vesicles. Phys. Rev. A. **43**, 4525–4528
Sackmann, E. (1990): Molecular and global structure and dynamics of membranes and lipid bilayers. Can. J. Phys. **68**, 1000–1012
Sackmann E. (1994): Intra- and extracellular macromolecular networks: physics and biological function. Macromol. Chem. Phys. **195**, 7–28
Seifert, U. (1990): Adhesion of vesicles. Phys. Rev. A. **42**, 4768–4771
Svetina, S., Zeks, B. (1989): Membrane bending energy and shape determination of phospholipid vesicles and red blood cells. Biophys. J. **17**, 101–111
Tanford, C. (1980): The Hydrophobic Effect. Wiley, New York.

Microtubule Dynamics

Henrik Flyvbjerg

Höchstleistungsrechenzentrum (HLRZ), Forschungszentrum Jülich, Germany
and
Department of Optics and Fluid Dynamics, Risø National Laboratory, Denmark

1 Introduction

Biological cell division, *mitosis*, is a very complicated process, and consequently error prone. Cells therefore employ special mechanisms which minimize, correct, or eliminate errors. A good example is the mother cell's partitioning of its duplicated chromosomes in a manner giving each daughter cell exactly one copy of each chromosome. This process must necessarily be carried out without errors. The error rate has been measured for yeast to be approximately one per 100,000 divisions (Alberts et al. 1994, 2nd ed., p. 762). This 99.999% precision is even more impressive when one considers that the process is carried out *blind-folded*, so to say:

Before mitosis, the chromosomes are duplicated to pairs of connected *sister chromatids*. These roll up on themselves in a complex manner which serves to prevent their entanglement with each other, something that obviously would hinder their separation. Thus rolled up, the chromosomes have their well-known appearance, easily seen in a light microscope; see e.g. Figs. 1 and 2. The cell does not have the same grand view of its interior as we do, however. Instead, it is equipped with a mechanism which can both *find* the chromatid pairs, can *assess* whether *all* of them have been found, and, when this has been confirmed, can *separate* the pairs in such a manner that one and only one sister chromatid ends up in each of the two daughter cells which are formed immediately after the separation.

This mechanism and its components are the subject of much current research, which is mostly carried out by biologist, biochemists, and medical researchers, for good reasons. But also by physicists, because the subject contains several physics problems. Modeling of non-linear and stochastic dynamical systems are examples treated in this chapter.

2 Microtubules

Eukaryotic cells, which are not about to divide, have next to their nucleus a center, the *centrosome*, from which thin, rigid tubes radiate in all directions towards the cell membrane, which they support, thus giving the cell its form, somewhat like tent poles supporting a tent—with the difference that the tubes

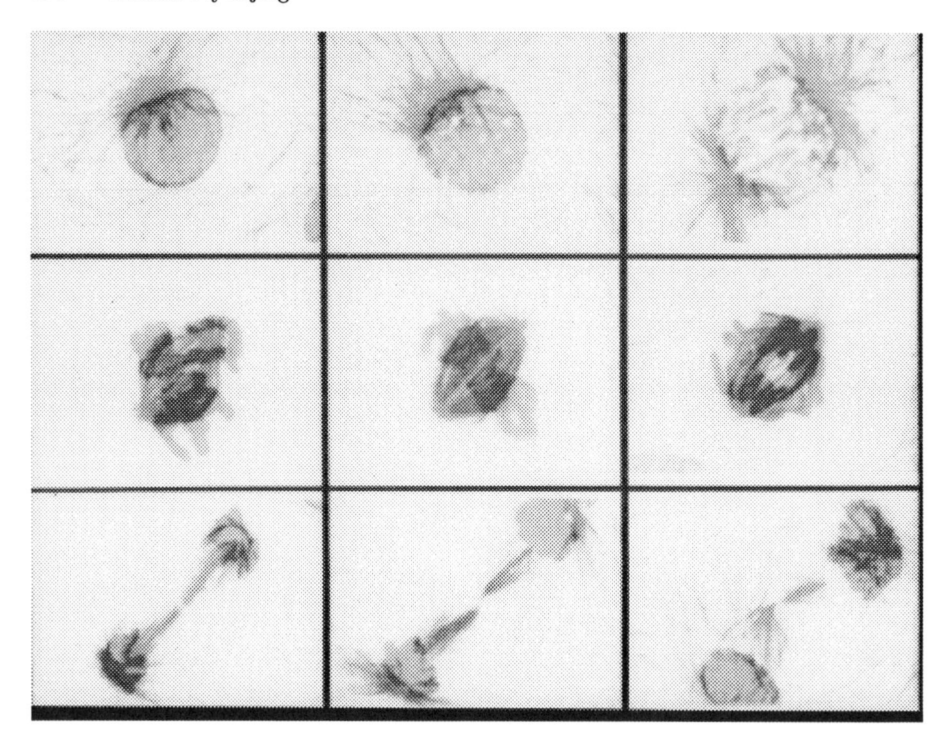

Fig. 1. Gray-scale rendition of a pseudo-color time series of a cell's division, starting in upper left frame (a), and progressing left to right, to finish in lower right frame (i). Only chromosomes and microtubules have been made visible by immuno-fluorescence. Compare with Fig. 2. (Merdes, Stelzer, and De Mey 1991)

in this *skeleton* of the cell are replaced continuously, as new ones form by growth from the centrosome, and old ones dissolve and their material is recycled.

These small tubes, called *microtubules,* are polymers formed from a protein, *tubulin.* The name is used for several, very closely related globular proteins with a diameter of approximately 4 nm. Two of them, α- and β-tubulin, bind well to each other, and form a stable *hetero dimer* which polymerizes to microtubules. Electron micrographs show that the tubulin in a microtubule forms a two-dimensional, cylindrical "crystal lattice" with a circumference of typically 13 hetero dimers. This cylinder has an outer diameter of approximately 25 nm, and an inner diameter of approximately 14 nm; see Fig. 3. So microtubules may, in a way, be regarded as being two-dimensional crystals, with one dimension rolled up to make the microtubule a cylinder.

From an engineering point of view, we note that if one wishes to form a beam with a maximum of rigidity from a minimum of material, the tubular shape is optimal for this purpose. And indeed, microtubules are the most rigid elements of construction found in cells. They serve as a universal design element inasmuch as everywhere in cells where a rigid construction is required, microtubules are

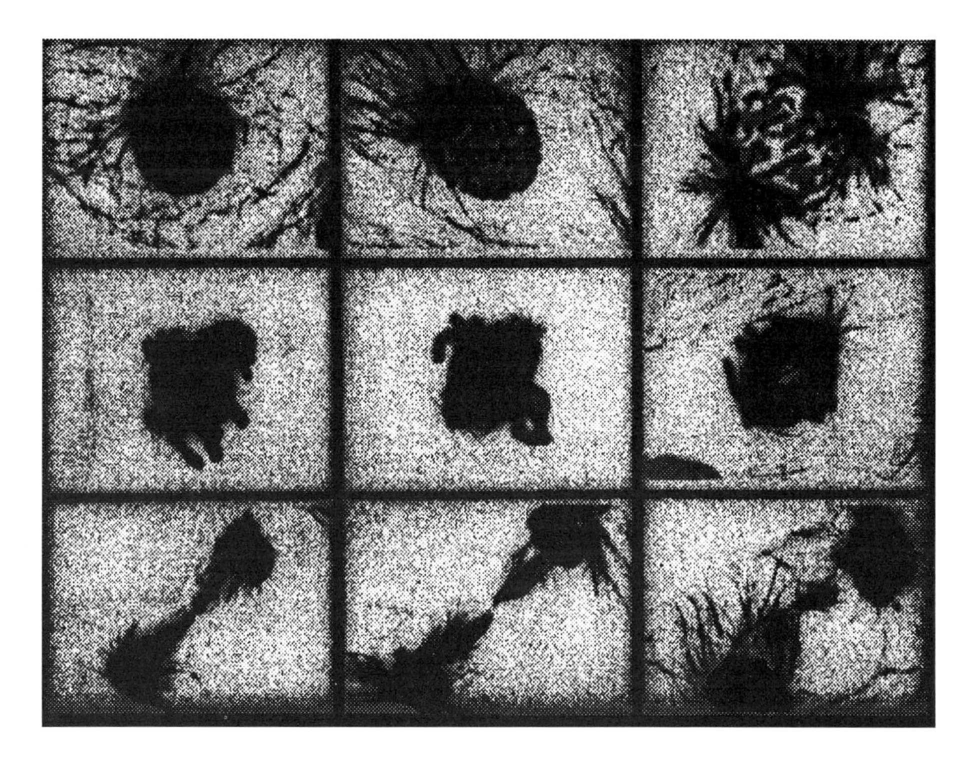

Fig. 2. Differently processed version of Fig. 1. (a) The almost circular dark object in the middle is the chromosomes filling the cell's nucleus. They are visible because they are fluorescent. (The grey-scale has been inverted in Figs. 1 and 2, so light appears dark and vice versa.) The "hairy" structures are fluorescent microtubules spanning the cell's body, most of which is within the frame. In the lower right corner of this frame, part of a neighboring cell is seen. In frame (a) the microtubules still form a cytoskeleton radiating from the *centrosome.* (b) The centrosome has become two centrosomes which are moving apart. (c) The nuclear envelope has disintegrated, and the microtubules polymerize and depolymerize rapidly from the two centrosomes, interacting with the chromosomes. (d) Some of the chromosomes—actually pairs of sister chromatids—have been found by microtubules from both centrosomes, and consequently been pulled to the plane midway between the centrosomes. Other chromosomes have been found only by microtubules from one centrosome, and have consequently been pulled over towards that centrosome. (e) All chromatid pairs are in the mitotic plane. (f) All chromatid pairs have simultaneously let go of each other, and have consequently been pulled towards the two centrosomes, one sister from each pair towards each centrosome. (g) Microtubules from the two centrosomes have interacted with each other, and pushed apart the two centrosomes with attached chromatids. A cell wall (invisible) has formed midway between them; it pinches (and hides) the interacting microtubules in the center of the frame. (h) Nuclear envelopes form around the chromosomes and new microtubules grow in all directions from the two centrosomes to form cytoskeletons of the two new daughter cells. (i) End of cycle: one cell has become two cells.

Fig. 3. A microtubule is a polymer of the protein *tubulin* which forms a cylindrical "crystal" of hetero dimers in a spiral arrangement. The Figure shows one kind of monomers as dark spheres and the other kind as light ones. Since the bond between the pair of tubulin units forming a hetero dimer is stronger than the bond between tubulin units in different, neighboring hetero dimers, the hetero dimers remain identifiable within the microtubule "crystal," and give a definite polarity to a microtubule. Consequently, its two ends differ. The more rapidly polymerizing end is referred to as the *plus-end,* while the other, more slowly polymerizing end is referred to as the *minus-end.* (Graphics by I. Jánosi)

used. Larger, more rigid structures are formed from many parallel microtubules, up to several hundreds aligned in regular arrays.

3 Cell division

Before an animal cell divides, its skeleton of microtubules dissolves, and its centrosome divides in two; see Fig. 2ab. From each new centrosome microtubules grow in all directions; see Fig. 2bc. It is believed that it is interactions between these microtubules which push the centrosomes apart. The pushing is believed to result from motor molecules like *kinesin* walking along microtubules from one centrosome while being attached to microtubules from the other centrosome, possibly by identical motor molecules, in an arrangement like that found in muscles (Howard and Gittes 1996).

Next, the nuclear envelope surrounding the chromosomes breaks down, and microtubules from the two centrosomes try to *find* the chromosomes—like a blind person's probing fingers—by polymerizing in different directions, and "returning" by depolymerization, when no chromosome is found in a given direction. That, at least, is what *appears* to be happening, when seen through the eyes of a rationalizing observer. Because, when the growing end of a microtubule by chance *does* find one of two specific spots called *kinetochores*, located at the middle of a pair of sister chromatids, this microtubule terminates its search, and pulls the chromatid pair towards the centrosome from which it polymerized; see Figs. 1cd and 2cd.

Since there are two kinetochores on a chromatid pair and they are placed on opposite sister chromatids, facing in opposite directions, microtubules from one centrosome can grab a pair of chromatids already held by a microtubule from the other centrosome. Not only *can* they do this, they *always* do it, sooner or later, because the blind search simply continues until the task has been done, and all pairs of sister chromatids have been found by microtubules polymerizing from both centrosomes. While this search process goes on, the chromatid pairs are violently pushed, pulled, and squeezed between each other, as they are pulled by microtubules attached from opposite sides, and possibly pushed by ends of unattached microtubules. Consequently all chromatid pairs end up in the middle between the two centrosomes; see Figs. 1de and 2de.

Once they are there (Fig. 1e), all pairs separate simultaneously, as if on command, and each half of a pair is quickly pulled over to the centrosome it is attached to (Fig. 1f), while the cell membrane contracts between the centrosomes, and pinches the cell in two, the two daughter cells (Fig. 1g). The membrane itself is invisible in Figs. 1g and 2g, but its effect can be seen on the microtubules, which are pinched together midway between the two clusters of chromosomes, and are invisible, shaded by the membrane, right at the midpoint.

Next, nuclear envelopes re-form around the chromosomes of the daughter cells, and each cell grows a skeleton of microtubules (Fig. 2hi). The cell cycle has been completed.

In rapidly dividing animal and plant cells, it takes only 1–2 hours to condense the chromosomes, find and separate them, and divide the cell in two, while the whole cell cycle lasts 16–24 hours. So the delicate process of partitioning the chromosomes correctly and error free seems to be efficient, and is not a bottle neck in the cycle.

We have just seen that the cell does not have to "know" or "see" where the chromosomes are located. It finds them in a *random search process,* which is efficient because it is done in parallel, many directions being searched simultaneously. But how does the cell "know" then, when all chromosomes have been found by both centrosomes, and which "command" coordinates the last phase of the division process, the separation of the sister chromatids, and the pinching off of the daughter cells?

It is believed that each kinetochore emits a chemical signal, unless it is under tension, being pulled at by microtubles. This chemical signal functions like input to a logical *and-gate:* As long as microtubules do not pull at *all* pairs of sister chromatids from *both* sides, kinetochores not under tension will emit the signal, and the search process continues. Only when all kinetochores have been "shut up," the resulting silence is interpreted as the command for the sister chromatids to let go of each other, with the result that they are pulled towards the centrosomes, exactly one to each.

An experiment demonstrates that tension is, indeed, a key factor: If, by micromanipulation, one breaks the connection between a chromosome and a centrosome just when all chromosomes are connected with both centrosomes, that chromosome is pulled over to the other centrosome which it is still connected to, and mitosis does not proceed, the sister chromatids do not separate. Instead, the microtubules keep searching until a new connection has been established. But if, by micromanipulation, one pulls at the piece of broken microtubule attached to the kinetochore, the cell is fooled into believing everything is okay, so the chromatids separate, and daughter cells with a wrong number of chromosomes form.

4 Two questions to be addressed here

Cell division is obviously a life process. Such processes are mostly deterministic, programmed in the genes, though designed with some leeway to allow them to unfold successfully in an environment which is not entirely predictable. On this background, the random search process carried out by microtubules during mitosis stands out by its entirely different nature. It is fully stochastic, not at all deterministic. It is easy to invent an explanation why it has to be so: The cell either would have to *know* where the chromosomes are located inside it—i. e., this information would have to be part of its genetic blue-print, as would a mechanism utilizing this information to divide the chromatid pairs evenly—*or* the cell must be able to *find* the pairs, one way or another, wherever they are, when it is time to separate them. The latter strategy was chosen, and the search mechanism is random. That does not make the outcome of cell division a random event. On the contrary: 100% determinism is achieved. We conclude that it seems that nature chose to encode a recipe for a random process in the cell's genetic blue-print. The cell will go through the various stages of its cycle in a deterministic manner, and, at a particular stage, will assemble a random number generator and turn it on, all in a deterministic manner. Once turned on, the random process does its job with much precision regarding outcome, and is then switched off again, deterministically, until it is needed again.

Is this scenario correct? It seems so, but no evidence could be more convincing than a description of the "random number generator" used. What is the source of the randomness in this macroscopic system? Where is the "random number generator" hidden?

Another mechanism is of interest because one would expect it to occur, but it does not, and that also serves a purpose: Cells contain tubulin in concentrations that result in continued—and in mitotic cells very rapid—polymerization of microtubules from the centrosomes. Why, then, do microtubules only grow from the centrosomes, and not nucleate and grow spontaneously everywhere in the cytoplasm, as *in vitro* experiments demonstrate they can?

Answers to these questions detail our understanding of the mechanisms of life. But the distance between fundamental research and practical application can be short: several anticancer drugs arrest cell division by specific targeting of microtubules' dynamics.

5 Microtubule growth—conflicting experiments?

Polymerization of microtubules is known to cost energy in the form of GTP (guanosine triphosphate). Only tubulin dimers liganded with two GTP molecules can polymerize. It is also known that shortly after a dimer has been incorporated into a growing microtubules' lattice structure, one of these GTP molecules is hydrolyzed to GDP (guanosine diphosphate). Tubulin-t is converted to tubulin-d is a brief way of saying the same. One has therefore assumed for many years that a polymerizing microtubule has a "cap" of newly added GTP which has not yet been hydrolyzed, and that the presence of this cap is the difference between the growing and the shrinking state of the microtubule end. Despite much experimental and theoretical work, this picture is controversial. The experiments seem to contradict each other.

The chance that a growing microtubule changes to its shrinking state within a given time interval is smaller the faster the microtubule grows; see Fig. 4. This change of state is called a *catastrophe,* and the decrease in the catastrophe rate accompanying an increase in the growth rate is explained as a consequence of the GTP cap's dynamics: It grows with the microtubule's polymerization from tubulin-t, and simultaneously shrinks by the hydrolysis of this tubulin-t to tubulin-d. Both processes are stochastic, so if the latter accidentally happens to catch up with the former, the cap is gone, and that is the "catastrophe," the microtubule end now consists entirely of tubulin-d, and consequently depolymerizes rapidly. The probability that the cap is lost by a fluctuation in its size decreases with its size, and the cap size is larger, the faster the microtubule grows by addition of tubulin-t. In this way the catastrophe rate's dependence on the growth rate seems explained, at least qualitatively.

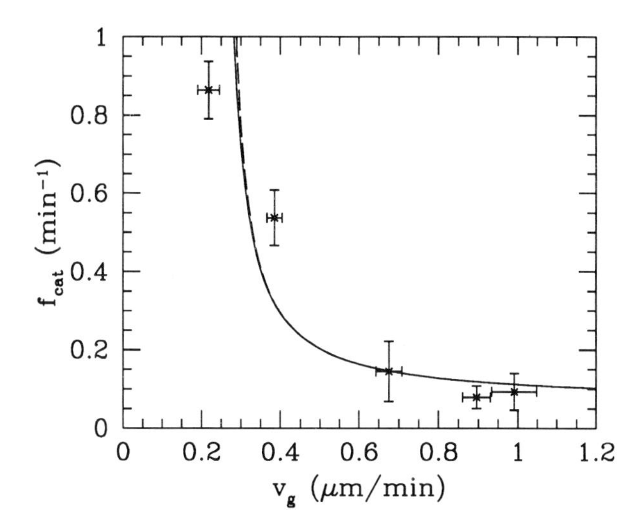

Fig. 4. The catastrophe rate vs the growth velocity. Experimental data points are from Drechsel et al. (1992). The full curve is the result of the full theory by Flyvbjerg, Holy, and Leibler (1994, 1996). The dashed curve is the approximation to this theory given in Eq. (5) below.

But if rapidly growing microtubules have larger GTP caps than slowly growing ones, then their larger caps must also take more time to disappear by hydrolysis in a situation where the experimenter suddenly halts the growth by flushing out the tubulin solution. So if one observes the length of an individual microtubule in a microscope, one would expect to observe a *delay* between the time when growth is halted by flushing, and the time when the microtubule starts shrinking because it lost its cap. One would also expect this delay time to depend on the velocity with which the microtubule grew before flushing, and expect it to increase with it.

This simple experiment has been carried out numerous times with different concentrations of tubulin-t, and consequently different growth velocities (Walker, Pryer, and Salmon 1991). The result was surprising: There was no correlation at all between the tubulin-t concentrations used, and the delay times observed! Moreover: the delay time turned out to be a stochastic quantity: at a given concentration of tubulin-t, many different delay times were observed, distributed with a standard deviation about as large as their mean.

Figure 5 shows the delay times measured. The same figure shows how a rather simple theory predicts the distribution of the delay times, and their dependence

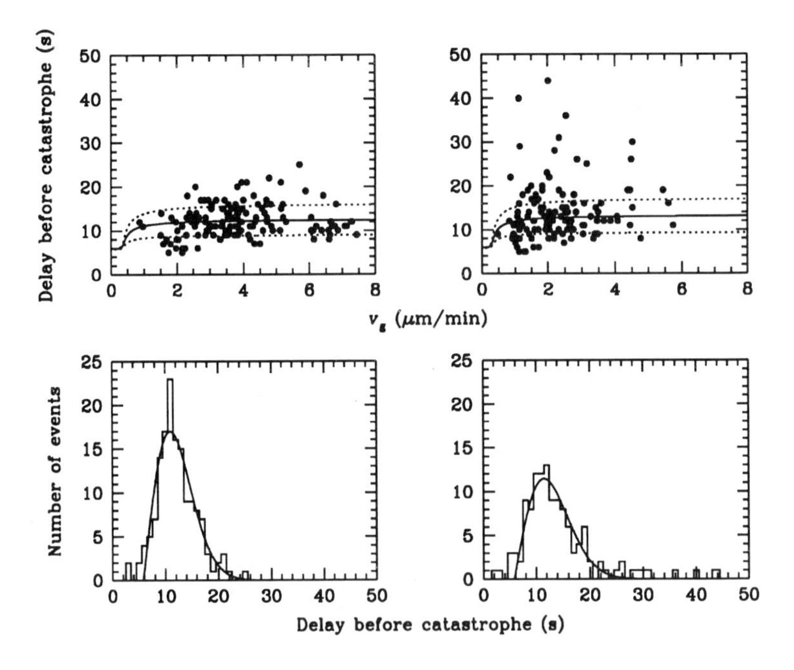

Fig. 5. Delays before catastrophe following dilution. Data taken from Walker, Pryer, and Salmon (1991). Left: plus-end; right: minus-end. Top: delay as a function of initial growth velocity. Each point represents a single measurement on a microtubule. Curves are theoretical mean (solid) and standard deviation (dashed) of the delay, from Eqs. (7) and (6) below. Bottom: histograms showing the experimental distribution of delays before catastrophe. The curves are fits of the theoretical distribution given in Eq. (6). Dilution was initiated at $t = 0$ and required some time for completion, so that our assumption of *instantaneous* dilution results in an *effective* time for the occurrence of this event. When treated as a free parameter, and fitted as shown here, this effective time is 6.1 sec (Flyvbjerg, Holy, and Leibler 1994, 1996).

on the tubulin concentration. Each of the two distributions is fitted to the data using just one parameter, the average of the distribution. In addition, one common, effective origin for the time axis was fitted. The rest, the shape of the distribution, was derived from the theory. The same simple theory also gave the curve through the data points in Fig. 4. So we see that it gives a single, unified explanation of experiments that seem to contradict each other.

A third type of experiment, which was to determine the cap size by using a radioactive marker, also gave results in apparent conflict with the idea of a GTP cap of a size depending on the growth velocity: GTP was doubly labeled, one marker labeling the phosphate that is hydrolyzed and released when GTP turns to GDP, the other marker being permanent. Tubulin liganded with this GTP was allowed to polymerize at a given concentration, and then the microtubules were quickly caught on a filter, washed, and both the amount of GTP and the amount of tubulin polymerized was measured. The amount of GTP was then related to the cap size. However, no matter how fast the microtubules were grown in this experiment, no trace of a GTP cap was detected! Whatever amounts of GTP had been present in the microtubules while they grew, was hydrolyzed in the 15–20 seconds that passed from growth was halted by filtering, and till the microtubules had been washed and measurements took place (Stewart, Farrell, and Wilson 1990).

Finding nothing is a result, even if it was an unexpected one in this case. This result is fully explained by the simple theory described in the next section, once its parameters have been determined by fitting the theory's predictions to the data in Figs. 4 and 5.

6 An effective theory

The theory giving the fits shown in Figs. 4 and 5 is a so-called *effective* theory. In this context, "effective" does not refer to the theory's effectiveness at modeling the data, but to the criteria used in formulating the theory. "Effective" means the opposite of "fundamental." When, in a particular context, one integrates out such parts of a theory which are irrelevant for the context, one refers to the resulting, context-bound theory as an effective theory, while the original theory is referred to as fundamental. A classical example is thermodynamics, which may be seen as an effective theory derived from statistical mechanics.

But the label "effective theory" is also used to add an air of respectability to phenomenological theories. By using this name, one indicates that one believes a fundamental theory can be found, which will have the phenomenological theory at hand as effective theory. Like thermodynamics was developed before statistical mechanics.

Obviously, this practice cannot be justified, but it is well established, and it is the way we use the label here, though we do do some "integrating out" of short length scales.

One makes it more plausible that the fundamental theory eventually can be found, by making the simplest possible assumptions and by introducing as few parameters as possible, when formulating the effective theory. While this always is a recommendable approach, it is especially useful when, as here, the data are limited in number and quality.

The theory assumes that a growing microtubule differs from a shrinking one by having a GTP cap. This cap grows at one end by the random addition of tubulin-t dimers to the growing microtubule. We denote the rate at which dimers are added by k_g. Each dimer added extends the length of the microtubule by an amount δx, where δx is known, $\delta x = 8\,\mathrm{nm}/13 = 0.6\,\mathrm{nm}$, since a dimer is 8 nm long, and it takes 13 of them, one for each protofilament, to extend the whole microtubule by 13 nm. With this notation, the *average* rate of growth of the tip of a microtubule is

$$v_g = k_g \delta x \ . \tag{1}$$

The cap shrinks from its other end by hydrolysis of tubulin-t which neighbors tubulin-d; see Fig. 6. This hydrolysis is assumed to occur at random, at a rate of k_h dimers per unit time, so shrinking is a random process, like growth, with average rate

$$v_h = k_h \delta x \ . \tag{2}$$

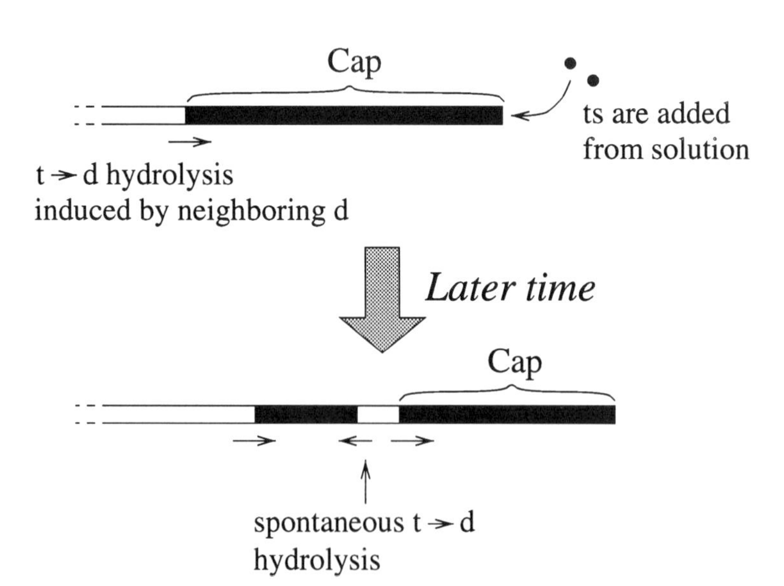

Fig. 6. A microtubule grows by polymerization from tubulin-t (black) which then hydrolyzes to tubulin-d (white) both at the interface between the tubulin-t cap and the tubulin-d body of the microtubule, and spontaneously in the interior of the cap.

Finally, it is assumed that a tubulin-t dimer *inside* the cap (as opposed to those at its trailing end) also can hydrolyze at random, but do so at a much slower rate, k'_h, than those neighboring tubulin-d at the cap's trailing edge. However, *when* such an event of *internal hydrolysis* occurs, the neighbors to the hydrolyzed dimer suddenly find themselves next to tubulin-d, and hydrolyze at a much faster rate. Thus an event of internal hydrolysis creates a spot of tubulin-d which spreads rapidly by hydrolysis of the surrounding tubulin-t. When this spot meets itself by spreading 360° around the microtubule, it has effectively divided the cap into two disconnected regions of tubulin-d: a *shorter* cap, *plus* a left-behind patch of tubulin-t which now shrinks by hydrolysis from both ends; see Fig. 6. This process of internal hydrolysis will cut the length, x, of a cap to any fraction of the length it had before it happened, because it occurs anywhere along the length of the cap with the same rate $r = k'_h/\delta x$ per unit length. Thus the rate at which a cap is shortened by this "cutting mechanism" is proportional to its length. Consequently, caps remain effectively limited in length, no matter for how long they have been growing.

For the cap's length x, the dynamics just described adds up to constant *growth* with the average velocity

$$v = v_g - v_h \ , \tag{3}$$

superposed with an unbiased random walk parametrized by a diffusion constant

$$D = \tfrac{1}{2}(k_g + k_h)\delta x^2 = \tfrac{1}{2}(v_g + v_h)\delta x \ , \tag{4}$$

and abrupt change of x to any fraction of x with equal probability, at a rate r per unit length of x. When x by chance vanishes—by a combination of abrupt change to an accidentally small value, followed by an accidental decrease to zero by random walk—then the cap is lost and we have a "catastrophe."

The theory just described is so simple that it can be solved analytically (Flyvbjerg, Holy, and Leibler 1994, 1996). The full analytical solution is rather complex and not at all handy. But it can be used to show that with the conditions in effect in the experiments that gave the data shown in Figs. 4 and 5, very simple analytical expressions are extremely good approximations to the exact analytical expressions describing the experimental results. Thus the catastrophe rate in Fig. 4 is given by

$$f_{\mathrm{cat}} = \frac{Dr}{v} = \frac{r\delta x(v_g + v_h)}{2(v_g - v_h)} \ . \tag{5}$$

Note that the graph of f_{cat} as function of v_g is just a hyperbola with asymptotes $f_{\mathrm{cat}} = r\delta x/2$ and $v_g = v_h$. The divergence of f_{cat} at v_h is an artifact of the approximation; the full analytical solution is finite. But the finite value of f_{cat} at $v_g \to \infty$ is a genuine result of this theory. While v_h may differ for the two ends of a microtubule, because it has an inherent polarity, r and δx cannot. So our expression (5) for the catastrophe rate predicts that both ends of a microtubule have *the same* finite asymptotic rate at large growth velocities v_g, a prediction which can be tested experimentally. It agrees with experimental findings, though the data leave something to be desired (Flyvbjerg, Holy, and Leibler 1994, 1996).

The waiting time for catastrophes induced by flushing out the tubulin solution is stochastic according to the theory, and distributed as

$$p_{\mathrm{cat}}(t) = \frac{\pi t}{2{t_{\mathrm{cat}}}^2} \exp\left(-\frac{\pi t^2}{4{t_{\mathrm{cat}}}^2}\right) , \tag{6}$$

where

$$t_{\mathrm{cat}} = \left(\frac{-2rv'}{\pi}\left(1 - \frac{v'}{v}\right)\right)^{-1/2} \simeq \left(\frac{\pi}{-2rv'}\right)^{1/2} \tag{7}$$

is the average waiting time for dilution from and to tubulin concentrations corresponding to cap growth velocities v and v', respectively. The last, approximate expression is typically a good approximation because the microtubule growth rate prior to dilution, v_{g}, typically is large compared to v_{h}, hence so is $v = v_{\mathrm{g}} - v_{\mathrm{h}}$, while dilution results in a small or vanishing concentration, resulting in $|v'| \sim v_{\mathrm{h}}$ and therefore $v \gg |v'|$. Equation (7) therefore predicts a waiting time distribution which is essentially independent of the growth rate prior to dilution. To the extent a dependence is predicted, the waiting time increases only slightly with the pre-dilution growth rate, v, except at small rates, $v \sim v_{\mathrm{h}}$.

The analytical expressions given in Eqs. (5) and (6) were fitted to the experimental data, as shown in Figs. 4 and 5, using the theory's parameters, v_{h}, r, and D, as fitting parameters. The fits are rather acceptable, so the simple effective theory seems to provide a single explanation of two different experiments which measure the same process, cap loss, but under circumstances so different that the registered rates differ by 1–2 orders of magnitude.

As mentioned in Sect. 5, a third type of experiment involving cap loss, that by Stewart, Farrell, and Wilson (1990), was also explained by this simple, effective theory.

An effective theory must agree with experiments to be of any interest. This is a necessary condition, but not a sufficient one: First of all it must be self-consistent. We have an internal consistency check on the theory: we assumed that spontaneous internal hydrolysis, parametrized by r or k'_{h}, is a rare event compared to the hydrolysis occurring at the edge of the cap, parametrized by v_{h} or k_{h}. This assumption is well satisfied: the fitted parameter values give $k'_{\mathrm{h}}/k_{\mathrm{h}} = 3 \times 10^{-4}$, which is a small number by most standards.

7 Spontaneous formation of microtubules

In cells microtubules grow from seeds, while *in vitro* spontaneous formation of microtubules is seen in the bulk of sufficiently concentrated tubulin solutions. So it is of some interest to understand the process, both to see how it is avoided in cells, and as a model-example of biological *self-assembly*, i.e. the process by which proteins aggregate to form larger structures with no use of templates or other information but their own characteristic shapes.

It is the initial stages of the process which are interesting. How do the tubulin dimers get the cylinder shaped crystal lattice started? After it has been initialized, its growth is just a repeated addition of dimers to the existing structure.

The latter we deduce from the well-established experimental result that the rate of growth of existing microtubules is simply proportional to the concentration of tubulin-t available for polymerization. This simplicity of the polymerization of existing microtubules is very helpful for the study of the initial nucleation of microtubules, because nucleation cannot be observed directly, experimentally. Only its consequences, the ensuing polymerization of microtubules, can be followed. So it is convenient that the latter is conceptually transparent, when we have to see through it, so to speak, to uncover the dynamics of nucleation. This is where the translational symmetry of the microtubule simplifies the task of describing its self-assembly in comparison with, e.g., the self-assembly of icosahedral procapsid shells (Prevelige, Thomas, and King 1993).

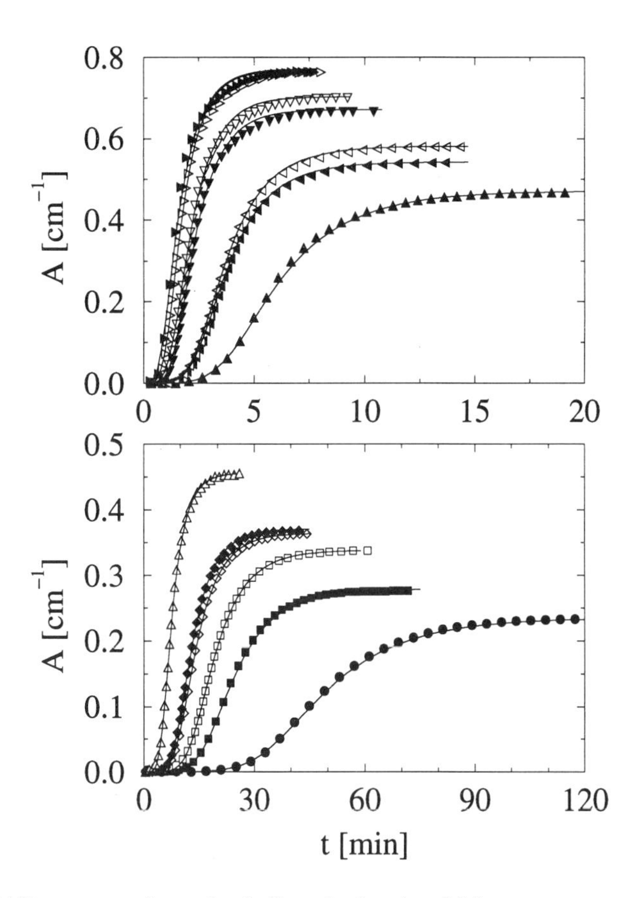

Fig. 7. Turbidity versus time of tubulin solution in which a temperature jump from 0 °C to 37 °C at time 0 has induced microtubules to self-assemble. Plotting symbols: turbidity time series for different initial concentrations of tubulin measured by Voter and Erickson (1994). Fully drawn curves: Two-parameter fits to experimental data of theoretical turbidity series derived in Flyvbjerg, Jobs, and Leibler (1996).

Figure 7 shows time series from 13 different experiments in which the turbidity of tubulin solutions was measured as a function of time at 350 nm. At this wave length the turbidity is proportional to the total amount of polymer formed. The only difference between the 13 experiments was the amount of tubulin present. In all 13 cases all tubulin was in solution at time zero. But after an abrupt increase in temperature from 0 °C to 37 °C at time zero, microtubules formed spontaneously. These microtubules grew by polymerization until they had used up all tubulin in solution, after which nothing happened any more. "Catastrophes" did not occur; this mechanism had been "switched off" by having glycerol in the solution; glycerol stabilizes microtubules.[1]

If one now assumes that microtubules always nucleate by the same mechanism, though at rates depending on the concentration, of course, then the time series shown must contain much information about this mechanism. Do they contain sufficient information to actually *determine* the mechanism? That would be an interesting situation, because chemical processes in solution cannot be monitored at the molecular level through a microscope. Neither can the processes of elementary particles, but that does not keep the high energy physicists from determining the processes from their products. Also, mathematical physicists know what it takes to determine a potential from its deflection of particles passing through it. So why not formulate a similar so-called *inverse problem* for non-equilibrium reaction kinetics?

8 Inverse problem

It *is* actually possible to determine a kinetic model for the spontaneous formation of microtubules from the experimental time-series shown. The result is shown as the curves through the the experimental data points in Fig. 7, and are seen to agree quite well with them. We sketch the line of arguments here. Details are given in Flyvbjerg, Jobs, and Leibler (1996).

The first step is entirely phenomenological; no theory whatsoever is involved. One simply observes that all the experimental time series have similar sigmoid shapes. It is then natural to ask whether they *scale,* and one finds that indeed they do to a good approximation—i.e. one can reproduce them all with *one* function, f, with a sigmoid graph by writing them as

$$A(t) = A_\infty f(t/t_0) \ . \tag{8}$$

Here $A(t)$ is the turbidity of a time series as a function of time t, and t_0 is a characteristic time for this series—e.g. the so-called *tenth time*, the time it takes for the turbidity to grow to 1/10th of its final value, A_∞. With this result

[1] Of the 16 time series shown in Voter and Erickson, (1984), Figs. 5 and 9, we have left out three with slowest rates of assembly here, because they were distinctly anomalous compared to the rest when analyzed as here. The properties of tubulin are known to change during experiments lasting as long as the slowest assemblies in Voter and Erickson (1984).

established, the task is reduced to that of finding a kinetic theory which gives the two functions f and $t_0(A_\infty)$, where the task initially was to find a theory which would give $A(t)$ as a different function for each time series, i.e. give $A(t; A_\infty)$.

The next step is equally phenomenological and void of theory. One plots the experimental values for the tenth time, t_0, for each time series against the final turbidity of the series, A_∞. One can argue that the final turbidity is simply proportional to the initial concentration, so for lack of precise knowledge about the latter, it is convenient to work with the former. One finds a surprisingly simple relationship,

$$t_0(A_\infty) \propto A_\infty^{-3} \, , \tag{9}$$

is valid with all precision possible, see Fig. 8.

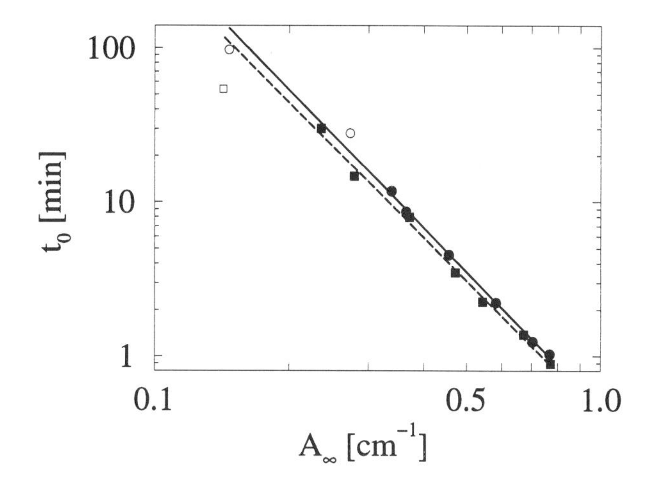

Fig. 8. Double-log plot of $t_0(A_\infty)$ vs A_∞. Circular symbols show results from Fig. 5 in Voter and Erickson (1984). The full straight line is a fit to the six filled round symbols, which correspond to the time series shown in Fig. 7 with open symbols. Its slope is $-2.97 \pm .05$. Square symbols show results from Fig. 9 in Voter and Erickson (1984). The dashed straight line is a fit to the seven filled square symbols, which correspond to the time series shown in Fig. 7 with filled symbols. Its slope is $-2.90 \pm .09$. The intercepts of the two straight lines with the second axis do not differ significantly, and give the constant of proportionality in Eq. (9) as $0.44 \pm .03\,\mathrm{min\,cm}^3$.

The simple form of these two 100% phenomenological results strongly indicates that the mechanism, we wish to uncover, is simple. The results are also precise mathematical requirements which any model for the mechanism must satisfy.

The third step consists in writing down a generic set of kinetic equations with as few built-in assumptions as possible. Then we demand from these equations that their solutions obey Eqs. (8) and (9). Note that we do not have to *solve* the equations in order to implement the demands on the solution! The demands on the solutions can be turned into demands on the form of the equations. The demand for scaling, for example, has as consequence that only equations which are dimensionless when rewritten in terms of the dimensionless scaling variables t/t_0 and $A(t)/A_\infty$ need be considered. Taken together with Eq. (9), this turns out to be a very strong demand. It defines a narrow class of equations which turn out to be analytically soluble up to one integral! Analytical solubility is a rare property among coupled non-linear differential equations, but then we are not considering a generic case when we demand that Eqs. (8) and (9) are valid.

Having solved the kinetic equations, the final step is to fit the solution to the experimental time series. One needs to fit an integer, the total number of equations, and two real rate constants to the data. Then one has the result shown in Fig. 7.

The model found this way provides a plausible explanation why microtubules can polymerize from seeds located in special centers in cells, so-called *centrosomes*, without new microtubules forming spontaneously in the cytosol from the available tubulin: The amount of tubulin which in a given time interval polymerizes onto existing microtubules is proportional to the tubulin concentration. But the amount of tubulin going into new microtubules nucleated in the same time interval is proportional to the tubulin concentration to the power 16, initially, and the power 7 in the steady state. The difference between these two powers and the power 1 is so high that it acts as an effective switch: Except in a very narrow range of concentrations, one of the two processes dominates.

The model also shows that the *size of the nucleus*—i.e. the number of dimers that go into the formation of a new microtubule before the addition of more dimers is a repetition of the same process—is 15. This number is intriguingly close to the typical number of protofilaments in microtubules. This suggests that the stable nucleus may be a single ring or proto-helix ("lockwasher") like those formed by the coating protein of tobacco mosaic virus (Durham, Finch, and Klug 1971, Voet and Voet 1990). Remarkably, γ-tubulin in centrosomes, where it participates in *in vivo* nucleation of microtubules forms ring-like structures of size similar to that of our nucleus, (Moritz et al. 1995). Though it is not known exactly how γ-tubulin contributes to microtubule growth from centrosomes, it has been demonstrated *in vitro* that γ-tubulin binds tightly and exclusively to the minus ends of microtubules in a saturable fashion with a stoichiometry of 12.6 ± 4.9 molecules per microtubule (Li and Joshi 1995). Taken together, these experimental results indicate that a ring of γ-tubulin with the size of our nucleus, nucleates microtubules in centrosomes. Hence it is natural to speculate whether the nucleus discussed in the present section has the same shape.

9 Conclusion

The stochastic element in microtubules' search for chromosomes can be explained as the result of a competition between two simple stochastic processes in the GTP cap: growth and hydrolysis. The suppression of spontaneous creation of microtubules at tubulin concentrations where existing microtubules polymerize willingly, can be explained by the extreme concentration dependence of the nucleation process.

The two problems discussed here exemplify well how molecular biology generates problems which biologists are not trained to address, while physicists are. This happens partly because biological systems are studied at length scales so short that the distinction between biology, chemistry, and physics becomes blurred, and partly because physical methods of measurement are used, so that the quality and quantity of data not only makes mathematical modeling possible, but also necessary, as we have seen, in order to interpret data.

Acknowledgment

The work presented here was done in collaborations with Timothy E. Holy, Elmar Jobs, and Stanislas Leibler. It was partially supported by *The Danish Natural Science Research Council,* grant 11-0244-1, by CONNECT, and by *Julie Damm's Studiefond.*

References

Alberts, B., Bray, D., Lewis, J., Raff, M., Roberts, K., Watson, J. D. (1994): *Molecular Biology of the Cell,* 3rd ed., Garland Publishing, Inc., New York

Drechsel, D. N., Hyman, A. A., Cobb, M. H., Kirschner, M. W. (1992): Modulation of the Dynamic Instability of Tubulin Assembly by the Microtubule-Associated Protein Tau, Mol. Biol. Cell **3**, 1141–1154

Drechsel, D. A., Kirschner, M. W. (1994): The minimum GTP cap required to stabilize microtubules, Current Biology **4,** 1053–1061

Durham, A. C. H., Finch, J. T., Klug, A. (1971): Nature London New Biol. **229**, 38

Flyvbjerg, H., Holy, T. E., Leibler, S. (1994): Stochastic Dynamics of Microtubules: A Model for Caps and Catastrophes, Phys. Rev. Lett. **73**, 2372–2375;

Flyvbjerg, H., Holy, T. E., Leibler, S. (1996): Microtubule Dynamics: Caps, Catastrophes, and Coupled Hydrolysis, Phys. Rev. E **54**, (1996) 5538–5560;

Flyvbjerg, H., Jobs, E., Leibler, S. (1996): Kinetics of Self-assembling Microtubules: An "Inverse Problem" in Biochemistry, Proc. Natl. Acad. Sci. (USA) **93**, 5975–5979

Kuchnir Fygenson, D., Flyvbjerg, H., Sneppen, K., Libchaber, A., Leibler, S. (1995): Spontaneous nucleation of microtubules, Phys. Rev. E **51,** 5058–5063

Howard, J., Gittes, F. (1996): *Motor Proteins,* Chapter in this book.

Li, Q., Joshi, H. C. (1995): γ-Tubulin Is a Minus End-specific Microtubule Binding Protein, J. Cell Biol. **131**, 207–214

Merdes, A., Stelzer, E. H., De Mey, J. (1991): The three-dimensional architecture of the mitotic spindle, analyzed by confocal fluorescence and electron microscopy, J. Electron Microsc. Tech. **18**, 16-73

Moritz, M., Braunfeld, M. B., Fung, J. C., Sedat, J. W., Alberts, B. M., Agard, D. A. (1995): Three-Dimensional Structural Characterization of Centrosomes from Early *Drosophila* Embryos, J. Cell Biol. **130**, 1149–1159

Prevelige, P. E., Thomas, D., King, J. (1993): Nucleation and growth phases in the polymerization of coat and scaffolding subunits into icosahedral procapsid shells, Biophys. J. **64**, 824–835

Stewart, R. J., Farrell, K. W., Wilson, L. (1990): Role of GTP Hydrolysis in Microtubule Polymerization: Evidence for a Coupled Hydrolysis Mechanism, Biochemistry **29**, 6489–6498

Voet D., Voet, J. G. (1990): *Biochemistry*, Chap. 13, pp. 339–340, John Wiley & Sons, Inc., New York

Voter, W. A., Erickson, H. P. (1984): The Kinetics of Microtubule Assembly, J. Biol. Chem. **259,** 10430–10438

Walker, R. A., Pryer, N. K., Salmon, E. D. (1991): Dilution of Individual Microtubules Observed in Real Time In Vitro: Evidence That Cap Size Is Small and Independent of Elongation Rate, J. Cell Biol. **114,** 73–81

Part IV

Neurons, Brains, and Sensory Signal Processing

A Physicist's Introduction to Brains and Neurons

William Softky[1] and Gary Holt[2]

[1] National Institutes of Health, Mathematics Research Branch, 9190 Wisconsin Ave. #350, Bethesda MD 20814, USA
[2] California Institute of Technology, Computation and Neural Systems Program, Pasadena CA 91125, USA

1 Introduction

We do not yet understand the function of a whole brain or of its neural components. These problems are very different from problems which physics has successfully solved in the past, in both complexity and experimental accessibility. But quantitative concepts—like information transfer, point-process statistics, and charge diffusion—can help us pose some crucial questions: Are neurons analog computing elements, or digital ones? Do they have high bandwidth, or low? Are they noisy, or deterministic? Does each neuron perform a simple computation, or a complex one? While answers await experimental data, the questions may give hints about whether Nature is a good engineer.

Our perceptions and behavior arise from electrochemical pulses sent between neurons in our brains. But we do not yet understand how. As a rough introduction to the problem, we can break it down into four basic questions:

1. What does one cell do with its many inputs to make an output pulse?
2. How are cells connected together[43]?
3. How do the connections change with learning[19]?
4. What is the whole circuit supposed to do[24]?

There has been a general consensus on answering the first question, but that consensus has several problems (as outlined below). The other three questions are still unanswered. As an indication of that disagreement—and as a way to appreciate the problems of measuring or modeling brains—consider the final question: What does the circuit do? Many proposals contain guiding principles which are equally plausible but diametrically opposed (table 1).

This confused situation may surprise physicists. We understood the basic properties of atoms and nuclei decades ago, so why can't we understand neurons, which are over 10^5 times larger and slower than single atoms? Because the properties which allowed single particles to be studied—their independence from one another, their identical characteristics, their stationary properties over time, their huge numbers, their absence of "design"—do not hold for brains. Brains are not atomic vapors or spin glasses.

1. **Neurons are macroscopic/classical.** They have complex shapes, with many (unknown) internal variables and no obvious simplifying principles.
2. **There are almost no fundamental "laws" of nervous systems** analogous to the laws of conservation, quantum mechanics, and thermodynamics.
3. **One cannot make large numbers of neurons to do exactly the same thing.**
4. **One can't extricate a neuron from its network,** either physically or functionally. A brain has electrically relevant structure covering five orders of magnitude in scale (10^{-6} m up to 0.1 m). Because of neurons' small size and packing density, it is very difficult even to see a single one in a live brain, much less to trace its connections to the others or to extricate it from them.
5. **Time-averaging $\neq$ population-averaging.** The ergodic principle does not apply. Even when a population of neurons seems to respond similarly to a controlled input (such as injected current), each neuron is "wired" differently in the network.
6. **Coupling is not simple.** The enormously tangled and complex connections between brain neurons cannot be realistically reduced to simple physical terms such as nearest-neighbor, isotropic, or common heat/radiation bath interactions.
7. **Brains evolved to have a function.** The best analogy for them is a special-purpose computer, not a lattice.
8. **Simplicity of structure is our own preference,** not biology's.

Some neurons, especially outside the brain proper, can be understood in a simple functional sense, as sensory or motor transducers. Researchers can provide a simple controlled sensory stimulus (such as an air puff) and measure the neuron's responses. Or researchers can compare the animal's motor output with the activity of a neuron thought to control it. This approach has worked spectacularly well for systems like the fly, in which we know that certain motor functions are controlled by only two neurons (see Bialek, this volume).

This approach has also been used with neurons farther from raw sensation. For instance, neurons in primary visual cortex (which receive fairly direct input from the eye) fire pulses rapidly in response to elongated patterns of light or dark, and neurons in other cortical areas respond best to motion, or to certain complex images. But these well-established results do not necessarily tell us either how the individual neuron functions or how the brain does, because brains are not simply sensory transducers—brains must also select input, interpret it, and act on it, and the role for single neurons in these broader functions is obscure. So it is not generally possible to infer a neuron's entire function from its stimulus-response properties[30]).

There are a number of considerations that make analysis of neurons in brains difficult:

1. **Cerebral cortex has massive feedback and interconnectivity between neurons**, so it is not clear which aspects of a response are caused by circuit interactions and which ones by a neuron's direct response to the input. Input and output are not conceptually distinct.

2. **The timescale of computation may be faster than the timescale of the input or output.** Even when both input and output signals are changing slowly (as often occurs in vision experiments), it is possible in principle that the timing of individual spikes within that slow envelope may have computational significance. (By analogy, a computer can compute faster than you can type.)
3. **The stimulus space is vast.** A circuit (like cortex) is capable of interpreting complex patterns, but it is impossible to test even a tiny subset of them. For example, a human might recognize patterns in a 30×30 pixel array[38]. Just static, black-white patterns in such an array number $2^{900} \approx 10^{300}$ possible combinations. In contrast, if a neuron is recorded for three hours with 10 stimuli/sec, only 10^5 stimuli can be tested.
4. **Experiments are biased towards high responses.** The most popular stimuli (such as isolated flashed or moving high-contrast patterns) are chosen primarily because they elicit strong neuronal responses, not because they are found commonly in natural scenes.

As a result of these constraints, it is very difficult to discuss the input–output properties of cortical neurons—whether in terms of correlations, filters, or information theory—without making strong simplifying assumptions about which stimuli to present and about how the network as a whole processes the stimulus.

Consider a concrete analogy. A typical "black-box" problem in an undergraduate physics class is to deduce the nature of a simple hidden circuit (perhaps a resistor, a capacitor, and a battery) based only on external measurements. But one cannot realistically apply the same approach to something like a computer's central processing unit (CPU) without a label or schematic, even if we were told how it works (e.g., by clocked binary computations using transistor gates).

Part of the problem is sheer complexity. For example, with a hundred-pin VLSI chip, there are almost 10,000 different ways of connecting just two of the power supply pins, and an almost unlimited number of ways to confuse the input and output lines. (The complexity of this problem is far greater than deciphering someone else's un-commented computer code, a problem we know to be difficult enough).

And if we decided to open up the CPU for internal inspection, we would encounter a measurement problem. A microprobe touched to its surface is typically larger (several μm) than a typical transistor gate (0.5 μm), so the probe would have difficulty touching just one gate. Furthermore, that probe has a much larger capacitance than a typical gate, and would capacitively load it, disrupting the circuit at typical operating speeds (e.g., 100 MHz).*

* These problems plague the microelectronics industry too. A typical CMOS transistor cannot drive an output pin directly; it requires an amplifier. So the method for measuring the output of a tiny transistor embedded in a working circuit is to re-design the circuit with a dedicated amplifier for each transistor of interest, then to re-fabricate the chip from scratch, and finally to measure the output of the amplifier on the new chip.

Because we know that both CPU's and brains compute, we can draw some tentative lessons from their analogy:

1. Devices which we know to be computationally powerful have as much internal structure as possible operating as fast as possible.
2. Computational efficiency can be the *opposite* of simplicity.
3. Those properties which confer computational efficiency (high bandwidth, close packing, and low power consumption) are almost by definition the hardest to measure externally. The brain has equally small anatomical elements (dendrites and spines), which we do not yet understand.
4. It is extraordinarily hard to deduce the operation of a sophisticated computing device unless you understand either its overall function or its design principles.

Because these problems arise from the fact that the brain was "engineered" by evolution, it helps to think like an engineer when trying to solve them. Such an engineering approach can have two flavors, which may be intermixed: reverse-engineering and forward-engineering.

Reverse-engineering[8] starts with assumptions about the system's components, and tries to deduce the system's overall function from their interactions. For example, one might assume a particular single-neuron model, and investigate how such neurons interact with various patterns of connectivity or learning rules. This approach has the powerful advantage that it builds directly upon experimental data. The difficulty lies in the multiplicity of theories: the number of possible combinations of simple components increases exponentially with the number of unknown parameters involved, and in a blind search one must either explore or explain away all the possible combinations.

On the other hand, the usual engineering approach ("forward" engineering) is to assume a function—such as redundancy-reduction, information transmission, or pattern retrieval—and deliberately design a system to implement it. There are two obvious disadvantages: the assumed function may be wrong, and/or its engineered implementation may have no biological relevance. On the other hand, the process of deliberate design is much more focussed than a broad search of parameter-space, and can invoke much more intelligence and intuition: each successive design is guaranteed to improve upon the previous one. Furthermore, the process of design can mimic evolution, and can locate and exploit useful combinations and features of elements, even if those arrangements at first seem rare and unlikely.

In the spirit of asking how we might design a brain, the following sections will outline some of the issues involved in transmitting information by spikes between neurons: how the information can be coded, how traditional cell models could deal with such codes, and how more realistic cell models can alter the traditional perspective.

2 Binary vs. Analog Coding

Suppose you want to design an "analog CPU", something which will compute using real-valued interconnections. And suppose furthermore that, in the absence of specifying the computation *a priori*, you only know that you want each unit to transmit the maximum rate of information to the units it feeds. Within preset voltage limits, you can choose the analog resolution of each unit to be whatever you want—a volt or a microvolt—and you then let the temporal resolution follow from that choice (for example, the inevitable noise in the system would make microvolt-sensitive units operate more slowly than millivolt ones). Alternatively, you could choose the temporal resolution, or clock rate, first, and let the voltage resolution follow.

What kind of choice should you make—slow and precise, or quick and dirty? If the total information I transmitted in some time interval is expressed as the number of distinguishable voltages (N_V) and the number of separate messages (N_T) in that interval, then

$$I = N_T \log_2(N_V). \tag{1}$$

Given the unavoidable tradeoff—that increasing N_T reduces the optimum N_V, and vice versa—then we have the choice of a linear benefit (higher message rate) with logarithmic penalty (degraded resolution), or a logarithmic benefit with linear penalty. The obvious choice (for reasonable tradeoff functions) is to maximize the linear term N_T: the information will peak when the code is "quick and dirty", using a high bandwidth and low resolution per message[7,48].**

3 Interpreting Cortical Spike Trains

We face a different kind of problem in understanding the brain's internal code. Here we know the nature of the signal—irregular, near-Poisson trains of spikes[50]—but we do not know what they mean. Do they constitute an average-rate code (like the signal from a Geiger counter), or a binary code (like a computer or a telegraph), or something else?

The experimental evidence for a rate-code is overwhelming. At least in the visual system, the image seen by the animal seems to influence only the average firing rate (and to a slight degree its temporal envelope), but not the exact timing of the spikes themselves[15] (see Hertz, this volume). If this effect is interpreted to mean that the spike code is an analog code—summing several spikes in some time-window—then one can estimate the maximum possible rate of information than can be transmitted with such a code. That upper bound (which depends on the time window, the firing rate, and so forth) is around five bits per second[49,51]. It is low because of the long time one must wait to transmit each analog signal, and because the spike train's irregularity acts as noise corrupting the signal.

** The marketplace seems to confirm this choice, since digital electronic circuits (very fast, with binary resolution) still vastly outsell analog ones, even for many analog signal-processing applications.

A competing hypothesis—that the timing of each spike in the brain carries important information in addition to the obvious rate-related signal—suffers from a lack of experimental evidence, although experiments do not exclude it either[2,48]. (The computational need for such a code is an open question, as is the nature of brainlike computations.[46])

But such a pulse code has a strong appeal to engineering efficiency: it would allow the same spike trains to carry a hundredfold more information. To estimate the peak information rate in such a code (without even knowing how the information is used!), one can assume the time-window for discrimination is a small fraction of the typical interspike interval, rather than several times it. Then a "message" in that narrow time-window becomes the presence or absence of a single spike. The irregularity of the spike train serves as information rather than noise, and the messages are sent much faster: a spike train could transmit hundreds of bits per second, rather than just a few.

4 Anatomical and Physiological Facts

The mammalian brain consists of a number of different tissue types, each of which appears to have significantly different microcircuitry[43]. As an example, we will focus here on mammalian neocortex (*cerebral cortex*, or *isocortex*), which in humans occupies almost the entire surface of the brain and much of its volume[13,17,18,55].

The cerebral cortex is a sheet, $1 - 3$ mm thick, which is often folded into the skull. This sheet can be subdivided into areas serving different functions. For example, visual signals first reach primary visual cortex (at the back of the head), then go to cortical areas like MT (for motion processing), V4 (color and shape), and IT (complex shapes and faces)[58,59].

The cortical sheet has six layers of distinct cell types with complex interconnections. For example, cells in layer IV receive input, and layers V and VI project output. But cells in any layer receive input from other cortical areas and send axons to other cortical areas, so input and output occur at every point of the circuit.

A cortical pyramidal neuron is tree-shaped, with a cell body (*soma*) and tens of branch-like *dendrites* (Fig. 1). Almost all of its roughly 10,000 synaptic inputs arrive on the dendrites. Most of these are on small ($1\,\mu$m) projections (*spines*[28]). A single output fiber (*axon*) near the cell body transmits the neuron's pulses to other neurons through another 10,000 or so synaptic connections. (This architecture is markedly different from digital computers, where a single gate may receive only a few different inputs and its output may contact only a few other gates.) Furthermore, we can know the circuitry only statistically. The exact details of the connections—nearly impossible to trace for even a few cells at a time—depend on the individual and its sensory experience.

When a pulse (*action potential* or *spike*) propagates along an axon and reaches a synapse, it releases a chemical *neurotransmitter*, which quickly diffuses to the

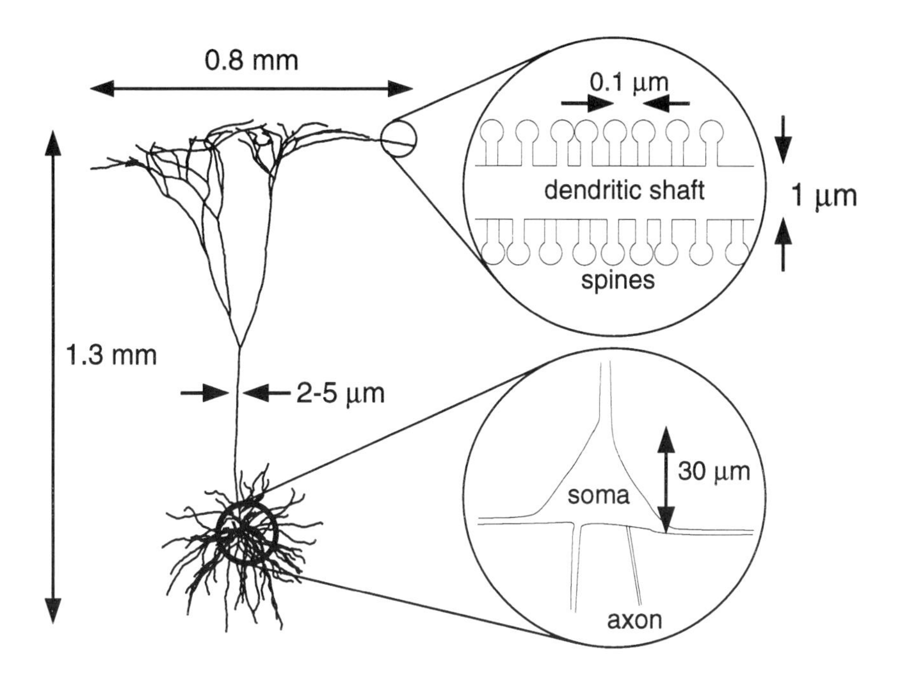

Fig. 1. The geometry of the dendrites of a large layer V excitatory neuron. Note that the thickness of the dendrites is not to scale.

recipient cell's membrane and opens tiny pores (*channels*) there[36,52]. Depending on the neurotransmitter and the ion channels, the effect may be *excitation* (raising the cell's voltage toward its firing threshold), *inhibition* (lowering it), or *shunting* (making other synapses less effective by increasing the membrane conductance). One almost-general principle is that a given axon's output synapses all use the same chemical neurotransmitter, and hence have the same "sign." Thus a given cell will be either excitatory or inhibitory, not both. Another principle is that (at least in cortex) inhibitory outputs remain relatively local, while only excitatory connections can be long range.

4.1 Simplified Models of Neurons

How does a cortical neuron actually compute something? We do not know the answer, but we know some of the cellular tools available.

On a time scale of tens or hundreds of milliseconds or longer, biochemical reactions can be used. For instance, calcium ions affect synaptic development and modification and many other processes as well. The concentration of Ca^{+2} can rise due to voltage changes or synaptic input, using both positive and negative feedback mechanisms. And many proteins—ion channels, enzymes, and even structural proteins—are sensitive to the calcium concentration or to its indirect effects[45]. Diffusable gases, such as NO and possibly CO, are also known to play an important role in neuronal signaling[10,14,40]. Other chemical messengers, such as inositol 1,4,5-trisphosphate, diacyl glycerol, and cyclic AMP (cAMP), are probably important as well. All of these substances interact in a complex, poorly-understood biochemical network[9].

Faster neuronal properties (at the millisecond scale) depend primarily on voltage. The membrane acts as a capacitor (about $0.6 - 1\,\mu\mathrm{F/cm}^2$), and specialized proteins (*channels*) allow ions to pass through it. Because of concentration differences, each ion species has a different electrochemical potential. For example, the concentration of potassium is higher inside than outside the cell, so opening potassium channels will cause potassium to flow outward and reduce the voltage inside. Conversely, sodium is concentrated outside the cell, so opening sodium channels lets sodium ions flow into the cell and raises the voltage. Channel opening may strongly depend on time and voltage, with complex feedback influences[22].

The most important channels are the *Hodgkin–Huxley conductances.* When the membrane voltage rises to a certain level (*threshold*), sodium channels open, and positive feedback quickly accelerates the voltage increase (by about 100 mV within 1 ms). The channels close (*inactivate*) after roughly 1 ms; then, other channels (for potassium ions) open, resetting the voltage below threshold. The whole voltage pulse constitutes one *action potential*; the neuron is unable to fire another one until after a $1-2$ ms dead time (*refractory period*) because potassium channels are still open and sodium channels are still inactivated.

A simple abstraction of this process is the integrate–and–fire model (Fig. 2A). Synaptic input is summed, or *integrated*, until the voltage reaches a threshold[23]. A spike occurs, the voltage is instantly reset, and the process begins again. Such a model performs temporal integration, because its output firing rate is linearly related to the time-integral of its input current.

A variety of modifications can make this model more realistic: a "leak" conductance (in parallel with the capacitor) represents the combined ohmic channels in the membrane*** (Fig. 2B), and a refractory period can prevent the cell from firing immediately after a previous spike (Fig. 2C).

If in fact the mean rate (averaged over many spikes) is the main signal, then one can eliminate spikes altogether and treat the mean rate as an analog value. Then the neuron becomes a simple real-valued function, integrating its inputs over time[21] and passing that sum through a nonlinearity. This is the basis of most of "neural network" theory. The simplicity of these neurons makes detailed mathematical analysis of the properties of networks possible[20]. This model is a reasonable approximation if the time constant τ_m is longer than the

*** This gives the cell membrane an intrinsic time constant $\tau_m = R_m C_m$ (usually $5-100$ ms). τ_m defines a time-limit to the temporal integration; after a few time constants, the cell "forgets" what its inputs were because they have decayed. For many years τ_m was also thought to be a measure of the time scale of a cell's response. In fact, we now know that it gives essentially no information about the cell's time scale[27].

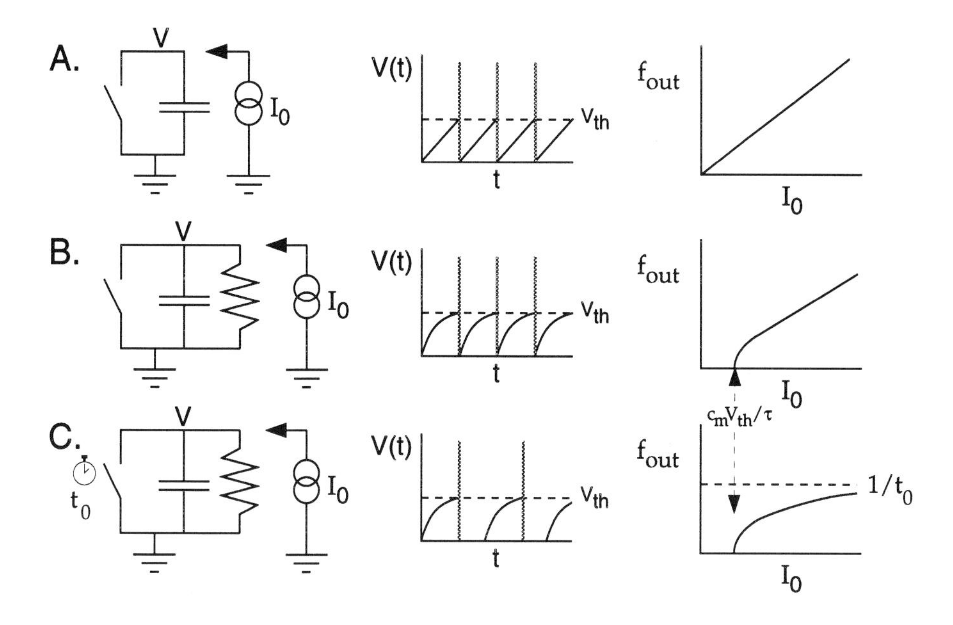

Fig. 2. Various integrate–and–fire models of a neuron. **A.** Synaptic input current charges up the cell membrane, modeled as a single capacitor C_m. When the voltage reaches a threshold V_{th}, a spike (grey) is fired, and the switch on the left closes briefly to reset the potential. The firing rate f_{out} of this simple model is a linear function of its input current I_0. **B.** A leak resistor represents the conductance of the cell membrane, and imposes a minimum current necessary for firing. **C.** During a refractory period (dead time) of $1 - 2$ ms immediately after a spike, the cell cannot fire another spike; this feature imposes a maximum firing rate.

mean interval μ_o between output spikes (Fig. 4.1A).

If, on the other hand, the signals of interest have a higher bandwidth than the average spiking frequency, then temporal integration (smoothing) makes little sense. A better strategy is to fire spikes in response to fluctuations (as would occur if $\tau_m \ll \mu_o$). Such a regime is often called "coincidence detection," because if each input pulse is small, then only several in coincidence can fire the cell (while the same number of pulses separated in time would not reach firing threshold[1]). Fluctuation– or coincidence–detection in cortex could potentially occur on the millisecond or even sub-millisecond time scale[47,48].

What evidence is there to help us decide between these two different kinds of computational strategies? Very little at present, at least for realistic cells with dendrites (see Sec. 4.2). But simple integrator-models can at least highlight some issues.

Suppose an integrate–and–fire neuron (with no leak; Fig. 2A) receives random synaptic input from many independent excitatory neurons, and that all

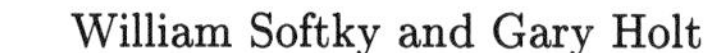

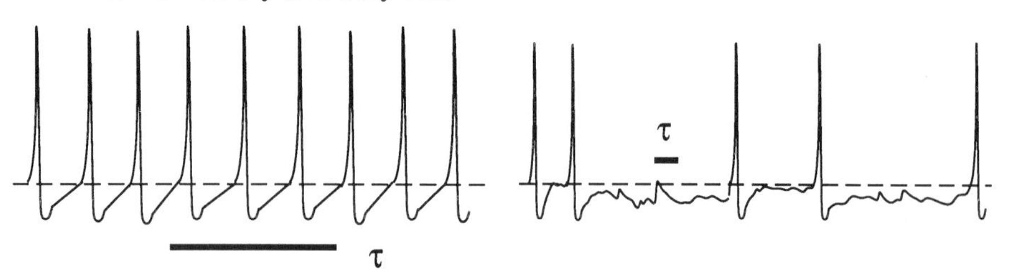

A. Temporal integration.
$\tau > \overline{\Delta t}$
weak inhibition
average input sets average dV/dt

B. Fluctuation detection.
$\tau \ll \overline{\Delta t}$
strong inhibition
average input sets average voltage

Fig. 3. Temporal integration vs. fluctuation detection.

synaptic amplitudes are equal. These assumptions mean that we can treat the superposition of the inputs as a single Poisson process, with mean interval μ_i and standard deviation σ_i. If it takes N_{th} input pulses to rise from reset to threshold, then the output spikes have $\mu_o = \mu_i/N_{\text{th}}$ and $\sigma_o = \sigma_i/\sqrt{N_{\text{th}}}$. So a dimensionless measure of the output variability, C_V (the Coefficient of Variation), is given by

$$C_V = \frac{\sigma}{\mu} = \frac{1}{\sqrt{N_{\text{th}}}} \tag{2}$$

Most researchers think that N_{th} is large (≥ 50) for neocortical cells, so the value of C_V should be low[50] (≤ 0.15). This is intuitively reasonable, since one would expect a neuron which sums many inputs to average out their irregularity, and fire more regularly as a result.

But for cortical neurons $C_V \approx 1$, the value expected for a completely random (Poisson) process. Thus there is a large contradiction between the predicted and observed values of C_V. Various refinements to the integrate–and–fire model can resolve the contradiction, but only by preventing temporal integration (e.g., by drastically reducing τ_m or N_{th}[50]).

Why do cortical neurons fire so irregularly? Some possible explanations are:

1. Neurons might be intrinsically noisy or chaotic devices. But there is no evidence for this. In fact, experiments have shown neurons to be quite reliable when their inputs can be quantified[11] or controlled[32].
2. Network effects (especially inhibition) may give rise to the irregularity through chaos or other simpler effects. For example, if both excitatory and inhibitory input arrive at a high rate, and the inhibitory input approximately balances the excitatory input, then integrate–and–fire neurons will fire irregularly[42]. Similarly, more sophisticated networks with a specific topography of excitation and inhibition have irregularly firing neurons[56]. In these cases, irregularity reflects responses to fast fluctuations rather than temporal summation.
3. Neurons are doing coincidence detection on a fast time scale[47,48,50].

4.2 "Realistic" Models in Space and Time

The simple and popular neuron models discussed above ignore the single most obvious aspect of real neurons in cortex: their intricately branched shapes, and the long, thin *dendrites* through which synaptic input passes on the way to generating output. While we have very little data on how those dendrites behave in animals[35,54], it is at least possible to numerically simulate how they might behave[34,41].

Each piece of dendrite can be modeled by a one-dimensional circuit element[39,44] (Fig. 4). Thus, each of the dozens or hundreds of dendrites branching out from the cell body is equivalent to a set of ohmic series resistances (the saline solution inside the cell), a set of ohmic parallel resistors (leak resistance through the membrane), and a set of parallel capacitors (the cell membrane itself). These three features all exist in real cells, and together constitute a linear, passive dendrite. (A more realistic active dendrite includes an additional set of parallel nonlinear, non-ohmic resistances, shown as the variable resistor in Fig. 4; the functional form of the nonlinearity affects the dendrite's behavior enormously.)

Let us review a passive dendrite first, approximated as a single isolated cylinder. Inside one such cylinder far from its ends, it can be shown that a steady-state voltage V_0 imposed at one point ($x = 0$) will decrease with distance as $V(x) = V_0 \exp(-|x|/\lambda)$ (Fig. 5A). The "electrotonic space constant" λ gives the scale of this voltage decay, and depends both on the axial and transmembrane resistances (but not on membrane capacitance). No signal—DC or time-varying—can exceed this envelope, so λ is often taken as a typical distance for signal propagation inside a dendrite, and synapses lying within λ of the cell body are often considered relatively unaffected by dendritic filtering.

This traditional view is misleading. First, space constants can change dramatically with synaptic activity[5] because synapses are conductances. Second, signals with strong high-frequency components (such as fast synaptic currents) are strongly influenced by the membrane capacitance, which is ignored in the calculation of λ. So $V(x)$ should really be $V(x,t)$. We can illustrate the main effects of the membrane capacitance in a simplified example (exact, formal solutions for passive cables are available elsewhere[22,39]).

At sufficiently fast timescales ($t \ll \tau_m$) and short distances along the dendrite ($x \ll \lambda$), we can ignore the leak resistors. So imagine that the resistors in Fig. 5A in parallel with the membrane capacitors have been eliminated: now no charge will leak outside the cell, but will instead diffuse across the inside of the membrane. This charge-diffusion—best visualized as packet of synaptic charge first piling up near the synapse, then broadening with time—introduces two crucial effects (Fig. 5B).

The first is that signals are delayed and smoothed at locations away from the synapse, so the dendrite acts as a kind of low-pass filter for signals passing through it *en route* to the cell body. A typical one-millisecond synaptic current

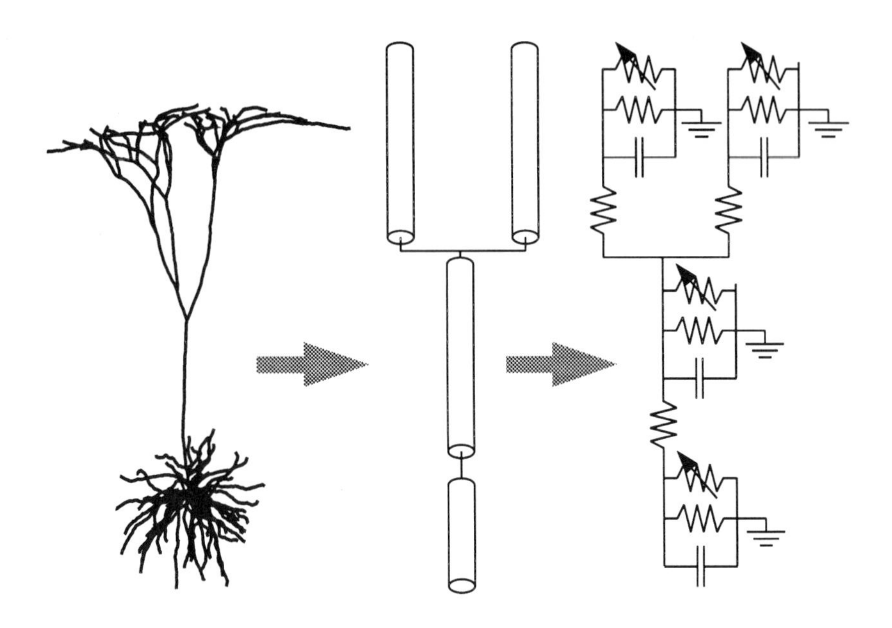

Fig. 4. How geometries are converted to finite elements for simulations or analysis. The cell is broken up into compartments, each of which contains a capacitor, a leak conductance, and possibly a conductance which depends on time, voltage, or calcium concentration. A realistic model cell may have to be broken up into several hundred compartments rather than just the four shown here.

could be smoothed to over ten milliseconds in a realistic situation. This smoothing effect, by itself, would make it very difficult for a cell to preserve the precise timing of input spikes arriving on far-away dendrites.

The other effect is that voltages *near* the synapse decay very rapidly in time ($V(t) \propto 1/\sqrt{t}$), much faster than the membrane time-constant (recall that $\tau_m = \infty$ with no leak). The physical process producing this effect—the rapid diffusion of charge away from the synapse while remaining inside the membrane—is fundamentally different from the slower process of membrane voltage-decay, in which charge leaks out through the membrane. Membrane capacitance prevents high-frequency signals from propagating down the dendrite, and localizes them near the synapse.

Thus, the passive properties of dendrites are ideally suited to localize signals tightly in both space and time. There are two ways a cell might make use of these properties. One is by locating a *shunting* inhibitory synapse nearer to the soma than an excitatory synapse. Then the inhibition can roughly cancel the excitation if the two are coincident, but the excitatory signal will reach the soma uncancelled if the synaptic conductances are non-overlapping[25,26].

The other mechanism is by nonlinear resistances which can amplify voltage

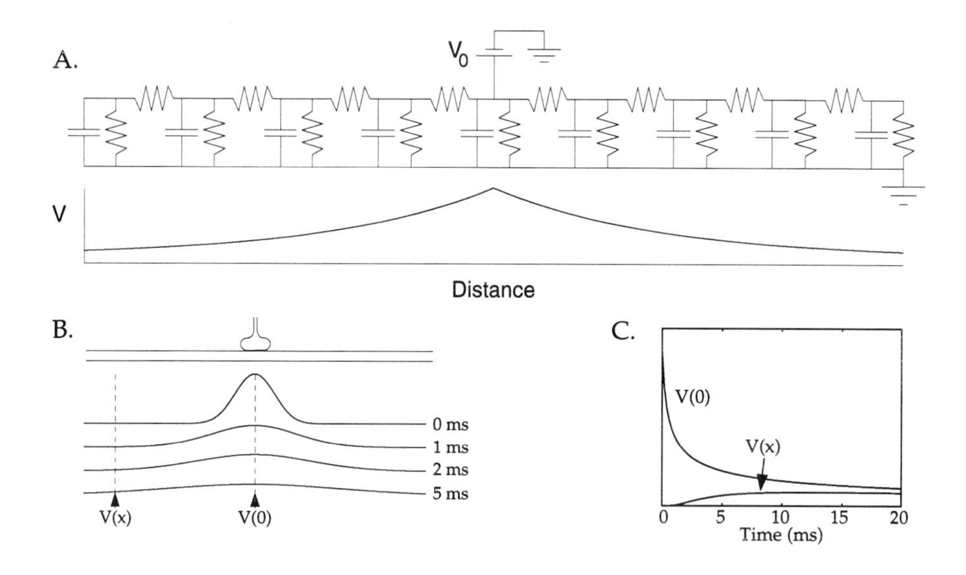

Fig. 5. A. Steady state voltage distribution in a long dendrite if the middle is held at V_0. **B.** Voltage distribution in a passive dendrite responding to a short pulse of synaptic input ending at $t = 0$. **C.** The time course of the response at the two marked locations: the voltage transient is very strong and brief near the synapse, but is weak and blurred further away.

fluctuations inside the dendrite (such as spike-generating sodium conductances[53]). Such local amplification means that two exactly coincident synaptic inputs might produce a stronger signal than if the same two were slightly offset. Furthermore, the Hodgkin-Huxley equations contain time-delayed responses (inactivation and rectification) which add high-pass properties to the amplification; as a result, dendrites might act more like temporal differentiators than like integrators.

These effects together can render a cell model sensitive to synaptic timing two orders of magnitude faster than the membrane time-constant[48]. However, the properties upon which such models depend—the existence of precise timing and location in a cell's input, and the existence of special membrane properties to make use of that precision—remain experimentally inaccessible. Properties of dendrites have a profound effect on a neuron's response to very brief signals, but the sign of the effect—amplifying vs. damping local fluctuations—is still unknown.

Nonlinear conductances in dendrites can perform computations at slower time scales as well. In some cases, the active nonlinearities may simply counteract such passive properties as saturation[6,12,29,37]. Other computationally interesting functions (AND, OR, and XOR) can also be generated in dendrites with simple combinations of known conductances[34].

5 Conclusions

We do not yet know even qualitatively how cortical cells process their inputs spikes to produce output spikes. The dominant simple models—single capacitors which temporally average over many small input events—are incapable of producing realistically irregular firing, except in networks where they functionally do not integrate over time. Simple models which do produce irregular firing do so by responding to fluctuations, rather than to averages. More complex models with biophysically realistic dendrites are capable of many diverse types of behavior, ranging from temporal integration to temporal differentiation, depending on unknown membrane properties. This confusion is enhanced by many persistent but false myths about the electrical behavior of single neurons: that λ is the shortest distance at which synapses can be distinguished; that τ_m is the shortest time at which two events can be distinguished; that temporal smoothing is inevitable in a realistic cell.

Why does this confusion matter? Because there are two opposing interpretations of spike codes in the brain, one which treats irregularity as noise, and the other which treats it as high-bandwidth information. We know that slow events (tens to hundreds of milliseconds) are important in the nervous system, but we do not know to what degree single spikes matter. With high speed would also come tight localization, giving a stark choice: between fast, precise, information-dense computations versus simple, slow, noisy ones. Engineers might prefer the former, but which did Nature choose?

References

1. M. Abeles. Role of cortical neuron: integrator or coincidence detector? *Israeli Journal of Medical Sciences* **18**:83–92, 1982.
   Although McCullogh and Pitts in 1943 proposed that cortical neurons perform fast coincidence-detection, this is the first paper to give a biophysical argument (based on assumptions about the relative timescales of spike-firing and subthreshold voltage fluctuations).
2. M. Abeles. *Corticonics*. Cambridge: Cambridge University Press, 1990.
   In order to account for an anomalous excess of precise firing patterns in cortical spiking, the author presents a highly innovative theory (*synfire chains*) of precisely-timed volleys of spikes which ricochet among interconnected neurons.
3. J. Atick and N. Redlich. Towards a theory of early visual processing. *Neural Computation* **2**:308–320, 1990.
   Starting with the hypothesis that the retina and optic nerve have evolved to carry visual information in a particularaly information-efficient form (*low redundancy*), the authors derive closed-form expressions for the expected spatiotemporal filter properties of those cells, and compare them to experiment.
4. H. B. Barlow. Single units and sensation: a neuron doctrine for perceptual psychology? *Perception* **1**:371–394, 1972.
5. Ö. Bernander, R. Douglas, K. Martin and C. Koch. Synaptic background activity determines spatio-temporal integration in single pyramidal cells. *Proceedings of the National Academy of Sciences USA* **88**:1569–1573, 1991.

A striking demonstration, using elaborate numerical simulations of anatomically-correct cells, that some cellular properties often thought of as constant (such as membrane time constants) in fact strongly depend on synaptic input.

6. Ö. Bernander, C. Koch and R. J. Douglas. Amplification and linearization of distal synaptic input to cortical pyramidal cells. *Journal of Neurophysiology* **72**:2743–2753, 1994.
7. W. Bialek, M. Deweese and F. Rieke. Bits and brains—information flow in the nervous system. *Physica A* **200**:581–593, 1993.
8. J. M. Bower. Reverse engineering the nervous system: an anatomical, physiological, and computer-based approach. In: *An Introduction to Neural and Electronic Networks*, pp. 3-24. Academic Press, 1990.
9. D. Bray. Protein molecules as computational elements in living cells. *Nature* **376**:307–312, 1995.
   A discussion of how properties of biochemical networks could be useful for computation. Weak on specifics, since so little is known about how biochemistry is actually used for computation.
10. D. S. Bredt and S. H. Snyder. Nitric oxide, a novel neuronal messenger. *Neuron* **8**:3–11, 1992.
11. W. Calvin and C. Stevens. Synaptic noise and other sources of randomness in motoneuron interspike intervals. *Journal of Neurophysiology* **31**:574–587, 1968.
   The relatively modest firing irregularity of a particular cell type outside the brain is consistent with the temporal summation of many small, random input pulses.
12. E. de Schutter and J. M. Bower. Simulated responses of cerebellar Purkinje cells are independent of the dendritic location of granule cell synaptic inputs. *Proceedings of the National Academy of Sciences USA* **91**:4736–4740, 1994.
13. R. J. Douglas and K. A. C. Martin. Neocortex. In: *The Synaptic Organization of the Brain*. G. M. Shepherd, ed., pp. 389-438. Oxford University Press, 1990.
   A good introductory review on the structure and circuitry of neocortex.
14. J. A. Galley, P. R. Montague, G. N. Reeke and G. M. Edelman. The NO hypothesis: possible effects of a short-lived, rapidly diffusible signal in the development and function of the nervous system. *Proceedings of the National Academy of Sciences USA* **87**:3547–3551, 1990.
15. T. Gawne and B. Richmond. How independent are the messages carried by adjacent inferior temporal cortical neurons? *Journal of Neuroscience* **13**:2758–2771, 1993.
16. A. P. Georgopoulos. Neural coding of the direction of reaching and a comparison with saccadic eye movements. *Cold Spring Harbor Symposium on Quantitative Biology* **55**:899–910, 1990.
17. C. D. Gilbert. Horizontal integration and cortical dynamics. *Neuron* **9**:1–13, 1992.
18. C. D. Gilbert. Circuitry, architecture, and functional dynamics of visual cortex. *Cerebral Cortex* **3**:373–386, 1993.
19. R. D. Hawkins, E. R. Kandel and S. A. Siegelbaum. Learing to modulate transmitter release: themes and variations in synaptic plasticity. *Annual Review of Neuroscience* **16**:625–665, 1993.
20. J. Hertz, A. Krogh and R. G. Palmer. *Introduction to the Theory of Neural Computation*. Addison–Wesley, 1991.
   The standard text on neural network theory.
21. J. Hopfield. Neurons with graded response have collective computational properties like those of two-state systems. *Proc. Nat. Acad. Sci. USA* **81**:3088–3092, 1984.
22. J. J. B. Jack, D. Noble and R. W. Tsien. *Electric Current Flow in Excitable Cells*. Oxford University Press, 1983.

The classic work on electrical properties of neurons with nonlinear conductances; highly mathematical.

23. C. Koch, Ö. Bernander and R. J. Douglas. Do neurons have a voltage or a current threshold for action potential initiation. *J. Computational Neuroscience* **2**:63–82, 1995.
24. C. Koch and J. L. Davis. *Large Scale Neuronal Theories of the Brain.* Cambridge, MA: MIT Press, 1994.
25. C. Koch, T. Poggio and V. Torre. Retinal ganglion cells: a functional interpretation of dendritic morphology. *Philosophical Transactions of the Royal Society B* **298**:227–264, 1982.
    An example of how to use shunting inhibition to compute direction selectivity.
26. C. Koch, T. Poggio and V. Torre. Nonlinear interaction in a dendritic tree: localization timing and role in information processing. *Proceedings of the National Academy of Science USA* **80**:2799–2802, 1983.
27. C. Koch, M. Rapp and I. Segev. A brief history of time (constants). *Cerebral Cortex* **in press**, 1996.
    A review of thinking about the passive membrane time constant, how it is measured, how it changes, and what (if anything) it means.
28. C. Koch and A. Zador. The function of dendritic spines: devices subserving biochemical rather than electircal compartmentalization. *Journal of Neuroscience* **13**:413–422, 1993.
    A proposal that the tiniest structures of a neuron serve to maintain strong, local calcium concentrations to aid learning functions at single synapses.
29. G. Laurent. A dendritic gain-control mechanism in axonless neurons of the locust, *Schistocerca Americana. Journal of Physiology* **470**:45–54, 1993.
30. S. R. Lehky and T. J. Sejnowski. Network model of shape-from-shading: neural function arises from both receptive and projective fields. *Nature* **333**:452–454, 1988.
31. R. Linsker. Self-organization in a perceptual network. *IEEE Computer* **21**:105–117, 1988.
    A proposal that the computations done by each cortical layer upon its input are in fact meant to preserve the maximum amount of information possible, subject to the inevitable noise added in transmission.
32. Z. F. Mainen and T. J. Sejnowski. Reliability of spike timing in neocortical neurons. *Science* **268**:1503–1506, 1995.
    Multiple presentations of a strong, quickly-fluctuating current input generate reproducible patterns out output spikes from a cortical cell, showing that such cells do not have an inherently stochastic firing mechanism.
33. K. A. C. Martin. A brief history of the feature detector. *Cerebral Cortex* **4**:1–7, 1994.
34. B. W. Mel. Information processing in dendritic trees. *Neural Computation* **6**:1031–1085, 1994.
    A fairly thorough review of the potential for computation, especially slower computation, using dendrites with nonlinear conductances such as NMDA or various calcium currents.
35. J. Midtgaard. Processing of information from different sources: spatial synaptic integration in the dendrites of vertebrate CNS neurons. *Trends in Neuroscience* **17**:166–173, 1994.
    A review of the physiology of active dendritic conductances, focussing mostly on the Purkinje cell in the cerebellum.

36. J. G. Nicholls, A. R. Martin and B. G. Wallace. *From Neuron to Brain: A Cellular and Molecular Approach to the Nervous System.* Sunderland, MA, USA: Sinauer Associates, Inc., 1992.
A good textbook introducing the visual system and neuronal biophysics.
37. A. Nicoll, A. Larkman and C. Blakemore. Modulation of EPSP shape and efficacy by intrinsic membrane conductances in rat neocortical pyramidal neurons in vitro. *Journal of Physiology* **468**:693–710, 1993.
38. B. A. Olshausen, C. H. Anderson and D. C. Van Essen. A neurobiological model of visual attention and invariant pattern recognition based on dynamic routing of information. *Journal of Neuroscience* **13**:4700–4719, 1993.
39. W. Rall. Cable theory for dendritic neurons. In: *Methods in Neuronal Modeling.* C. Koch, I. Segev, eds., pp. 9-62. MIT press, 1989.
A detailed mathematical introduction to cable theory as applied to passive dendrites.
40. E. M. Schuman and D. V. Madison. Nitric oxide and synaptic function. *Annual Review of Neuroscience* **17**:153–184, 1994.
41. I. Segev. Single neurone models: oversimple, complex, and reduced. *Trends in Neuroscience* **15**:414–421, 1992.
42. M. N. Shadlen and W. T. Newsome. Noise, neural codes and cortical organization. *Current Opinion in Neurobiology* **4**:569–579, 1994.
A review of firing variability studies in neocortex, concluding that the variability is of no significance for coding—that it just represents noise. See the response to this in reference 48 for the opposite conclusion from exactly the same data and simulations.
43. G. M. Shepherd. *The Synaptic Organization of the Brain.* Oxford University Press, 1990.
An excellent, readable introduction to a number of different areas of the mammalian brain, focussing on computational issues—synaptic connectivity and electrical properties of each kind of neuron.
44. G. M. Shepherd and C. Koch. Dendritic electrotonus and synaptic integration. In: *Synaptic Organization of the Brain.* G. M. Shepherd, ed., pp. 439-473. Oxford University Press, 1990.
A readable introduction to cable theory as applied to passive dendrites.
45. P. B. Simpson, J. Challiss and S. R. Nahorski. Neuronal Ca^{+2} stores: activation and function. *Trends in Neuroscience* **18**:299–306, 1995.
46. W. Singer. Putative functions of temporal correlations in neocortical processing. In: *Large-Scale Neuronal Theories of the Brain.* C. Koch, J. L. Davis, eds., pp. 201-237. Cambridge, MA: MIT press, 1994.
Temporal modulation can multiplex information into the average firing rate of neurons, and such information can in principle aid in computations necessary for perception, such as "binding" and feature-segmentation.
47. W. Softky. Sub-millisecond coincidence detection in active dendritic trees. *Neuroscience* **58**:13–41, 1994.
An exploration of possible mechanisms by which thin dendrites might perform fast, localized computations, using both approximate physical scaling principles and numerical simulations.
48. W. R. Softky. Simple codes versus efficient codes. *Current Opinion in Neurobiology* **5**:239–247, 1995.
A response to reference 42.

49. W. R. Softky. Fine analog coding minimizes information transmission. *Neural Networks* **in press**, 1995.
50. W. Softky and C. Koch. The highly irregular firing of cortical cells is inconsistent with temporal integration of random EPSPs. *Journal of Neuroscience* **13**:334–350, 1993.
A detailed analysis of variability in neocortical cells, including both measured responses (from awake monkeys) and the failure of standard theories to account for those results.
51. R. Stein. Some models of neuronal variability. *Biophys. J.* **7**:37–68, 1967.
52. C. F. Stevens. The neuron. *Scientific American* **241 (Sep)**:55–65, 1979.
A good non-technical introduction to most of the basic biophysics about neurons and synapses. Our understanding of this material has not changed much since this article was published.
53. G. Stuart and B. Sakmann. Active propagation of somatic action potentials into neocortical pyramidal cell dendrites. *Nature* **367**:69–72, 1994.
54. G. Stuart and N. Spruston. Probing dendritic function with patch pipettes. *Current Opinion in Neurobiology* **5**:389–394, 1995.
A review of experimental techniques for recording from dendrites and results from such studies.
55. A. M. Thomson and J. Deuchars. Temporal and spatial properties of local circuits in neocortex. *Trends in Neuroscience* **17**:119–126, 1994.
56. M. Usher, M. Stemmler, O. Zeev and C. Koch. Network amplification of local fluctuations causes high spike rate variability, fractal firing patterns and oscillatory local field potentials. *Neural Computation* **6**:795–836, 1994.
57. D. C. Van Essen, C. H. Anderson and D. J. Felleman. Information processing in the primate visual system: an integrated systems perspective. *Science* **255**:419–423, 1992.
A brief review of the different areas and their possible functions.
58. D. C. Van Essen, D. J. Felleman, E. A. Deyoe, J. Olavarria and J. Knierim. Modular and hierarchical organization of extrastriate visual cortex in the macaque monkey. *Cold Spring Harbor Symposium on Quantitative Biology* **55**:679–696, 1990.
A more detailed but less up-to-date review of different visual cortical areas and processing streams.
59. D. C. Van Essen and J. L. Gallant. Neural mechanisms of form and motion processing in the primate visual system. *Neuron* **13**:1–10, 1994.

Cerebral cortex (the largest part of human brains) has evolved to transmit as much information as possible inward from the senses to higher brain areas (**information maximization**[31]).	In practice, our visual systems throw away the vast majority of their input information, winnowing about 10^8 bits/s of photoreceptor signals down to about 1 bit/s that we can recall from long-term memory (**information minimization**[57]).[a]
Computation occurs by **stable equilibrium and attractor dynamics**[21].	Computation occurs by **non-attractor dynamics**, allowing fast processing and quick responses to changing inputs.
Neurons are **slow** devices to average signals over time.	Neurons are **fast** to allow high bandwidth and computational power.
Neurons are approximately **linear** devices, to preserve maximum information.	Neurons are strongly **nonlinear** devices; nonlinearity is essential for making decisions.
Neurons **ignore the spatial location** of synaptic input[12] to simplify the process of wiring during development.	Neurons **use the spatial location** of synaptic input to augment memory capacity and computational power[34].
Neural systems are **robust**, giving the same output regardless of small input variations	Neural systems are **sensitive**, preserving or amplifying small changes.
Neurons act as **feature detectors**, signaling the presence of a specific property in the real world[4,33].	The brain's neurons use a **combinatorial code**[3]; a single neuron does not code for any particular recognizable feature in the real world. In motor control, the motion is coded by **an average of many correlated neurons**[16].

[a] Obviously, neither maximizing information nor minimizing it can yield a very interesting model of brains; the best way to preserve information is merely to copy it without modification, and the best way to remove information is to throw all of it away immediately.

Table 1. Contradictory guiding principles in neurobiology.

Statistical Mechanics and Sensory Signal Processing

William Bialek

NEC Research Institute, 4 Independence Way, Princeton, New Jersey 08540 USA

1 Introduction

The sensory systems have long held a special fascination for physicists, since these are the instruments with which we make measurements on the physical world around us. These instruments are really very impressive, often approaching the fundamental physical limits of thermal and quantal noise (de Vries, 1956; Bialek, 1987). This performance poses important challenges for how we think about the molecular and cellular mechanisms of transduction and amplification in the receptor cells.*

Even unicellular organisms are "instrumented" with an array of sensors, including chemoreceptors which allow the cell to measure gradients in the concentrations of nutrients and toxins. On the $\sim 1\,\mu$m scale of a bacterium, chemical sensing is severely limited by concentration fluctuations, while the motor system is strongly constrained by the peculiarities of low Reynolds number hydrodynamics, so that one has a chance at understanding the whole sensory-motor loop in terms of fundamental physical principles. This is the program laid out nearly twenty years ago by Berg and Purcell (1977), and there are still interesting open problems. But in big complicated animals (like flies ... or even us) it is hard to believe that attention to physical principles will carry us all the way from sensory input to motor output – or, more subtly, to our internal understanding of the sensory world. Surely at some point in the nervous system some more "biological" considerations must take over. Perhaps thinking like a physicist forces us to stay near the periphery of the sensory systems, nibbling at the edges of the brain but never really getting to the essence of what brains do.

There is, however, a hint that physics does not stop abruptly once we leave the receptor cell. In the same way that receptor cells perform near the limits of physical noise sources, so the brain as a whole reaches decisions and makes estimates which are near the limits imposed by these same noise sources (Barlow, 1981; Bialek, 1992). This means that "thinking like a physicist" cannot stop at the receptor, but must be pushed through the processing of sensory information far enough to see if we understand this impressive reliability. From a theoretical point of view, making accurate estimates and reliable decisions is not merely a matter of avoiding extraneous noise sources. Maximally accurate estimates,

* See, for example, Lagnado and Baylor (1992), or the historical snapshot provided by Delbruck (1969).

for example, are possible *only* if one processes the input data in a particular way – the "optimal estimator" is a *unique* function of the sensory inputs. If real brains reach the physical limits to perception, then they *must* evaluate these functions. If we can take seriously the notion of the brain as an optimal processor of sensory information, then the theory of these optimal devices provides us with a predictive theory of what the brain should be computing. We will see that this problem of optimal estimation is in turn a statistical mechanics problem, and hence the title of this chapter.**

2 Is there a Physics Problem?

You should be skeptical of any claim that "brains do **X**." The history of experiments on animal and human behavior shows that behaviors which seem to involve brilliant abstract reasoning are often the result of applying simple and, in retrospect, rather stupid rules, while behaviors which look trivial are highly adaptable, sophisticated and learned rather than hard-wired. All this tells us that we don't have much intuition about the problems which brains have evolved to solve. There is one area where, as physicists, we do have some intuition, and this is in the processing of sensory information. We understand these problems because they are the same problems as those which are solved in the laboratory when we make measurements. We know that a well designed particle physics experiment begins with the construction of sensitive and reliable detector elements (like the receptor cells of our sense organs), and the design must continue through the construction of algorithms which extract meaningful events from the vast array of detector outputs. We know that at each step there are physical limits: Detectors of a certain size have some irreducible noise level, cables have limited capacity for transmitting information, and so on. We also know that our understanding of the rules which govern the microscopic "world" of the collision region is essential in separating interesting events from the random background. So the questions are clear: Does biology build good detectors? Is the processing of the detector outputs efficient? How extensive is the brain's "knowledge" of the rules which govern the sensory world, and how is this knowledge applied to aid in the interpretation of the raw sense data?

2.1 Real brains approach the physical limits

In this section I try to assemble some of the evidence that animals can perform signal processing tasks down to the limits of precision imposed by fundamental physical noise sources. There is no way that I can do justice to generations of beautiful experiments, and I encourage you to read the original literature.

** I gave very similar lectures at the School of Biophysics in Erice (May 1995) and at the Princeton Lectures on Biophysics (June 1995). The present text is roughly what I would have liked to have said (or wish I had managed to say) at each school, rather than a precise record of what I said in Humlebæk, Denmark. With the permission of the publishers, essentially the same text appears in the proceedings of the Erice School (Conti and Torre, 1996).

Photon counting. Imagine sitting in a dark room. Someone flashes a dim light, just a short pulse. Do you see it? If the flash is bright, you see it easily. But if the flash is very dim, maybe you don't see it at all. In fact whether you see any particular flash seems almost random. How does this randomness arise? Leaving aside the biology, we know that with dim (conventional) light sources the number of photons counted by an ideal photon counter is a random number with Poisson statistics. Indeed, the random responses of human observers under dark-adapted conditions are an almost perfect reflection of this inevitable physical randomness in photon arrivals (Hecht, Shlaer and Pirenne, 1942; van der Velden, 1944; Sakitt, 1972). With modern optical techniques one can manipulate the statistics of photon arrivals, and Teich et al. (1982) have observed the resulting changes in the statistics of seeing. Even the detection of single photons involves discrimination against a background of dark noise, and careful analysis of human psychophysical experiments gives an estimate of this dark noise level; see, for example, Sakitt (1972). Recordings from individual photoreceptor cells in toads (Baylor, Lamb and Yau, 1979; Baylor, Matthews and Yau, 1980) and later in monkeys (Baylor, Nunn and Schnapf, 1984) showed that these cells produce clear single photon responses but also a random background of spontaneous photon-like events. These discrete events are due, almost certainly, to the thermal activation of the photopigment rhodopsin, and because there are $\sim 10^9$ rhodopsin molecules in the receptor cell, the observed event rate of one per minute (in toad rods at 25°C) means that the half-life of one molecule is several thousand years. The rate of these events matches the estimated noise levels computed from psychophysical data (Barlow, 1988). This agreement was tested further in experiments by Aho et al. (1988), who studied the reliability of photon counting by toads who must strike at a target illuminated by a dim flash of light. Again one can estimate dark noise levels from the behavioral data, and again they agree with the dark noise measurements in single receptor cells, but now one can change the temperature (this is harder with people) and show that the two measurements track each other over a large range of noise levels. This means that when the toad is colder it behaves more reliably, down to the temperature where it does not behave at all. To summarize, receptor cells can count single photons and the brain can register, accumulate and interpret these counts with a reliability limited only by the dark noise of the detector itself.

Echo delay estimation by bats. Bats navigate by echolocation, producing high frequency acoustic pulses and listening for the returning echoes. Man-made sonar or radar systems typically operate with a narrow beam which is scanned, but bats do all the imaging in "software" – their pulses spread over a wide angular range and they uses the waveforms of echoes at their two ears to compute an image of the three-dimensional world around them. Several generations of experiments have given some insight into the nature of this image (Griffin, 1958; Simmons, 1989). In particular, Simmons and co-workers have measured the accuracy with which bats can discriminate changes in the distance of a target, or more precisely changes in the delay of simulated echoes. Early experiments (Simmons, 1979) showed that bats could achieve reliable discriminations with delay differences of order 1 μsec or a bit less, and more surprisingly that the bat could be confused by delay differences of $\sim$30 μsec which match the fudamental period of the acoustic waveform. This last effect is qualitatively what one expects from an optimal receiver which correlates the echo against a perfect copy of the emitted pulse. But given the typical signal-to-noise ratios in the bat's environment, such an optimal system should be capable of discriminating delay differences of order 10 *nano*seconds. If you remember that we are talking about behavioral experiments on a living, walking and potentially flying animal, it takes some courage even to attempt an experiment which searches for such precision. But this is exactly what Simmons et al. (1990) did, and they found that bats could make reliable discriminations at the predicted 10 nsec. More importantly they could manipulate the signal-to-noise ratio and show that the bats' performance tracks the optimum over a range of noise levels. Given that the basic time scale for the dynamics of individual neurons is $\sim$msec, the observation that bats can make temporal discriminations down to tens of nsec is surely one of the most remarkable quantitative results about the nervous system.

Frequency discrimination and pitch estimation. Humans can detect reliably differences of 3 Hz among pure tones near 1000 Hz. Is this precision comparable to the limits imposed by noise in the inner ear itself, or more directly to the limits imposed by the randomness in the activity of the auditory neurons which carry information from ear to brain? This issue has a long history, starting in the papers of Siebert (1965, 1970) which in turn were motivated by the early quantitative experiments on auditory nerve activity (Kiang et al., 1965). Siebert's work provided a rather sophisticated mathematical view of many questions which are debated in the current literature on neural coding (Rieke et al., 1996). A second question, emphasized by Goldstein (1973), is whether the brain can make judgements about more complex sounds (not just single sine waves) with the precision that one would expect from the measurements on pure tones. We now know the following: First, the precision of single-tone frequency discrimination is consistent with the performance of a device which makes optimal use of the information available from the intervals between action potentials in the auditory nerve (Goldstein and Srulovicz, 1977; Srulovicz and Goldstein, 1983). Second, if we ask listeners to make judgements about the pitch of complex sounds (e.g., the fundamental frequency in a harmonic complex), then the reliability of discrimination and identification is consistent with the optimal processing of the component frequencies (Goldstein et al., 1978; Beerends and Houtsma, 1986). The optimal processor theory of pitch perception also predicts systematic shifts of pitch for almost-harmonic sequences, a point to which we return below.

'Higher' vision – symmetries, objects, ... Many of the interesting features of the visual world are at best probabilistically related to the images which appear on our retinae. Thus the shading of an object tells us something about its three dimensional structure, but there is also some chance that the object has a surface with non-trivial variation in reflectance. This element of randomness in the cues for different features means that there are statistical limits on our ability to extract these features from the raw sense data, and of course these limits are made more severe by any true noise (such as photon shot noise). One can try to isolate these issues by the systematic construction of images with different statistical properties. In this spirit, Barlow (1980) constructed images which had a statistical tendency toward bilateral symmetry and asked observers to distinguish these from images which were generated without correlations. The reliability of discrimination was (in appropriate units ...) within a factor of two of the limit which comes from the spontaneous near-symmetries of the random patterns. These results were confirmed and substantially extended by Tapiovaara (1990). More recently Blake, Bulthoff and Sheinberg (1993) as well as Liu, Knill and Kersten (1995) have shown that human observers make efficient use of several different sources of information which give cues about the identity and structure of three dimensional objects. This work is still in its early stages, and the analysis is difficult, but I think it is important because it represents a transfer of the optimal processor ideas to problems which have been viewed as more "cognitive" in character.

Aside: Limits to information transmission. In addition to estimating what is going on in the outside world, the nervous system needs to move information from one place to another. From your retina to the first processing station in the thalamus, for example, is a distance of several centimeters, and there is roughly one meter between the touch receptors (or motor nerve endings) in your fingers and and their contacts in the spinal cord. If the electrical properties of the nerve cell membrane consisted just of capacitance and resistance, then, with typical values for these parameters, electrical signals would dissipate over about one millimeter. In fact the current flow across the cell membrane is controlled by ion channels, protein molecules which provide a pore through which ions can pass. These channel molecules can open and close in response to changes in voltage; because the cell maintains a difference in ionic concentration across the membrane, this opening and closing of channels causes currents to flow which can amplify and regenerate a propagating electrical signal. Furthermore, because the coupled dynamics of the channels and the membrane voltage are nonlinear, it turns out that there are unique stable propagating pulses (like solitons) which are called *action potentials* or *spikes*.*** This reliance on discrete pulses is an

*** This picture was worked out by Hodgkin and Huxley, in a beautiful series of experimental papers followed by a phenomenological theory (Hodgkin and Huxley, 1952a–d). It is worth appreciating that they found stable nonlinear pulses by numerical means, using what would now be called a "shooting" method. The molecular foundations for the Hodgkin-Huxley model were solidified by Neher and Sakmann, who were able to record the currents flowing through individual ion channel using their "patch-clamp" technique (Neher, 1992; Sakmann, 1992). These single channel recordings are now a common part of the biologist's technique, so it is easy to forget that noise levels in these recordings have been pushed below a femtoamp in a almost universal feature of the sensory systems, and much of the internal communication among neurons in the brain proceeds in this way as well. But this means that sensory signals are, at a very early stage, *encoded* in these pulse sequences, and this will bc important in the analysis of the fly's visual system discussed below. For now I want to introduce the idea that this encoding places limits on information transmission, an idea originally discussed by MacKay and McCulloch (1952). Imagine that the nervous system observes these pulses in bins or time slices of some duration $\Delta\tau$, which sets the resolution of the system. Because neurons have a minimum "refractory period" separating successive spikes, we can choose this bin size so that we never observe more than one spike per bin. The spikes then form a long binary sequence, or alternatively an Ising chain where spin up (down) represents the presence (absence) of a spike in a given time bin. If there is a deterministic mapping from input signals to output spike trains, then the entropy of this Ising chain is exactly the information, as defined by Shannon (1948), which the spike train provides about the sensory stimulus, and more generally this entropy is an upper bound to the transmitted information. Sensory neurons have information rates which are within a factor of two of this fundamental limit (Rieke, Warland and Bialek, 1993). In the frog auditory system more naturalistic input signals – with spectra shaped to match those of frog calls – are coded even more efficiently, with information rates reaching 90% of the entropic limit (Rieke, Bodnar and Bialek, 1995). There is a long history behind the idea that the early stages of sensory processing must provide an efficient representation of incoming data; certainly these results encourage us to think that some sort of information theoretic optimization principle is at work in the coding strategies of sensory neurons. For recent theoretical work on efficient coding see Atick (1992), and for the development of these ideas at the level of spike trains see DeWeese (1995). For a general overview of neural coding see the forthcoming book by Rieke et al. (1996).

2.2 Optimal estimation is a compromise

It is tempting to think that our estimates about what is happening in the world are based on data collected by an array of sensors. This would seem to be true both in the physics lab and in our own sensory systems. But in both cases this view is misleading.

An example from the laboratory. The simplest signal processing strategy is just averaging or integrating over time. This is used in the lab every day, perhaps so often that we ignore why it works (except when it doesn't). When we try to measure, for example, the optical absorption cross-section of a material, we set up the experiment so that this quantity varies slowly in time if at all. Thus if we observe high frequency fluctuations in the output of our photodetector – which we will, if we look closely – we can be sure that these are just noise and can be

one Hertz bandwidth, and that it is a single molecule whose transitions are being monitored as the current flickers on and off.

averaged away without losing the real signal. Indeed, nobody would ever report that the absorption profile of a rock fluctuates on a microsecond time scale, even though the current at the output of the photodetector fluctuates in this way. Thus what we call real properties of the world involve a compromise between what we measure and what we expect.

An example from language. We have all had the experience of trying to carry on a conversation in a noisy environment. Although you may not have realized this, fairly comfortable conversation is possible even at noise levels where we would fail to discriminate among the basic speech sounds (phonemes) if they were presented in isolation. This is essentially the same phenomenon which we can illustrate in written text: You -an r-ad t-is e-en t-oug- I ha-e de-ete- eve-y fo-rth -ett-r. We say colloquially that we make sense out of the broken text by its "context." I encourage you to think that this is really the same as what happens in the time averaging of the photomultiplier outputs – although many different letters could have been behind the blanks, we know something about what to expect and we use this knowledge to suppress noise. Again our best estimate of what really happened (in this case, what I really meant to type) is a compromise between what we measure and what we know a priori. Better knowledge of what to expect yields better noise suppression and hence more reliable estimates, as the non-native English speakers in the audience will testify.

Toads do it too. Let us return to the experiments of Aho and co-workers, who studied the ability of toads to strike (and hit) dimly illuminated moving targets (Aho et al., 1993). At moderate light levels, the photodetector cells and the underlying retinal circuitry respond fairly quickly. But as the light dims the system adapts to increase its integration time, presumably to achieve better averaging and noise suppression. This adaptation also increases the latency of the retinal response – as the lights get dimmer, the time required for *anything* to leave the retina and be sent to the brain gets noticeably longer. Thus the toad cannot know where the target is at this moment, only where it was some time ago, and this delay gets longer as it gets darker outside. One might expect this delay to show up as part of the behavioral reaction time, and it does, so that the toad strikes where the target used to be, and this spatial displacement between the location of the target and the strike point gets larger as it gets darker. We might say that the toad experiences an illusion that the target is slightly displaced, and acts upon this illusion. Fortunately the typical targets (and those in the experiment) are reasonably large, and it doesn't matter if you hit the leading edge of a worm or the center of its body. But if this effect continued down to the very lowest light levels, where retinal delays exceed one second, it would be a disaster and the toad would miss every time. In the experiments the toad simply does nothing on the first few trials at the lowest light level, as if it were trying to learn something about the typical trajectories of the target, and after this learning period it strikes well ahead of the location signaled by the retinal output. It seems clear that the toad is using its knowledge of the "world" (as created by the experimenter) to correct for systematic distortions introduced by the retinal circuitry. Again the actions of the toad are based on a compromise between the retinal data and the internally stored expectations.

Bayesian formulation. The notion that estimation involves a compromise between what we know and what we observe has a precise mathematical formulation. If our observations do not uniquely determine the state of the world, then there is an element of randomness or ambiguity in the inferences we shall be able to draw. Usually we think about the world being in some definite state, and this state generates data in our instruments which is drawn from some probability distribution which we write schematically as $P(\text{data}|\text{world})$. But the states of the world are themselves drawn from some (presumably quite complex) probability distribution $P(\text{world})$. By the usual rules of probability we can write the joint probability for the occurrence of some state of the world and some set of data as

$$P(\text{world}, \text{data}) = P(\text{data}|\text{world}) \times P(\text{world}) \ . \tag{1}$$

But we can also decompose the joint distribution in another way: The data are chosen from some distribution $P(\text{data})$ and from these data we can infer the probability of the world being in some state, which is given by the conditional distribution $P(\text{world}|\text{data})$, so that

$$P(\text{world}, \text{data}) = P(\text{world}|\text{data}) \times P(\text{data}) \ . \tag{2}$$

These two different factorizations refer to the same joint distribution, so they must be equal:

$$P(\text{data}|\text{world}) \times P(\text{world}) = P(\text{world}|\text{data}) \times P(\text{data}). \tag{3}$$

$$\Rightarrow P(\text{world}|\text{data}) = \frac{1}{P(\text{data})} P(\text{data}|\text{world}) P(\text{world}). \tag{4}$$

This last formula, (4), is called Bayes' rule,† and it shows us how our prior expectations about what we will find in the world, represented by $P(\text{world})$, must be combined with the data to draw inferences about what is really happening. In simple cases, for example, one can use this formulation to show that signals in a background of noise should be filtered to suppress those frequency bands where we expect the signal-to-noise ratio will be low, and so on. Going back to the analogy at the start of this section, $P(\text{data}|\text{world})$ describes the physics of our detectors, while $P(\text{world})$ describes the physics of the world we are observing.

† The mathematics here is so elementary that one wonders why the mention of Bayes rule can still trigger passionate discussion, with partisans displaying a nearly religious zeal. I think the problem lies with the claim that all of our prior expectations about the world can be encapsulated in a probability distribution. Thus when Pauli postulated the existence of the neutrino, he was trying to avoid a breakdown in conservation of energy. But would he have been willing to give a probability that energy is not conserved? I think that in the discussion below the probabilistic description of expectation is not controversial, but we should be careful.

Caution: What do we know about natural ensembles? There is a problem in the background of this discussion, and in some way it may be the most important problem of all. If optimal estimation requires a statistical model of the signals which can occur in the world, it is natural to ask whether we (as physicists, thinking abstractly; let's leave the hardware of neurons out for now ...) can produce such a model. We know that the objects in the world around us obey the laws of physics, but as soon as we get into a reasonably realistic situation we really don't know how to translate this knowledge into statements about the relative likelihood of different configurations or trajectories. In the olfactory system signals are carried by turbulent air flow, and understanding the statistics and dynamics of these turbulent plumes is still an important physics problem (Shraiman and Siggia, 1994); for experiments on the 'interpretation' of these turbulent signals by the nervous system see Mafra-Neto and Cardé (1994) and Atema (1995). In vision, observations on the statistics of natural images reveal a hierarchy of structures on all angular scales (Ruderman and Bialek, 1994), but as in turbulence the low-order correlation functions do not characterize the essential structures. In the auditory system, signals such as speech have long-range, apparently scale-invariant correlations in their amplitude and frequency modulations (Voss and Clarke, 1977), and presumably these structures connect to the long-range correlations in language itself (Shannon, 1951; Ebeling and Poschel, 1994). Although we can write down statistical models which capture some aspects of the structure in each sensory modality, it is clear that in each case we are missing the essential features. It is tempting to think that we are missing just one deep idea which would tie together all of these "natural" structures, and one can even argue that this speculation is supported by some of the similarities in statistical structure found in the different domains.

2.3 Optimal estimation is a statistical mechanics problem

Some of the important ideas in estimation theory can be illustrated by a simple example. A friend picks a number, x, at random from a distribution, $P(x)$. Someone else then adds some "noise," η, where η is Gaussian-distributed with zero mean and variance σ^2. Then you are told the number $y = x + \eta$, and you are supposed to come up with your "best estimate," x_{est}, for x. Let me emphasize that x_{est} is really a function of y, and in the case where x and y are functions of time (or space in the visual system) the estimator is a functional. Thus we need to find, in general, functionals of our input data which optimize some measure of closeness between our estimate and the right answer.

What's optimal? Obviously, the first problem to is to say what we mean by a good estimate. One conventional choice is to measure χ^2, the mean-square deviation between our estimate and the right answer, $\chi^2 = \langle |x_{\text{est}}(y) - x|^2 \rangle$. You might think that if we adopt this measure, then for each estimation problem we have to do the variational calculation of finding the functional $x_{\text{est}}(y)$ which minimzies χ^2. This is not necessary since one can give a general formal solution.

From the discussion of Bayes' rule above, we know that everything we can say about x by observing y is contained in the conditional distribution $P(x|y)$. The estimator which minimizes χ^2 is then the conditional mean,

$$x_{\text{est}}(y) = \langle x \rangle_y = \int dx\, x P(x|y) \ . \tag{5}$$

This is a well known result and I leave the proof as an exercise for the reader.

Constructing $P(x|y)$. To compute the conditional mean we need to build $P(x|y)$. For that, we use Bayes' rule, $P(x|y) = P(y|x)P(x)/P(y)$. In the statement of this model problem, I assume that you know $P(x)$. $P(y|x)$ describes the noise, which we take to be Gaussian, so that

$$P(y|x)\frac{1}{\sqrt{2\pi\sigma^2}} \exp\left[-\frac{(y-x)^2}{2\sigma^2}\right] \ . \tag{6}$$

Finally $P(y)$ serves as a normalization factor which we can obtain in the usual way by integrating over x. I will push all such normalizations into a term $Z(y)$, anticipating the analogy with statistical mechanics. Putting the various factors together, x_{est} is given as

$$\begin{aligned} x_{\text{est}}(y) &= \frac{1}{P(y)}\frac{1}{\sqrt{2\pi\sigma^2}} \int dx\, x \exp\left[-\frac{(y-x)^2}{2\sigma^2} + \ln P(x)\right] \\ &= \frac{1}{Z(y)} \int dx\, x \exp\left[\ln P(x) - \frac{x^2}{2\sigma^2} + x\frac{y}{\sigma^2}\right] \ . \end{aligned} \tag{7}$$

Let us pause for a moment to remind ourselves of the connection with sensory processing. The external data you receive, like y in this example, represents some information about the world, x, but is corrupted by noise, η). Your prior knowledge of the world, $P(x)$, tells you what sorts of features to expect.

The analogy. Equation (7) has a worthwhile analogy with the motion of a Brownian particle in a potential $V(x)$ and an additional constant applied force F. In that case, the distribution of particle positions at temperature T is

$$P(x|F) = \frac{1}{Z(F)} \exp\left[-\frac{V(x)}{k_{\text{B}}T} + x\frac{F}{k_{\text{B}}T}\right] \ . \tag{8}$$

Comparing (7) and (8), we see that the force corresponds to your data y, and the potential to $\ln P(x) - x^2/2\sigma^2$. Forces are measured in units related to the temperature, and data are measured in units related to the noise level, making explicit the intuitive connection between temperature and noise. External data acts like a forcing term on your prior knowledge, at least for Gaussian noise. This analogy tells us that the calculation of the optimal estimator in (7) is exactly the problem of computing the response of our particle to an applied force. When the noise level is high, the temperature in the statistical mechanics problem is high, and we should be able to compute the response in perturbation theory. When the noise level is low the temperature is also low and we should search for minimum energy configurations, or equivalently evaluate the integrals by a saddle point approximation. When we generalize to think about estimating functions of time, these integrals become path integrals, and it may be more useful to think about the analogy to quantum mechanics; low noise is then the semi-classical limit. Finally, statistical mechanics involves more than just computing expectation values. At some level one would like to understand the structure of the whole distribution, and this understanding will allow us to go beyond the choice of χ^2 as a metric for measuring the quality of our estimates.

3 Motion Estimation in Fly Vision

One small part of the fly's visual system provides an excellent testing ground for the ideas of optimal estimation. As a theorist, my interest in flies was sparked by having the office next door to Rob de Ruyter, and this resulted in a theory/experiment collaboration which is now (embarrassingly) a decade old. Personal history aside, I can try to give a rational justification for why flies (or more generally invertebrates, animals without backbones) provide an ideal system for physicists interested in the nervous system. First, the nervous system of a typical invertebrate simply has fewer neurons than found in a mammal or even a 'lower vertebrate' such as a fish or frog. The fly's visual brain has roughly 5×10^5 cells, while just the primary visual cortex of a monkey has $\sim 10^9$. In addition, many of the cells in the invertebrate nervous system are *identified.* This means that cells of essentially the same structure occur in every individual, and that if one records the response of these cells to sensory stimuli (for example) these responses are quantitatively reproducible from individual to individual. Thus the cells can be named and numbered based on their structure or function in the neural circuit, and there is now a serious attempt to collect these data into an on-line atlas of the fly's brain (Armstrong et al., 1995). Finally, the overall physiology of invertebrates allows for very long stable recordings of the electrical activity of their neurons. I have had the pleasure of collaborating in experiments on the fly visual system in which we studied the responses of a particular cell, H1 (more about this below), and Rob routinely records from this one cell for periods of up to five or six days, with occasional pauses for feeding the fly. In short, experiments on invertebrate nervous system look and feel like the physics experiments with which we all grew up – we have nice stable 'samples' with quantitatively reproducible behavior. In his undergraduate lectures on physics Feynman discussed the relations among the sciences, and pointed out that while psychology aims at understanding humans, it would be impressive if we understood dogs (Feynman et al., 1963). I hope to convince you that flies aren't such a bad place to start.

3.1 Why do we think that flies estimate angular velocity?

I have already cautioned you to be suspicious when someone tells you that brains perform a particular computation. I will now try to overcome your suspicions and argue that the fly's brain estimates the fly's angular velocity relative to the world, $\dot{\theta}(t)$, using the data provided by the array of receptor cells in the compound eye.

Behavior. If you watch a fly flying around in a room or outdoors, you will notice that flight paths tend to consist of rather straight segments interrupted by sharp turns. These observations can be quantified, both through the measurement of free-flight trajectories (Land and Collett, 1974; Wagner, 1986) and in experiments where the fly is suspended from a torsion balance (Heisenberg and Wolf, 1984). Given the aerodynamics for an object of the fly's dimensions, even flying straight is tricky. In the torsion balance one can demonstrate directly that motion across the visual field drives the generation of torque, and the sign is such as to stabilize flight against rigid body rotation of the fly. Indeed one can close the sensory-motor feedback loop by measuring the torque which the fly produces and using this torque to (counter-)rotate the visual stimulus, creating an imperfect 'flight simulator' for the fly in which the only cues to guide the flight are visual; under natural conditions the fly's mechanical sensors play a crucial role. Despite the imperfections of the flight simulator, the tethered fly will fixate small objects, thereby stabilizing the appearance of straight flight toward the object. This sort of fixation and tracking is related to the more natural chasing behavior which ultimately leads to mating (Land and Collet, 1974). The combination of free-flight and torsion balance (or 'flight simulator') experiments strongly suggests that flies can estimate their angular velocity from visual input alone, and then produce motor outputs based on this estimate (Reichardt and Poggio, 1976).

Physiology and anatomy. Flies have compound eyes. In large flies (like the blowfly *Calliphora* which I discuss below) there are about 5,000 lenses in each eye, with approximately 1 receptor cell behind each lens.[‡] The lens focuses light on the receptor, which is small enough to act as an optical waveguide. Each receptor sees only a small portion of the world, just as in our eyes. One difference is that

[‡] This is the sort of sloppy physics-speak which annoys biologists. The precise statement is different in different insects. For flies there are actually eight receptors behind each lens. Two provide sensitivity to polarization and some color vision, but these are not used for motion sensing. The other six receptors look out through the same lens in different directions, but as one moves to neighboring lenses one finds that there is one cell under each of six neighboring lenses which looks in the same direction. Thus these six cells are equivalent to one cell with six times larger photon capture cross-section, and the signals from these cells are collected and summed in the first processing stage (the lamina). One can even see the expected six-fold improvement in signal-to-noise ratio (de Ruyter van Steveninck and Laughlin, 1995).

diffraction is much more significant for organisms with compound eyes – because the lenses are so small, flies have an angular resolution of about 1°, while we do about 100× better. There is a beautiful literature on optimization principles for the design of the compound eye; the topic even makes an appearance in the Feynman lectures (Feynman et al. , 1963). Voltage signals from the receptor cells are processed by several layers of the brain, each layer having cells organized on a lattice which parallels the lattice of lenses visible from the outside of the fly. After passing through the lamina, the medulla, and the lobula, signals arrive at the lobula plate. Here there is a stack of about 50 cells which are are sensitive to motion (Hausen, 1984). The cells are all specialized to detect different kinds of motion, and they are identified in the sense mentioned above. If one kills individual cells in the lobula plate then the simple experiment of moving a stimulus and recording the flight torque no longer works (Hausen and Werhahn, 1983), strongly suggesting that these cells are an obligatory link in the pathway from the retina to the flight motor. If one lets the fly watch a randomly moving pattern, then by filtering the action potentials from a motion sensitive cell it is actually possible to reconstruct the time-dependent angular velocity signal (Bialek et al., 1991). Taken together, these observations support a picture in which the fly's brain estimates angular velocity and encodes its estimate in the activity of these few neurons.

Aside: Learning vs. hard-wiring .... At the beginning of this section I sung the praises of flies as a 'brain of choice' for physicists. Now for the bad news. Perhaps invertebrates don't do the things that we find so interesting about brains. Thus these 'simpler' nervous systems are often described as 'hard-wired,' meaning that the patterns of connectivity and the resulting computations are determined by genetically specified rules rather than through learning and experience as in the mammalian brain. This impression of pre-determined circuitry also leads to the belief that insects, for example, respond to sensory stimuli with a set of simple reflexes; the fly's motor response to movement across the visual field has been held up as a classic example of this reflex view. One might therefore argue that experiments on the insect brain will, at best, reveal these rules, and this does not take us very far toward our goal of understanding brains in general. While I believe (like everyone else) that our brains do some things which insects cannot, the belief that insects are hard-wired automata turns out to be wrong. In a beautiful series of papers, Heisenberg and co-workers have revisited the flight simulation experiments on flies and shown that the fly's behavior in this setting exhibits a great deal of plasticity (Heisenberg and Wolf, 1988; Wolf and Heisenberg, 1990). To begin, the torque generated in response to the same trajectory of motion across the visual field is different if the feedback from torque to motion is open or closed – it is as if the fly 'knows' whether its actions are having an effect on what it sees, and if there is no effect then the motor responses to sensory stimuli are much reduced. Furthermore, the fly can adjust its behavior in responses to changes in the feedback within a very short time, perhaps a few tens of wing beats. In related experiments on locusts, Möhl

(1988, 1989) used the activity of particular pairs of motor neurons to control the motion of images across the visual field, and he found that the pattern of neural activity becomes sensitive to precisely which neurons are in the feedback loop. In particular, neurons which are out of the loop drift and show none of the synchronization between the left and right wings which is required for stable flight, and again this plasticity of behavior can be triggered very rapidly. Finally, the flight simulator apparatus has been used to explore the fly's memory for visual patterns (Dill, Heisenberg, and Wolf, 1993) and these experiments led to the discovery that flies have a spontaneous preference for 'looking at' novel patterns (Dill and Heisenberg, 1995). These results (and I have not even touched the question of learning and memory in bees ...) are a bit off the main point I am trying to make in these lectures, but they are fascinating experiments. I suspect that we are far from reaching the limits of what insect brains can do, and that the boundaries between what they do and what we do are less clear than one might have liked.

3.2 How accurately do they do it?

To introduce the idea of optimal processing for motion estimation, I have to convince you that estimation in real flies is, after all this, optimal or at least close enough to motivate the discussion. This is not a trivial problem, in part because the output of the motion computation is encoded in sequences of action potentials in the motion sensitive neurons. We will have to understand enough about the structure of this code that we don't confuse a complex code for an error or random noise in the computation. But first we need to get an order of magnitude feel for the problem.

Hyperacuity and the physical limits. Each receptor in the fly's eye sees a region of about $\phi_0 \sim 1.3°$ in arc, a basic angular scale which is set by diffraction. Alternatively, images are blurred by this amount as they pass through the array of lenses. If the fly is looking at a scene where the mean photon counting rate is R and the typical variations from point to point are given by $\Delta R \sim CR$ (which defines the contrast C), then when the fly rotates by a very small angle $\delta\theta$ the change in photon count rate is typically $\delta R \sim (\Delta R)\delta\theta/\phi_0$. If we are allowed to average for a time τ, then the mean count in each cell is $\bar{N} = R\tau$, the variance of the counts is equal to the mean, and the change in the mean count is $\delta N = \delta R\tau$. Putting the factors together, we find that the signal-to-noise ratio for detecting the displacement $\delta\theta$ is just

$$\mathrm{SNR} \sim (\delta N)^2/\bar{N} = (\delta\theta)^2 \frac{C^2 R\tau}{\phi_0^2} \ . \tag{9}$$

The SNR should improve by a factor of N_{cells} if we can average over this many receptor cells, which means that to reach a signal-to-noise ratio of unity we need to see a displacement of

$$\delta\theta \sim \phi_0 \left[N_{\mathrm{cells}} C^2 R\tau\right]^{-1/2} \ . \tag{10}$$

This is basically the right answer, up to the usual (and sometimes dangerous) constant factors of order unity; a detailed account of direct relevance to the discrimination experiment discussed below is given an earlier set of lecture notes (Bialek, 1992). Typical values for the experiments on H1 are $R \sim 10^4\,\mathrm{s}^{-1}$, $N_{\mathrm{cells}} \sim 10^3$, and $C \sim 0.1$, while we know from behavioral observations that relevant integration times are $\tau \sim 30\,\mathrm{ms}$. Plugging in the numbers, it should be possible (at the optimum) to detect a motion of $\delta\theta \sim 0.1°$. Now the spacing between the lenses is also about 1°, which means that if the fly processes signals optimally, it should be able to discriminate motions about 10 times smaller than the elementary spacing of the retinal lattice. *We* can do this in a variety of situations, which are collectively termed *hyperacuity* (Westheimer, 1981). Can the fly do it?

A discrimination task. One way to characterize the precision of neural computation is in a discrimination experiment (de Ruyter van Steveninck and Bialek, 1995). In these experiments, the fly is immobilized and positioned to face an oscilloscope screen. An electrode is inserted in the back of the head and positioned so that it senses the output of H1. The fly is then shown "movies" on the screen, and the action potentials or spikes produced by H1 are recorded. At some moment in time the pattern on the screen is displaced by an angle θ_0 or θ_1, and we ask if by looking at the spikes produced immediately after this motion we (as observers) can decide if the step was of size θ_0 or θ_1. Again, from the behavioral data we know that the time window for making this decision should be $\sim$ 30 ms, during which time H1 can produce at most a few spikes. This means that the decision rule will come down to measuring the arrival times of these few events. When the dust settles, it turns out that one can make 75% correct discriminations between motions which differ by $|\theta_0 - \theta_1| = \delta\theta = 0.12°$. This is in excellent agreement with our rough estimate for the physical limit. A more careful analysis of the dependence of discrimination performance on integration time shows that the agreement between measured and optimal performance is best for smaller time windows, and in these small windows there is typically only one spike. Thus the fly's visual system can extract an estimate of motion down to the limits imposed by diffraction and photon shot noise, and then encode this estimate in the timing of just one spike.

Continuous estimation. The discrimination experiment gives us a very detailed view of the fly's motion computation, but in a very limited (and admittedly somewhat unnatural) setting. As an alternative, we can let the fly look at a pattern which moves randomly along some continuous angular trajectory $\theta(t)$. If we believe that the fly's brain is estimating this trajectory (or more precisely the angular velocity $\dot{\theta}(t)$) then we can try to extract this estimate from the sequence of spikes produced by H1. This amounts to *decoding* the cell's response. In principle this could be a very complex task, and it might not work at all – the fly could perhaps control its flight only by knowing about some limited features of the velocity waveform, and hence the motion sensitive neurons might not contain enough information to reconstruct the velocity waveform itself. In fact it turns out that one can decode the output of H1, and that the decoding algorithm can be very simple (Bialek et al., 1991): We construct an estimate of the angular velocity which is a linearly filtered version of the spike train itself,

$$\dot{\theta}_{\text{est}}(t) = \sum_i F(t - t_i) \ , \tag{11}$$

where t_i are the arrival times of the spikes. Initially, F is some unknown function, but it can be optimized so that our reading of the code is as accurate as possible, minimizing the mean-square error

$$\chi^2 = \langle |\dot{\theta}(t) - \dot{\theta}_{\text{est}}(t)|^2 \rangle \ . \tag{12}$$

In a typical experiment, some fraction of the data is used to construct F, and once we have settled on this "understanding" of the code we use the rest of the experiment to measure the quality of the reconstructions. Notice that errors could arise simply because we have failed to understand the code, but if our reconstructions match the real signal then this must reflect the accuracy of the fly's own computation. The errors in the reconstruction are themselves a function of time, and we quantify this 'noise' by its power spectrum. Over a broad range of frequencies one finds that the effective angular displacement noise $N_\theta \sim 10^{-4}$ deg^2/Hz. The physical limit is $N_\theta^{\text{opt}} \sim (\delta\theta)^2\tau$, with the parameters from above, and we see again that the fly is very close to optimal performance. Theory predicts that if we increase the contrast of the patterns on the screen one should increase the accuracy of the estimates, and this is observed. Although it would be nice to check on the range of conditions over which optimality is maintained, the agreement between the discrimination and the reconstruction experiments strongly suggests that the fly is very close to being the optimal motion processor, limited by the physics of the visual input, and not by the hardware of its tiny brain. This raises a sharp theoretical question: What is the structure of this optimal estimator?

3.3 Finding the optimal estimator

We now turn to the problem of finding the optimal motion estimator, in a sense the estimator which the fly would "like" to build if it could. It is important to realize that our "model" is a model of the problem the fly is solving, not a model of the neural circuitry which implements the solution. Thus we need not concern ourselves with the details of fly neuroanatomy; what is important is that we capture the essential features of the fly's environment which make the problem of motion estimation non-trivial. The discussion here follows the work of my student M. Potters and is very schematic; a more detailed account can be found in Potters and Bialek (1994).

Formulation and prior knowledge. For simplicity, let's look at a one dimensional version of the problem in which the visual world is parameterized by a single azimuthal angle ϕ. The "real world," as seen in fly-centered coordinates, is represented by a contrast pattern $C(\phi - \theta(t), t)$, where θ is the orientation of the fly, and the extra t dependence arises because the scene may change over time. The fly's retina turns these contrast patterns into a set, $\{V_n\}$, of photoreceptor voltages, and this transformation involves spatial averaging (because of diffraction, as discussed above), temporal filtering (which we ignore), and the addition of noise (mostly photon shot noise). Each of these aspects of the receptor response has been quantified experimentally, so when we write down a model relating the $\{V_n(t)\}$ to $C(\phi, t)$ we are just summarizing the experimental facts.[§] The essential observation is that the response is linear over the range of signals used in these experiments, so we write

$$V_n(t) = \eta_n(t) + \int dx\, C(x - \theta(t), t) f(x - x_n) = \eta_n(t) + \bar{V}_n(t) \ , \tag{13}$$

where η is the noise, and $f(x - x_n)$ describes the region of the visual field seen by the nth receptor. If the noise is dominated by shot noise from photons striking each photoreceptor at an average rate R, then it is spectrally white and, unless we look at very short times (comparable to $1/R$), the noise is also Gaussian, so that

$$P\left[\{V\}|C, \theta\right] \propto \prod_n \exp\left[-\frac{R}{2}\int dt\, |V_n(t) - \bar{V}_n(t)|^2\right] \ . \tag{14}$$

We would like to turn these equations around and estimate the trajectory $\theta(t)$ from the voltages $\{V\}$. One crucial point is that the fly has no independent knowledge of the pattern dynamics $C(\phi, t)$ – *all* she has to work with are the receptor voltages. In principle this leads to massive ambiguity, even in the absence of noise: There are an infinite number of ways of interpreting the data in terms of some combination of motion and intrinsic dynamics. Similarly, if we visualize the flow of an incompressible fluid, the fundamental quantity is the vector velocity at each point, but visualization gives us only scalar data (concentration of a tracer, for example). The only principled way to resolve these ambiguities (especially in the presence of noise) is the Bayesian formulation given above. In the present case, we imagine that the distribution of possible patterns $C(\phi, t)$ is known, and then we can proceed through the algebra to find

$$P[\theta|\{V\}] = \frac{P[\{V\}|\theta]P[\theta]}{P[\{V\}]} = \frac{\int \mathcal{D}C P[\{V\}|C, \theta]P[C]P[\theta]}{P[\{V\}]} \tag{15}$$

where as usual $\int \mathcal{D}z$ denotes the integral over all possible functions z. Then our best estimate $\dot{\theta}_{\rm est}(t)$ for the angular velocity becomes

$$\dot{\theta}_{\rm est} = \int \mathcal{D}\theta\, \dot{\theta} P[\theta|\{V\}] \ . \tag{16}$$

§ As an example, see the measurements of noise levels by de Ruyter van Steveninck and Laughlin (1995)].

All of the remaining work is in evaluating these functional integrals. Guided by our previous example, we know that at low signal-to-noise we should use perturbation theory, and that at high SNR we should do a saddle point evaluation.

Low SNR: Perturbation Theory. A perturbation calculation of (16) to lowest order yields the estimator

$$\dot{\theta}_{\mathrm{est}}(t) = \sum_{n,m} \int d\tau\, d\tau'\, V_n(t-\tau) V_m(t-\tau') K_{mn}(\tau,\tau') \ . \tag{17}$$

Note that this is second-order in the Vs; the first order term drops out because reversing the contrast cannot change the optimal estimate of the velocity. The functions $K_{mn}(\tau,\tau')$ embody the compromise between suppressing noise (which encourages longer averaging) and time resolution. Therefore the Ks depend on the noise, the contrast, and on the typical trajectories $P[\theta]$. Note that we are computing a *correlation* between the parts of the visual field at different times; this agrees with a previously-proposed model of motion estimation. This analysis shows that such a model is optimal in the low SNR limit, and gives a systematic way of choosing the kernels K_{mn}.

High SNR: Saddle Point Evaluation. One has to make certain assumptions about typical trajectories ($P[\theta]$) and contrasts ($P[C]$) in order to compute the functional integrals. For a class of choices, the optimal estimator has the general structure

$$\dot{\theta}_{\mathrm{est}} \sim \frac{\sum_n \dot{V}_n (V_n - V_{n+1})}{\# + \sum_n (V_n - V_{n+1})^2} \ . \tag{18}$$

Here # is some number which depends on the SNR. In the case that the V_ns are nearly noiseless and that the whole field translates uniformly, this is something like

$$\dot{\theta}_{\mathrm{est}} \sim \frac{\partial_t V}{\partial_x V} \ . \tag{19}$$

Equation (19) is the first thing you might think of in trying to compute the velocity, and it works fine if there is essentially no noise. For "real" data, the original version in (18) is much more robust.

3.4 Comparison with experiment

This analysis shows that thc statistics of the images determines the best algorithm for estimating motion changes. If the fly is estimating motion optimally, then it must be changing its computation as its environment changes. One can look for adaptation in the fly's visual system. A typical experiment consists of showing movies to the fly. First there is a fairly long "adaptation movie," in which you control the statistics (e.g. the contrast, light level, etc.) and hope that the fly adapts to these types of scenes. Then, after a brief rest (at most a few seconds) you show a short "test movie" and record the response of H1 to the motion of the scene. Different adaptation movies are followed by a common test movie; if the responses to the test differ, then the fly's reactions must be influenced by its memory of the adaptation movie.

Correlation. Consider the case when the test movie consists of a pattern moving at a fixed velocity $\dot{\theta}$. Suppose we adjust the contrast C of both the adaptation and test movies, and leave all other parameters fixed. We decode the output of H1 to measure the fly's estimate $\dot{\theta}_{\rm est}$ of $\dot{\theta}$. At low test contrasts, $\dot{\theta}_{\rm est}$ rises quadratically, in accord with the correlator method of motion detection (17). As the test contrast rises, $\dot{\theta}_{\rm est}$ rises to the actual value $\dot{\theta}$ and saturates, as predicted by (18). Furthermore, the estimate depends on the contrast of the adaptation movie, indicating that the fly really does learn from its recent past history and uses that information in making its estimates.

Getting the "right answer." Other experiments which probe the time-response of motion estimation show that adaptation movies with rapid motion encourage the fly to make rapid (if noisier) estimates of motion, while adapting to slower-moving scenes teaches the fly to respond more leisurely (and more accurately).

Dynamics. These experiments show that the fly, rather than being "hard--wired", really does adapt to its environment and uses what it knows to adjust its computations.

4 Ambiguous Stimuli

In the problems discussed so far, "estimation" has been a well defined problem because there is some unique interpretation of the data which provides the best compromise between our measurements and our expectations. It is possible that this is not the case, and hence that there are multiple interpretations of the same sensory experience. From the statistical mechanics point of view there will be a symmetry or degeneracy, and this might be spontaneously broken or not. These situations are surprisingly easy to generate just by making drawings on a sheet of paper, and most of you have seen the Necker cube or the vases/faces examples.[¶] Much of the literature on these perceptual phenomena is concerned with very "high level" matters, such as shifts in our attention and the problem of visual awareness. But I think that it is profitable to go back and ask a simple question, namely whether we understand what a completely objective, optimal processor would do with such ambiguous sensory data. This is what M. DeWeese and I did recently, and the results were at least a little surprising (Bialek and DeWeese, 1995).

[¶] For a brief historical review of the different ambiguous figures see Fisher (1968).

4.1 Introduction to the phenomena

When we view ambiguous figures, such as the Necker cube, we have the subjective impression that our perception switches at random between two equally plausible interpretations of the input data. This impression of randomness has been quantified in psychophysical experiments (Borsellino et al., 1972). Ambiguities are not restricted to static figures, and one can give multiple interpretations to pairs of moving dots as well as to non-visual stimuli such as multi-tone complexes which can have ambiguous pitch; both examples will be discussed below. If you view different images with each eye, you will see "binocular rivalry" in which you are aware first of one eye's view, then this view dissolves into the other eye's view, and so on seemingly at random. Although each example has some special features, it seems reasonable to suppose that there is a general class of problems where the sense data are truly ambiguous, providing equal support (even in the rigorous Bayesian sense) for each of two estimates of what is really happening in the world. Under these conditions it would seem that our perceptions flip at random, despite the fact that the input signals are static and apparently of rather high signal-to-noise ratio – you have no trouble seeing the lines which make up the Necker cube, or discriminating the tones which make up the ambiguous pitch. We would like to understand the origin of the random switching, as well as identify the parameters which set the time scales for the switching events.

4.2 Bayesian formulation and the random field model

We consider for simplicity that the interpretation of an ambiguous stimulus can be reduced to the problem of estimating a single variable or feature which may vary in time, $f(t)$. To arrive at this estimate the brain makes use of some sense data which we can collect into an array $\mathbf{d}(t)$. Given these data, what can the brain (or any machine, for that matter) conclude about $f(t)$? Since the data are noisy, all one can state is the relative likelihood that the data were generated by different features, which is the conditional probability of $f(t)$ given $\mathbf{d}(t)$. Bayes' theorem tells us that this probability can be written as

$$P[f(t)|\mathbf{d}(t)] = \frac{1}{P[\mathbf{d}(t)]} \cdot P[\mathbf{d}(t)|f(t)]P[f(t)] \ , \tag{20}$$

where $P[f(t)]$ is the a priori probability of the time variation $f(t)$. This prior distribution embodies the observer's knowledge that rapid variations in the feature f are unlikely in the natural world or in a given experimental setup. The distribution of the data averaged over all trajectories of the feature $f(t)$, $P[\mathbf{d}(t)]$, serves just as a normalization factor.

Setting up the distributions. If we look at one instant of time, we can define some notion of "goodness of fit" between the data $\mathbf{d}$ and some possible value of the feature f. We will call this goodness of fit $\chi^2[\mathbf{d}(t); f(t)]$, since under sufficiently strong assumptions the conventional χ^2 statistic is the relevant measure.

For simplicity we assume that the fluctuations in the data are effectively white noise, so that we can write

$$P[\mathbf{d}(t)|f(t)] \propto \exp\left(-\frac{1}{2N}\int dt \chi^2[\mathbf{d}(t); f(t)]\right) , \qquad (21)$$

where N is the noise level. The fact that we are viewing an ambiguous figure means that there are two distinct values of f, f_+ and f_-, separated by a difference $|f_+ - f_-| = \Delta f$, which minimize χ^2. In particular if we ignore the noise in $\mathbf{d}$ and set $\mathbf{d}(t) = \bar{\mathbf{d}}$, then $\chi^2[\bar{\mathbf{d}}; f(t)]$ has two degenerate minima so that the two alternative interpretations of the data are equally likely. If we try to change f continuously from one stable interpretation to the other, we must surmount a barrier in χ^2, and let's call the height of this barrier as χ^2_{max}. To summarize our knowledge that features vary slowly, we assume that the time derivative of f is chosen independently at each instant of time from a Gaussian distribution. This means that our a priori distribution corresponds to a random walk of the feature, with effective diffusion constant D,

$$P[f(t)] \propto \exp\left(-\frac{1}{4D}\int dt \dot{f}^2(t)\right) . \qquad (22)$$

Putting these terms together and assuming that the noise $\delta\mathbf{d}(t)$ in the data is small, we have

$$\begin{aligned} P[f(t)|\mathbf{d}(t)] \;\propto\; & \exp\left(-\frac{1}{4D}\int dt \dot{f}^2(t) - \frac{1}{2N}\int dt \chi^2[\bar{\mathbf{d}}(t); f(t)]\right) \\ & \times \exp\left(-\frac{1}{2N}\int dt \left.\frac{\partial \chi^2[\mathbf{d}(t); f(t)]}{\partial \mathbf{d}(t)}\right|_{\mathbf{d}(t)=\bar{\mathbf{d}}} \cdot \delta\mathbf{d}(t)\right) . \qquad (23) \end{aligned}$$

Mapping to the Ising model. If we leave aside (for the moment) the term involving the noise $\delta\mathbf{d}$, the probability distribution for $f(t)$ is exactly the (imaginary time) path integral for a quantum mechanical particle moving in a double potential well. The most likely trajectory $f(t)$ is a constant value which sits at one or the other minimum of χ^2. In the estimation problem this means that the most likely estimate of the feature f is one which is constant at one or the other of the two possible stable interpretations of the ambiguous figure. But from the quantum-mechanical analogy we know that there are also instanton trajectories which "tunnel" from one interpretation to the other (Coleman, 1985). These switching events occur in a small time τ_0, and the mean time between switching events is $\tau_{\mathrm{switch}} \sim \tau_0 \exp(S_0)$, where

$$\tau_0 \sim |\Delta f| \sqrt{\frac{N}{2D\chi^2_{\mathrm{max}}}} \qquad S_0 \sim \frac{|\Delta f|}{2}\sqrt{\frac{\chi^2_{\mathrm{max}}}{2DN}} . \qquad (24)$$

Thus if we ignore dynamics on times faster than τ_0, the particle can be in only one of two states, which we identify with the states ± 1 of an Ising spin σ_n; the index n counts time in bins of size τ_0. The fact that flips are rare means that spins in adjacent time bins tend to be parallel, and indeed the ferromagnetic coupling between neighbors is just S_0.

Adding back the noise. Small white noise fluctuations in the data can favor either stable interpretation, producing an equivalent magnetic field h_n which is an independent random variable in each bin. We emphasize that our estimate of the feature $f(t)$ is now approximated as the sequence of spins, while the random field arises from a particular instance of noise in the input data; thus the random field must be viewed as quenched disorder. The probability for a configuration of spins $\{\sigma_n\}$ is then given by the one dimensional random field Ising model,

$$P[\{\sigma_n\}] \propto \exp\left[S_0 \sum_n \sigma_n \sigma_{n+1} + \sum_n h_n \sigma_n\right] , \tag{25}$$

where the random field has a variance

$$\langle h_n^2 \rangle = \frac{\tau_0}{4N} \left(\left. \frac{\partial \Delta\chi^2[\mathbf{d}(t); f(t)]}{\partial \mathbf{d}(t)} \right|_{\mathbf{d}(t)=\bar{\mathbf{d}}} \right)^2 , \tag{26}$$

$$\Delta\chi^2[\mathbf{d}(t); f(t)] = \chi^2[\mathbf{d}(t); f(t)]\Big|_{f(t)=f+} - \chi^2[\mathbf{d}(t); f(t)]\Big|_{f(t)=f-} . \tag{27}$$

Qualitative solution. The long-distance behavior of the random field Ising model was found by Imry and Ma (1975). Their argument (which was controversial but ultimately shown to be correct) tells us that at low noise level (so that both S_0 and $\langle h_n^2 \rangle$ are large) the configuration of spins breaks into domains of spin up and spin down, with the typical domain size

$$\xi = S_0^2 / \langle h_n^2 \rangle . \tag{28}$$

In the present context this means that estimates of the stimulus parameters will flip between the two stable configurations with a typical switching time $\tilde{\tau}_{\text{switch}} \sim \xi \tau_0$ rather than τ_{switch} from above. Putting the various factors together we find

$$\tilde{\tau}_{\text{switch}} \sim \frac{2|\Delta f|^2}{D} \chi^2_{\max} \left(\left. \frac{\partial \Delta\chi^2[\mathbf{d}(t); f(t)]}{\partial \mathbf{d}(t)} \right|_{\mathbf{d}(t)=\bar{\mathbf{d}}} \right)^{-2} . \tag{29}$$

At low noise levels, the optimal interpretation of ambiguous incoming data thus switches randomly at a rate *independent* of the noise level. Indeed, the switching rate is proportional to the a priori expected drift rate between the two ambiguous interpretations, $|\Delta f|^2/D$, but is suppressed in (linear) proportion to the χ^2 "barrier" between these interpretations. This is very different from the prediction of network models with internal noise, where we expect that the jump from one stable state to the other is some sort of Kramers' problem (Kramers, 1940), so that the switching rate must depend exponentially on the noise level.

4.3 Implications, comparison with experiment, predictions

The idea that, at low noise levels, the switching rate becomes independent of the noise level seems intriguing. As promised, it is at least a little surprising in the context of signal processing. In statistical mechanics, however, we know that seemingly robust collective ordering can in fact be unstable to the introduction of arbitrarily small amounts of disorder, and the essence of the argument is that optimal estimation with ambiguous data is an example of this effect. Now it would be nice to make contact with some data ....

Ambiguous pitch. When we hear a harmonic sequence, e.g., 1000 Hz, 1200 Hz, 1400 Hz, we assign a pitch equal to the fundamental even if it is not present in the physical signal (de Boer, 1976). This search for the missing fundamental continues if the signals are slightly anharmonic, as in a sequence $f_n = nf_1 + \delta$, and the perceived pitches can be predicted as those f which minimize

$$\chi^2 = \sum_{\mu}(f_\mu - n_\mu f)^2 \ . \tag{30}$$

Let me emphasize that this formulation is an example of Goldstein's optimal processor view (Goldstein, 1973). Note that there are multiple minima corresponding to different assignments of the integers $\{n_\mu\}$. When the signals are maximally anharmonic, $\delta = f_1/2$ and there are two near degenerate minima of χ^2; if we choose the set $\{n_\mu\}$ which minimizes χ^2 at fixed f we find that the resulting $\chi^2(f)$ has the standard double-well form. Human observers hear both pitches, and the percept switches at random, as for ambiguous figures in vision. In the context of our analysis above, the parameter we are trying to estimate is the pitch f, and the data are the representations of the individual components f_μ in the sensory nerves. In naturally occurring sounds, such as speech, frequency modulations of this magnitude occur on times scales of several tens of msec, so we human listeners probably have an a priori distribution for pitch fluctuations with $|\Delta f|^2/2D \sim 0.01$–$0.1$ sec. The remaining factor in Eq. (9) is dimensionless and determined entirely by the choice of integers $\{n_\mu\}$, and for typical experimental parameters we predict $\tau_{\text{switch}} \sim 1.2$–$12$ sec, in reasonable agreement with experiment.

Expectation. The predicted switching times depend on the parameters of the stimulus *and* on the a priori assumptions of the observer. Thus individual differences in mean switching time could be substantial, although the distribution of switching times should be more reproducible, in agreement with experiment (Borsellino et al., 1972). If the human brain adjusts its "prior" expectations in response to recent sensory experience in the manner discussed for the fly, it should be possible to manipulate the observable switching times by having the observer listen to sounds with different statistical properties. Certainly human observers adjust their expectations in relation to the instructions given at the start of the experiment, so that instructing the observer to expect changing signals should increase the reversal rate, as observed (Girgus et al., 1977; Rock and Mitchener, 1992).

Arrays of figures. In vision one can have not just one ambiguous figure but a whole array, in one or two spatial dimensions. Ramachandran and Anstis (1985) have studied arrays of alternating dots which can be seen as moving either horizontally or vertically. In this case the feature $f(t)$ is the orientation of the local motion vector, and it is clear that a reasonable a priori distribution will include not only the penalties for temporal variation discussed above but also a term which penalizes spatial gradients of the velocity. The problem of making optimal estimates for the array then maps onto a two or three dimensional random field Ising model. At zero noise of course both models have a true phase transition to ferromagnetic order, and this ordering survives the random field perturbation in the 3D case. A ferromagnetic phase corresponds to perceiving all of the dot pairs moving in the same direction, or a coherent motion percept. For two-dimensional arrays, then, it is clear that one should observe coherent motion, and switching between the two possible directions of motion should be very slow, infinitely slow in the limit of large arrays. This is in excellent agreement with experiment. In an array of Necker cubes there is no obvious a priori distribution which would couple the relative depth parameters in neighboring cubes, so we expect a paramagnetic phase where different cubes switch at random, as observed (Long and Toppino, 1981).

New tests? A coherent percept across the entire 2D array is predicted to occur even when the prior distribution has only local terms and hence the optimal processor can be constructed entirely from local operations. True "magnetization" in the 1D array would require long range interactions, but it may be difficult to distinguish this from the marginal case. Controlled manipulation of the noise level, perhaps by jittering the dots which give rise to ambiguous motion, might make it possible to test the prediction that the transition to coherent motion is indeed a phase transition. One can even fantasize about studying the scaling behavior of our perceptions in the neighborhood of such a transition, but this may be asking too much of the experiments.

4.4 What have we learned?

Optimal estimation of a *potentially* time-varying feature leads, inevitably, to random perceptual switching in response to ambiguous signals. This feature of our perceptions may thus be a feature of the problem we are solving – and solving well! – rather than an artifact of our neural mechanisms. The predicted switching rate is independent of the (small) noise level in the sense data, but does depend on the observer's a priori hypotheses. If the observer can assume that features are truly static ($D \to 0$), then the predicted switching rate vanishes, in agreement with experiments on carefully instructed human observers. It is attractive that this combination of randomness and apparent subjectivity emerges from an objective theory of optimal estimation.

5 Conclusions, Directions, Speculations

What we admire about the brain, and biology more generally, is that it works. Evolution has selected mechanisms at the level of molecule, cell and system which solve real problems. One might have thought that these problems are solved in some sort of 'quick and dirty' compromise, doing the best one can with limited hardware, but time and again we see that the nervous system functions at the limits of what the laws of physics allow. In fact the molecular biologists have taught us that the hardware even in a single neuron is not so limited, with an enormous repetoire of channels, receptors and messengers available, all regulated by complex networks of biochemistry. The deep question is then how the cell, and ultimately the system, chooses among these myriad possibilities. We have seen that to solve the problems of sensory signal processing, the computations carried out by the nervous system must embody a statistical model of the signals in the outside world, and it is very pleasing that we can begin to 'read out' the parameters of this model in the experiments on H1. There is also the hint that this notion of model building provides a bridge from the very physical problems of photon counting to the more 'cognitive' problems of perception. But I have been completely silent on the question of how the system builds these models. Perhaps more seriously, I have talked about model building for the sake of solving signal processing problems, but I haven't said why the organism chooses these problems over all others. Clearly we are missing some organizing principles here.

6 Acknowledgements

It was great fun to give these lectures and I thank the organizers for their efforts and the students for their patience. Tim Holy acted as 'scientific secretary' and was a great help in producing a (hopefully) comprehensible version of these ideas; all obscurities which remain merely reflect those in the original lectures. Finally, my thinking on these matters has evolved largely through my enjoyable collaborations with M. DeWeese, M. Potters, F. Rieke, R. de Ruyter van Steveninck, and D. Warland, and it is a pleasure to thank them here.

References

Aho, A.-C., Donner, K., Helenius, S., Larsen, L. O., Reuter, T. (1993). Visual performance of the toad (*Bufo bufo*) at low light levels: Retinal ganglion cell responses and prey-catching accuracy, *J. Comp. Physiol. A* **172,** 671-682.

Aho, A.-C., Donner, K., Hydén, C., Larsen, L. O., Reuter, T. (1988). Low retinal noise in animals with low body temperature allows high visual sensitivity, *Nature* **334,** 348-350.

Armstrong, J. D., Kaiser, K., Muller, A., Fishbach, K.-F., Merchant, N., Strausfeld, N. J. (1995). Flybrain, an on-line atlas and database of the *Drosophila* nervous system, *Neuron* **15,** 17-20.

Atema, J. (1995). Chemical signals in the marine environment: Dispersal, detection and temporal signal analysis, *Proc. Nat. Acad. Sci. USA* **92,** 62-66.

Atick, J. J. (1992). Could information theory provide an ecological theory of sensory processing?, in *Princeton Lectures on Biophysics,* W. Bialek, ed., pp. 223-289 (World Scientific, Singapore).

Barlow, H. B. (1980). The absolute efficiency of perceptual decisions, *Philos. Trans. R. Soc. Lond. Ser. B* **290,** 71-82.

Barlow, H. B. (1981). Critical limiting factors in the design of the eye and visual cortex, *Proc. R. Soc. Lond. Ser. B* **212,** 1-34.

Barlow, H. B. (1988). Thermal limit to seeing, *Nature* **334,** 296.

Baylor, D. A., Lamb, T. D., Yau, K.-W. (1979). Responses of retinal rods to single photons, *J. Physiol. (Lond.)* **288,** 613-634.

Baylor, D. A., Matthews, G., Yau, K.-W. (1980). Two components of electrical dark noise in toad retinal rod outer segments, *J. Physiol. (Lond.)* **309,** 591-621.

Baylor, D. A., Nunn, B. J., Schnapf, J. F. (1984). The photocurrent, noise and spectral sensitivity of rods of the monkey *Macaca fascicularis, J. Physiol. (Lond.)* **357,** 575-607.

Beerends, J. G., Houtsma, A. J. (1986). Pitch identification of simultaneous dichotic two-tone complexes, *J. Acoust. Soc. Am.* **80,** 1048-1056.

Berg, H., Purcell, E. M. (1977). Physics of chemoreception, *Biophys. J.* **20,** 193-219.

Bialek, W. (1987). Physical limits to sensation and perception, *Ann. Rev. Biophys. Biophys. Chem.* **16,** 455-478.

Bialek, W. (1992). Optimal signal processing in the nervous system, in *Princeton Lectures on Biophysics,* W. Bialek, ed., pp. 321-401 (World Scientific, Singapore).

Bialek, W., DeWeese, M. (1995). Random switching and optimal processing in the perception of ambiguous signals, *Phys. Rev. Lett.* **74,** 3077-3080.

Bialek, W., Rieke, F., de Ruyter van Steveninck, R. R. and Warland, D. (1991). Reading a neural code, *Science* **252,** 1854-1857.

Blake, A., Bulthoff, H. H., Sheinberg, D. (1993). Shape from texture – Ideal observers and human psychophysics, *Vison Res.* **33,** 1723-1737.

de Boer, E. (1976). Pitch review.... *Handbk. Sens. Physiol.*

Borsellino, A., De Marco, A., Allazetta, A., Rinsei, S., Bartolini, B. *Kybernetik* **10,** 139 (1972).

Borst, A., Egelhaaf, M. (1989). Principles of visual motion detection, *Trends. Neurosci.* **12,** 297-306.

Chittka, L., Menzel, R. (1992). The evolutionary adaptation of flower colours and the insect pollinators' colour vision, *J. Comp. Physiol. A* **171,** 171-181.

Coleman, S. *Aspects of Symmetry* (Cambridge University Press, Cambridge, 1985).

Conti, F., Torre, V. (1996). *Neurobiology: Proceedings of the 1995 School of Biophysics* (to be published).

Delbrück, M. (1970). A physicist's renewed look at biology: Twenty years later, *Science* **168,** 1312-1315.

DeWeese, M. (1995). *Optimization Principles for the Neural Code* (Dissertation, Princeton University).

Dill, M., Heisenberg, M. (1995). Visual pattern memory without shape recognition, *Phil. Trans. R. Soc. Lond. Ser. B* **349,** 143-152.

Dill, M., Wolf, R., Heisenberg, M. (1993). Visual pattern recognition in Drosophila involves retinotopic matching, *Nature* **365,** 751-753.

Ebeling, W., Poschel, T. (1994). Entropy and long-range correlations in literary English, *Europhys. Lett.* **26,** 241-246.

Feynman, R. P., Leighton, R., Sands, M. (1963). *The Feynman Lectures on Physics, Volume I.* (Addison-Wesley, Reading MA).

Fisher, G. H. *Perception & Psychophysics* **4,** 189 (1968).

Franceschini, N., Riehle, A. and le Nestour, A. (1989). Directionally selective motion detection by insect neurons, in *Facets of Vision,* R. C. Hardie and D. G. Stavenga, eds. pp. 360-390 (Springer-Verlag, Berlin).

Girgus, J. J., Rock, I., Egatz, R. *Perception & Psychophysics* **22,** 550 (1977).

Goldstein, J. L. (1973). An optimum processor theory for the central formation of the pitch of complex tones, *J. Acoust. Soc. Am.* **54,** 1496-1516.

Goldstein J. L., Srulovicz, P. (1977). Auditory-nerve spike intervals as an adequate basis for aural spectrum analysis, in *Psychophysics and Physiology of Hearing* E. F. Evans and J. P. Wilson, eds., pp. 337-346.

Goldstein, J. L., Gerson, A., Srulovicz, P., Furst M. (1978). Verification of the optimal probabilistic basis for aural processing in pitch of complex tones, *J. Acoust. Soc. Am.* **63,** 486-497.

Griffin, D. R. (1958). *Listening in the dark: The acoustic orientation of bats and men* (Yale University Press, New Haven). Dover edition, 1974 (Dover, New York).

Hassenstein, S., Reichardt, W. (1956). Systemtheoretische analyse der zeitreihenfolgen und vorzeichenaauswertung bei der bewegungsperzeption des rüsselkäfers *Chlorophanus, Z. Naturforsch.* **11b,** 513-524.

Hausen, K. (1984). The lobular complex of the fly: Structure, function, and significance in behavior, in *Photoreception and vision in invertebrates,* M. Ali, ed., pp. 523-559 (Plenum, New York NY).

Hausen, K., Wehrhahn, C. (1983). Microsurgical lesion of horizontal cells changes optomotor yaw responses in the blowfly *Calliphora erythrocephala, Proc. R. Soc. Lond. B* **21,** 211-216.

Hecht, S., Shlaer, S., Pirenne, M. H. (1942). Energy, quanta, and vision, *J. Gen. Physiol.* **25,** 819-840.

Heisenberg, M., Wolf, R. (1984). *Vision in Drosophila: Genetics of Microbehavior* (Springer-Verlag, Berlin).

Heisenberg, M., Wolf, R. (1988). Reafferent control of optomotor yaw torque in *Drosophila melanogaster, J. Comp. Physiol. A* **163,** 373-388.

Hodgkin, A. L., Huxley, A. F. (1952a). Currents carried by sodium and potassium ions through the membrane of the giant axon of *Loligo, J. Physiol.* **116,** 449-472.

Hodgkin, A. L., Huxley, A. F. (1952b). The components of membrane conductance in the giant axon of *Loligo*, *J. Physiol.* **116,** 473-496.

Hodgkin, A. L., Huxley, A. F. (1952c). The dual effect of membrane potential on sodium conductance in the giant axon of *Loligo*, *J. Physiol.* **116,** 497-506.

Hodgkin, A. L., Huxley, A. F. (1952d). A quantitative description of membrane current and its application to conduction and excitation in nerve, *J. Physiol.* **117,** 500-544.

Imry, Y., Ma, S. K. *Phys. Rev. Lett.* **35,** 1399 (1975).

Kiang, N. Y.-S., Watanabe, T., Thomas, E. C., Clark, L. F. (1965) *Discharge Patterns of Single Fibers in the Cat's Auditory Nerve* (The MIT Press, Cambridge MA).

Klein, S. A., Levi, D. M. (1985). Hyperacuity thresholds of 1 sec: Theoretical predictions and empirical validation, *J. Opt. Soc. Am. A* **2,** 1170-1190.

Kramers, H. A. *Physica* **7,** 284 (1940).

Lagnado, L., Baylor, D. (1992). Signal flow in visual transduction, *Neuron* **8,** 995-1002.

Land, M. F., Collett, T. S. (1974). Chasing behavior of houseflies (*Fannia canicularis*): A description and analysis, *J. Comp. Physiol.* **89,** 331-357.

Liu, Z. L., Knill, D. C., Kersten, D. (1995). Object classification for human and ideal observers, *Vision Res.* **35,** 549-568.

Long, G. M., Toppino, T. C. *Perception* **10,** 331 (1981).

MacKay, D., McCulloch, W. S. (1952). The limiting information capacity of a neuronal link, *Bull. Math. Biophys.* **14,** 127-135.

Maddess, T., Laughlin, S. B. (1985). Adaptation of the movement sensitive neuron H1 is generated locally and governed by contrast frequency, *Proc. R. Soc. Lond. Ser. B* **225,** 251-275.

Mafra-Neto, A., Cardé, R. T. (1994). Fine-scale structure of pheromone plumes modulates upwind orientation of flying moths, *Nature* **369,** 142-144.

Marwan, W., Hegemann, P., Oesterhelt, D. (1988). Single photon detection by an archeabacterium, *J. Mol. Biol.* **199,** 663-664.

Möhl, B. (1988). Short-term learning during flight control in *Locusta migratoria, J. Comp. Physiol. A* **163,** 803-812.

Möhl, B. (1989). 'Biological noise' and plasticity of sensorimotor pathways in the locust flight system, *J. Comp. Physiol. A* **166,** 75-82.

Neher, E. (1992). Ion channels for communication between and within cells, *Science* **256,** 498-502.

Poggio, T., Reichardt, W. (1976). Visual control of orientation behavior in the fly. Part II. Towards the underlying neural interactions, *Q. Rev. Biophys.* **9,** 377-438.

Potters, M., Bialek, W. (1994). Statistical mechanics and visual signal processing, *J. Phys. I France* **4,** 1755-1775.

Ramachandran, V. A., Anstis, S. M. *Perception* **14,** 135 (1985).

Reichardt, W., Poggio, T. (1976). Visual control of orientation behavior in the fly. Part I. A quantitative analysis, *Q. Rev. Biophys.* **9,** 311-375.

Rieke, F., Bodnar, D., Bialek, W. (1995). Naturalistic stimuli increase the rate and efficiency of information transmission by primary auditory neurons, *Proc. R. Soc. Lond. Ser. B* in press.

Rieke, F., Warland, D., Bialek, W. (1993). Coding efficiency and information rates in sensory neurons, *Europhys. Lett.,* **22,** 151-156.

Rieke, F., Warland, D., de Ruyter van Steveninck, R. R., Bialek W. (1996). *Spikes: Exploring the Neural Code* (MIT Press, Cambridge MA).

Rock, I., Mitchener, K. *Perception* **21,** 39 (1992).

Ruderman, D. L., Bialek, W. (1994). Statistics of natural images: Scaling in the woods, *Phys. Rev. Lett.* **73,** 814-817.

de Ruyter van Steveninck, R. R., Bialek, W. (1995). Reliability and statistical efficiency of a blowfly movement-sensitive neuron, *Phil. Trans. R. Soc. Lond. Ser. B* **348,** 321-340.

de Ruyter van Steveninck, R. R., Bialek, W., Potters, M., Carlson, R. H. (1994). Statistical adaptation and optimal estimation in movement computation by the blowfly visual system, *Proceedings of the 1994 I. E. E. E. Conference on Systems, Man and Cybernetics,* pp. 302-307.

de Ruyter van Steveninck, R. R., Laughlin, S. B. (1995). Information transmission at the first synapse in fly vision, preprint.

de Ruyter van Steveninck, R. R., Zaagman, W. H., Mastebroek, H. (1986). Adaptation of transient responses of a movement-sensitive neuron in the visual system of the blowfly *Calliphora erythrocephela, Biol. Cybern.* **54,** 223–236.

Sakitt, B. (1972). Counting every quantum, *J. Physiol. (Lond.)* **223,** 131-150.
Sakmann, B. (1992). Elementary events in synaptic transmission revealed by currents passing through single ion channels, *Science* **256,** 503-512.
Shannon, C. E. (1948). A mathematical theory of communication, *Bell System Tech. J.* **27,** 379-423 & 623-656. Reprinted in Shannon and Weaver (1949).
Shannon, C. E. (1951). Prediction and entropy of printed English, *Bell Sys. Tech. J.*
Shannon, C. E., Weaver, W. (1949). *The Mathematical Theory of Communication* (University of Illinois Press, Urbana).
Shraiman, B. I., Siggia, E. D. (1994). Lagrangian path integrals and fluctuations in random flow, *Phys. Rev. E* **49,** 2912-2927.
Siebert, W. M. (1965). Some implications of the stochastic behavior of primary auditory neurons, *Kybernetik* **2,** 206-215.
Siebert, W. M. (1970). Frequency discrimination in the auditory system: Place or periodicity mechanisms?, *Proc. I. E. E. E.* **58,** 723-730.
Simmons, J. A. (1979). Perception of echo phase information in bat sonar, *Science* **204,** 1336-1338.
Simmons, J. A. (1989). A view of the world through the bat's ear: The formation of acoustic images in echolocation, *Cognition* **33,** 155-199.
Simmons, J. A., Ferragamo, ., Moss, C. F., Stevenson, S. B., and Altes, R. A. (1990). Discrimination of jittered sonar echoes by the echolocating bat, *Eptesicus fuscus*: The shape of target images in echolocation, *J. Comp. Physiol. A* **167,** 589-616.
Srulovicz, P., Goldstein, J. L. (1983). A central spectrum model: A synthesis of auditory-nerve timing and place cues in monoaural communication of frequency spectrum, *J. Acoust. Soc. Am.* **73,** 1266-1276.
Tapiovaara, M., (1990). Ideal observer and absolute efficiency of detecting mirror symmetric random images, *J. Opt. Soc. Am. A* **7,** 2245-2253.
Teich, M. C., Prucnal, P. R., Vannucci, G., Breton, M. E., McGill, W. J. (1982). Multiplication noise in the human visual system at threshold. III: The role of non-Poisson quantum fluctuations, *Biol. Cybern.* **44,** 157.
Valbo, A. B. (1995). Single afferent neurons and somatic sensation in humans, in *The Cognitive Neurosciences,* M. Gazzaniga, ed., pp. 237-252 (MIT Press, Cambridge MA).
van der Velden, H. A. (1944). Over het aantal lichtquanta dat nodig is voor een lichtprikkel bij het menselijk oog, *Physica* **11,** 179-189.
Voss, R. F., Clarke, J. (1975). 1/f noise in music and speech, *Nature* **258,** 317-318.
de Vries, Hl. (1956). Physical aspects of the sense organs, *Prog. Biophys.* **6,** pp. 208-264.
Wagner, H. (1986). Flight performance and visual control of flight in the free-flying house fly *(Musca domestica L.).* I: Organization of the flight motor, II: Pursuit of targets, III: Interactions between angular movement induced by wide- and small-field stimuli, *Phil. Trans. R. Soc. Lond. Ser. B* **312,** 527-595.
Westheimer, G. (1981). Visual hyperacuity, *Prog. Sens. Physiol.* **1,** 1-30.
Wolf, R., Heisenberg, M. (1990). Visual control of straight flight in *Drosophila melanogaster, J. Comp. Physiol. A.* **167,** 269-283.

Part V

Evolution, Micro- and Macro-Scale

Molecular Evolutionary Biology

From Concepts to Technology

Peter Schuster, Jacqueline Weber, Walter Grüner, and Christian Reidys

Institut für Molekulare Biotechnologie e.V. Jena, Germany

1 Evolutionary Dynamics

Current biology is facing a grand synthesis combining knowledge from three different disciplines: molecular biology, developmental biology, and evolutionary biology. The first step in this direction was taken already in the late sixties by the pioneering works of Sol Spiegelman (1971) who developed new biochemical methods of evolution in the test tube. About the same time Manfred Eigen (1971) conceived a kinetic theory of evolution at the molecular level. Since then the study of the evolution of molecules in laboratory systems has become a research area in its own rights. It contrasted and complemented conventional studies in *molecular evolution* by adding a dynamical component to the essentially phylogenetic issues of sequence data comparisons as initiated and scholarly developed, for example, by Margaret Dayhoff and Park (1969). Meanwhile, experiments with replicating molecules in the test tube have shown that evolution in the sense of Charles Darwin's principle of variation and selection is no privilege of cellular life: optimization of properties related to the *fitness* of replicating molecules is observed readily *in vitro* with *naked* ribonucleic acid (RNA) molecules in evolution experiments.

Dynamics of evolutionary optimization characterized as Darwinian dynamics is visualized as a hill climbing process on a *fitness landscape* that assigns fitness values to genotypes or polynucleotide sequences (DNA or RNA). In terms of dynamical systems theory Darwinian dynamics is simple in the sense that it follows a gradient and eventually reaches a (local or global) fitness maximum. Darwinian dynamics, however, was found to be just one feature of evolutionary systems. Others being, for example, suppression of optimization of individual fitness by (mutualistic) interaction through catalysis, predator-prey or host-parasite interactions. Evolutionary dynamics then needs not approach a steady state but may also give rise to complex dynamical phenomena like oscillations, spatial pattern formation or deterministic chaos in space and time. Spatiotemporal patterns cover only one aspect of evolutionary phenomena others being, for example, historical like the reconstruction of phylogenies, or genetic like the fixation of allels in populations. We introduce here a comprehensive model that tries to account for most of the relevant features.

Evolutionary dynamics is considered as a highly complex process described in three different abstract metric spaces which are used for convenient and illustrative projections of biological evolution:

(1) the *sequence space* of genotypes being DNA or RNA sequences,
(2) the *shape space* of phenotypes, and
(3) the *concentration space* of biochemical reaction kinetics.

The sequence space is a metric point space containing all sequences. Relations or distances in sequence space are commonly expressed by the number of single nucleotide exchanges or point mutations converting a given sequence into another and accordingly the metric is the Hamming metric (Hamming (1986)). The sequence space of binary sequences is simply a hypercube of dimension n with n being the chain length of the genotype (Wright (1932)). The shape space is an abstract space covering all possible structures under consideration. Notions of structure are largely context dependent: meaningful comparisons of active sites of enzymes (or ribozymes) require atomic resolution whereas studies of phylogenetic conservation of structures can be done much better on the coarse-grained level of *ribbon* or *wire* diagrams. As we shall see later on, some levels of coarse-graining are not only of physical relevance but also suitable for mathematical modeling. Secondary structures of RNA molecules may serve as an example. Similarity and dissimilarity of RNA structures can be expressed by means of quantitative measures with metric properties. Concentration space, finally, is the conventional space in which chemical reaction kinetics is usually described. It was formalized and put into precise mathematical terms by Martin Feinberg (1977).

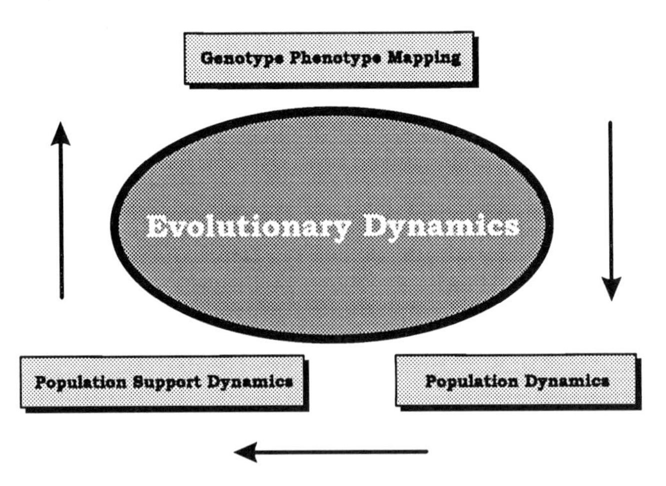

Fig. 1. Evolutionary dynamics in the spaces of sequences, shapes, and concentra tions.

Projections of the complex evolutionary scenario onto each of the three abstract spaces highlights one out of several dynamical aspects of evolution (figure 1): (population) support dynamics visualizes the migration of populations in sequence space, genotype-phenotype mapping induces the evolutionary relevant structure in shape space, and population dynamics is the projection of evolution onto concentration space.

- Genotype-phenotype mapping assigns a phenotype to every genotype. In molecular evolution this is tantamount to folding biopolymer sequences into structures. Accordingly the *shape space* is most appropriate for description. Precisely, it is the mapping from sequences into structures and further into functions that is relevant for evolution: genotype-phenotype mapping provides all internal (kineti c and thermodynamic) parameters for population dynamics.
- Population dynamics describes the temporal evolution of population variables (particle numbers, genotype frequencies or concentrations). It is properly described in the conventional *concentration space* of chemical reaction kinetics (population dynamics of sexual species is population genetics). The number of possible genotypes is hyperastronomically large and thus the majority of them will neither be materialized in an evolution experiment nor in nature. Only a small subset of all possible genotypes can be present at a given instant in the population. Whenever a new variant is formed by mutation or some genotype dies out the concentration space changes, a new dimension housing the variable describing the frequency of the new variant is added or the obsolete dimension is removed, respectively.
- Population support dynamics is the process taking place in *sequence space* since it deals with migrating sets of genotypes. The (population) support highlights the genotypes that are currently present in the population. Moreover, support dynamics is fundamental for genotype-phenotype mapping since it defines the regions in sequence space where new genotypes might appear.

The three projections of evolutionary dynamics onto the three abstract spaces form a conceptual cycle in the sense that each of them provides the input for the next one: genotype-phenotype mapping provides the parameters (fitness, for example, being most relevant for evolution) for population dynamics. Population dynamics deals with temporal alterations in concentrations and thus hands information on arriving new and disappearing old genotypes over to the support dynamics. Support dynamics in turn transfers changes in the population support to genotype-phenotype mapping in order to make new parameters available for population dynamics and thereby closes the cycle.

2 Molecular Phenotypes

In the past most molecular biologists considered molecular evolution as a topic for outsiders. It turned out, however, that this new discipline provides handles to deal with several issues of evolutionary theory that were previously unaccessible. Phenotypes are commonly understood as organisms which are formed in a highly complex unfolding process by instructions from the genotype. In case of higher multi-cellular organisms this occurs through the process of embryonic

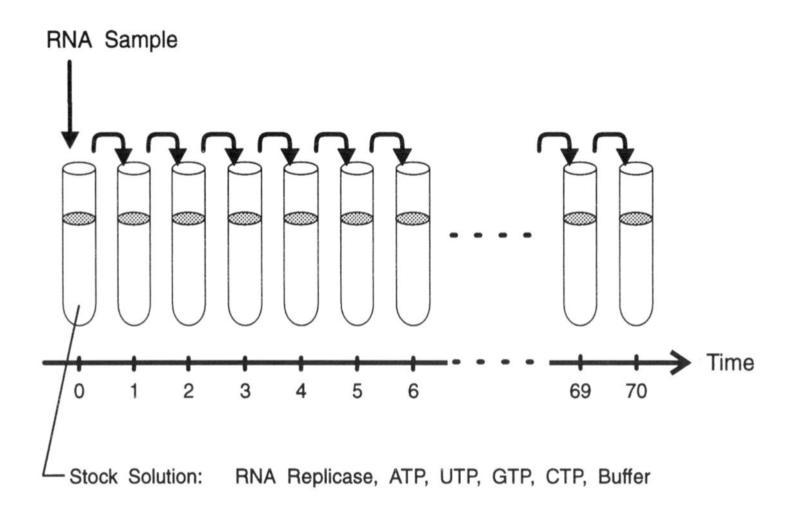

Fig. 2. Serial transfer experiments with RNA molecules.

development. At present molecular developmental biology reveals the genetic details of embryonic morphogenesis but it is still far away from an understanding of entire process. In molecular evolutionary biology unfolding of the phenotype is reduced to replication and folding of biopolymers. Sol Spiegelman (1971) interpreted the spatial structures of replicating RNA molecules as their phenotypes. In RNA based molecular evolution the genotype and the phenotype thus are two different features of one and the same molecule, the genotype being the nucleotide sequence that contains the genetic information and the phenotype being the spatial structure. In complete analogy to macroscopic biology the phenotype provides the kinetic and thermodynamic parameters that are relevant for the fitness of a replicating molecule.

Evolution of RNA molecules in the test tube has been studied first by means of serial-transfer experiments figure 2. These investigations have shown that evolution is not dependent on the existence of cellular life. Meanwhile RNA replication has been studied extensively and its molecular mechanism is well understood. Currently it represents the best known example of complementary copying of molecules (Biebricher and Eigen (1988)).

RNA replicating enzymes, so called RNA replicases, bind to the 3'-end of the template and produce a complementary strand step by step through incorporation of this activated mononucleotide triphosphte (ATP, UTP, GTP or CTP) which forms a Watson-Crick base pair (**AU** and **GC**) with the corresponding digit (U, A, C or G, respectively) on the template. Then the replicase takes care of separating the two strands in order to avoid formation of long double helical stretches that would not readily separate under the conditions of replication (figure 3). Double stranded RNA is no template for bacteriophage specific RNA replicases. In the case of $Q\beta$-replicase isolated from *Escherichia coli* cells

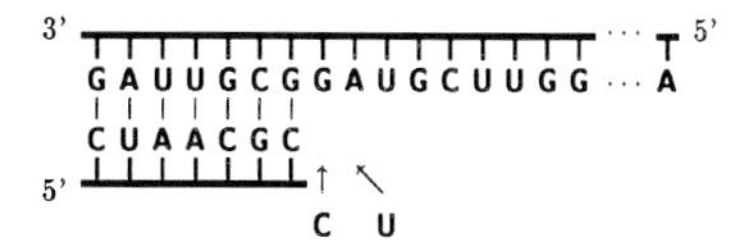

Incorporation of next base into the growing chain

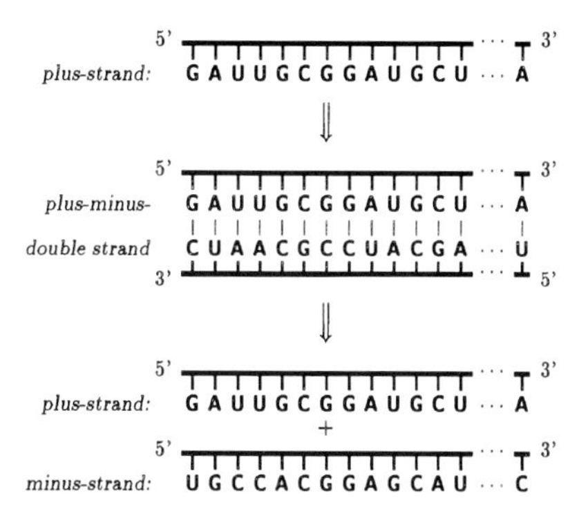

Replication via complementary strands

Fig. 3. Molecular principles of complementary RNA replication. The upper part presents a sketch of the incorporation of individual complementary mononucleotides (in form of their triphosphates ATP, UTP, GTP or CTP) into the growing chain of the newly synthesized strand. The lower part shows (schematically) the two tasks that the replicating enzyme, the RNA replicase, has to fulfil in complementary replication: completion of the (plus) template strand to a (plus–minus) double strand and separation of the double helix into separate (plus) and (minus) strands. (Note that the minus-strand in the double helix runs from 3' to 5'-end which is opposite to the conventional representation in molecular biology.)

infected by the RNA phage $Q\beta$, minimum requirements which make sure that they are accepted as templat es by the enzyme and then replicated (Biebricher and Luce (1993)) are known for RNA molecules. The smallest templates that have been isolated and sequenced are only about 25 nucleotides long and both strands have hairpin loops at their 5'-ends.

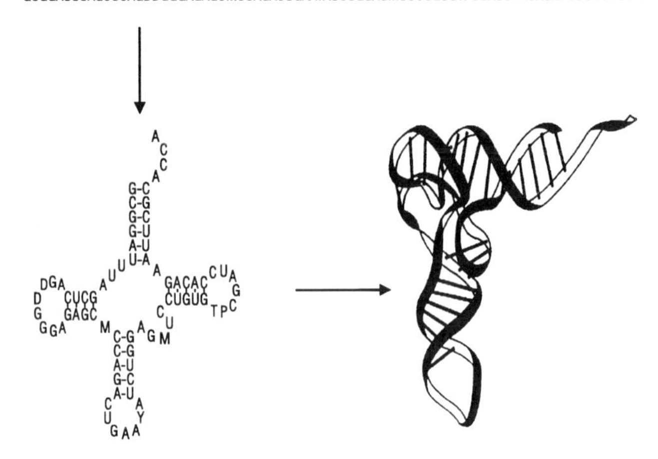

Fig. 4. Folding of RNA sequences into secondary structures and spatial structur es.

3 Statistics of Genotype-Phenotype Mappings

Genotype-phenotype mapping is the problem in the core of molecular evolution. Despite its general importance predictions of phenotypic structures and properties from known sequences commonly fail because of the enormous complexity of the relations between genotypes and phenotypes. Higher multicellular organisms provide many formidable challenges and unsolved problems for a molecular theory of evolution. For example, the molecular mechanisms of development are just now being explored by molecular geneticists. Even in case of the simplest known organisms, viroids, viruses or bacteria, we are not yet in a position to guess the consequences of mutations from known changes in the polynucleotide sequences.

One of the primary difficulties in the prediction of phenotypic properties is concerned with the very nature of cellular metabolism: it is a complex multifunctional and highly connected network of biochemical reactions and the results of changes in the genotypes and associated alterations in protein structures and enzymatic functions are impossible to be predicted reliably. Given metabolic network dynamics were fully understood the situation would be not much better, however, since the relations between amino acid sequences, protein structures and protein functions are also not yet understood sufficiently well. The same is true for polynucleotides, in particular for RNAs. Understanding of biopolymer structures and properties is certainly a main issue in current molecular biophysics, but progress is slow.

The most simple systems showing the essential features of evolution are revealed by RNA molecules replicating in the test tube: as said previously, genotypes and phenotypes of RNA molecules are two features of the same molecule. Unfolding of the phenotype is then tantamount to folding the RNA sequence into a three-dimensional structure. Full three-dimensional structures of RNA molecules are still to complex for straightforward modeling and, despite impressive progress in the field, they are not sufficiently well understood yet. So far high resolution X-ray structures are available only for two classes of molecules, transfer-RNAs and hammerhead ribozymes (Pley et al. (1994)). Secondary structures (figure 4) serve as a tractable and relevant coarse grained versions of full RNA structures. In essence, they represent listings of Watson-Crick and **GU** base pairs. They are conserved in evolutionary phylogenies and they were found to be useful for the prediction of RNA properties throughout more than thirty years of biochemistry.

The present state of knowledge allows detailed analysis and theoretical approaches to genotype-phenotype mappings only in this admittedly highly reduced case. Nevertheless, molecular evolution experiments provide a solid experimental basis for this theoretical approach. In addition, RNA molecules show all the essential features of a Darwinian scenario in test tube evolution experiments. Interpretation of experimental data, hence yielded and steadily yields conclusive answers to several open questions of evolutionary biology.

In conventional biophysics considers the science of biopolymers mainly concerned with the

$$\text{(one) sequence} \Longrightarrow \text{structure} \Longrightarrow \text{function}$$

prediction problem. Statistical problems, for example the question, "How many sequences form the same structure?", are never addressed by the standard concept. The approach to evolutionary dynamics followed here, however, requires precisely such global informations on the relations between genotypes and phenotypes. These relations are understood as mappings from the space of genotypes into the space of phenotypes. In constant environment every genotype leads to a defined phenotype. The mapping, however, need not be invertible: commonly there are (many) more genotypes than evolutionarily distinguishable phenotypes and thus many genotypes must form the same phenotype.

In order to study the mapping of RNA sequences into secondary structures one needs the concept of sequence space: every sequence is represented as a point in sequence space and the distance between sequences is the Hamming distance, the (minimal) number of point mutations converting two sequences into each other. In mathematical notation $\mathcal{Q}_\kappa^n$ is the space of sequences of chain length n on an alphabet with κ letters (There are κ^n different sequences. For binary alphabets, $\kappa = 2$, the sequence space is a hypercube in n-dimensional space). For natural polynucleotides, DNA or RNA we have $\kappa = 4$. The mapping of RNA sequences into (secondary) structures can be formally written as

$$f_n : \quad \mathcal{Q}_\kappa^n \longrightarrow \mathcal{S}_n$$

where $\mathcal{S}_n$ is the set of RNA (secondary) structures. The notion of compatible sequences turned out to be essential for the understanding of evolutionary dynamics because they are intermediates in transitions between structures. It will be briefly introduced here. An RNA (secondary) structure (figure 4) is tantamount to a set of contacts $[i, k]$ (i and k being two positions on the structure). In a (real) RNA molecule every contact corresponds to a base pair (Watson-Crick

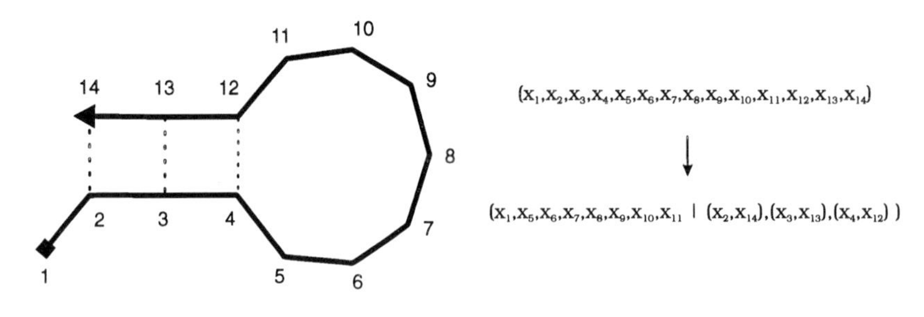

Fig. 5. Partition of a sequence according to structure. Natural RNA sequences are assembled from a four letter alphabet and form six base pairs. The 5'-end and the 3'-end of the RNA sequence are denoted by a square or an arrow, respectively. The symbolic notation assigns a dot, "·", to every unpaired base and a left or right parenthesis, "(" or ")", to every base in a base pair depending on whether it pairs upstream or downstream (in the 5'-end to 3'-end direction).

or **GU**, respectively). A sequence $(x_1, \cdots, x_n)$ is said to be compatible with a reference structure s if all base duplets (x_i, x_k) can form a base pair where $[i, k]$ is a contact of s (figure 5). A compatible sequence thus may, but need not, form the reference structure under minimum free energy conditions. Then the reference structure, however, is always contained in the set of non-optimal foldings of the sequence. Every structure induces a set of compatible sequences, $\mathbf{C}[s]$, that is smaller than the set of all sequences (There is one trivial exception: $\mathbf{C}[s]$ is the set of all sequences if and only if s is the fully open structure with no base pair). (Compatible) sequences can be decomposed with respect to the reference structure s into sets of unpaired and paired digits which will be addressed as unpaired and paired segments (figure 5). The two segments may be considered as elements in two different combinatorial (sequence) spaces, $\mathcal{Q}^{n_u}_{\alpha_u}$ and $\mathcal{Q}^{n_p}_{\alpha_p}$ (Here α_u and α_n denote the multiplicities of bases and base pairs, respectively, and n_u and n_p are the numbers of unpaired bases and base pairs. Hence, $\alpha_u = 4$ and $\alpha_n = 6$ for natural polynucleotides as shown in figure 5, and $n = n_u + 2n_p$). All compatible sequences form a graph in sequence space that can be expressed as the Cartesian product

$$\mathcal{C}[s] \;=\; \mathcal{Q}^{n_u}_{\alpha_u} \times \mathcal{Q}^{n_p}_{\alpha_p} .$$

Two sequences are nearest neighbors if they differ either

- in a single position i which is unpaired in s, or
- in two positions i and j which form a contact $[i, j] \in s$.

In natural RNA molecules the exchange of an admissible base pair may lead to a shift of Hamming distance one (for example **GC**$\rightarrow$**GU**) or two (for example **GC**$\rightarrow$**AU**).

All sequences that are mapped into a secondary structure s are certainly contained in the graph of compatible sequences $\mathcal{C}[s]$. Remembering the definition of the entire sequence to structure mapping, $f_n : \mathcal{Q}^n_\kappa \longrightarrow \mathcal{S}_n$, the *neutral network* of structure s, $\Gamma_n[s]$ is the subgraph induced by $f_n^{-1}(s)$ in $\mathcal{C}[s]$.

The generic properties of sequence to structure mappings of biopolymers, $f_n : \mathcal{Q}^n_\kappa \longrightarrow \mathcal{S}_n$, can in fact be studied by means of a mathematical model. Preimages are constructed as random subgraphs of the compatible sequences by means of the following procedure: let s be some secondary structure and $\mathbf{C}[s]$ be its set of compatible sequences, then a random graph is assigned to the neutral network $\Gamma_n[s]$ by choosing unpaired segments from $\mathcal{Q}^{n_u}_{\alpha_u}$ and paired segments from $\mathcal{Q}^{n_p}_{\alpha_p}$ with probabilities λ_u and λ_p, respectively. The procedure yields two disjoint random graphs, $\Gamma_{n_u}[s] \subseteq \mathcal{Q}^{n_u}_{\alpha_u}$ and $\Gamma_{n_p}[s] \subseteq \mathcal{Q}^{n_p}_{\alpha_p}$, which are formally combined to the neutral network with respect to s via the Cartesian product

$$\Gamma_n[s] \;=\; \Gamma_{n_u}[s] \times \Gamma_{n_p}[s].$$

The complete sequence to structure mapping is then obtained by iterating the random graph construction with respect to a given order of the secondary structures. Mappings derived by this procedure are not totally random since they have the property: $f(v) = s \Longrightarrow v \in \mathbf{C}[s]$.

Mathematical modeling of sequence to structure maps yields complete probability spaces of mappings. Generic stability properties can be detected by searching for those features that are fulfilled with probability one. Several properties which proved to be essential for evolutionary optimization were in fact found to be generic (see section 6). Random maps designed as outlined above do not take into account the biochemical and biophysical properties of RNA except the base pairing logics. Consequently, all statistical features of RNA folding that are due to specific molecular properties can be derived from comparisons of random maps with those obtained by RNA folding.

Sequence to structure mapping of real RNA molecules has been investigated in great detail [10-14]. In order to determine the sequences that fold into a given structure as well as to make an estimate on their numbers we conceived an algorithm for inverse folding of RNA secondary structures, implemented it (Hofacker et al. (1994)) and applied it to several common (for example t RNA) and randomly chosen structures (Schuster et al. (1994),Schuster and Stadler (1994)). The results were essentially the same for all cases: thousands of sequences forming the same structure were easily found and, in essence, they are randomly distributed in sequence space.

The development of a fast prediction algorithms for RNA secondary structures (Hofacker et al. (1994)) made also large scale statistical studies possible which revealed several other global features of RNA sequence to structure mappings as well (Fontana et al. (1993)). In summary sequence to secondary structure mappings of RNA molecules have the following properties:

- The number of sequences is much larger than the number of RNA secondary structures.
- Individual structures differ largely with respect to their frequencies.
- Sequences folding into the same secondary structure are distributed randomly in sequence space.
- Structural and evolutionary properties of RNA molecules depend strongly on the base composition (**A**:**U**:**G**:**C** ratio) of the sequences.

These global properties of RNA sequence to structure mappings provide important insights into the structure of shape space and have led to the formulation of the shape space covering concept to be discussed in the next section.

RNA secondary structures may be considered as a kind of toy universe where we have computational access to genotype-phenotype mappings. Are the results obtained in this model system representative for more complex systems or not? At the present state of our knowledge it seems very likely that extensive studies on three-dimensional structures of nucleic acids or proteins will provide similar results. Extensions of the approach presented here to the most simple organisms, viroids and simple RNA viruses seem to be possible. In the following sections we consider two important features of the RNA toy world that were revealed by our studies on RNA secondary structures, *shape space covering* and the existence of *neutral networks*, which have direct implications for evolutionary dynamics.

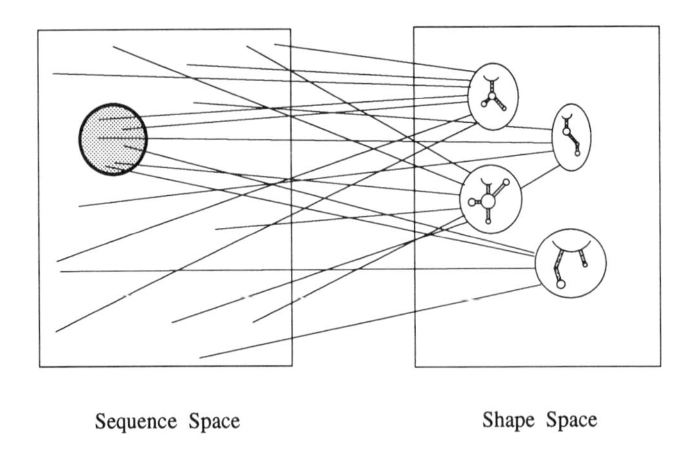

Fig. 6. Covering of RNA shape space by a small fraction of sequence space.

4 Shape Space Covering

The results obtained for the mapping of RNA sequences into secondary structures were combined into the general principle of *shape space covering* describing a regularity in sequence space which governs evolutionary dynamics (Schuster et al. (1994)). Only a small fraction of all possible sequences which are found in the environment of any arbitrarily chosen reference sequence has to be searched in order to find a sequence that folds into a given structure (figure 6). Shape space covering applies to all common structures.

To define *common* we divide the number of sequences by the number of distinct structures to obtain the average number of sequences folding into the same structure. Every structure is common when it is formed by more sequences than the average. The radius of the shape space covering sphere can be estimated for any given structure ($\mathcal{S}_k$) by simple statistics (Schuster (1995)):

$$r_{cov}(\mathcal{S}_k) \doteq \min\big(h \text{ with } B_h > f_k^{-1}\big),$$

where B_h is the number of sequences in a ball of radius h in sequence space and f_k is the frequency of structure $\mathcal{S}_k$: $f_k = N_k \;/\; \kappa^n$, N_k being the number of sequences forming structure $\mathcal{S}_k$, κ the number of monomer classes ($\kappa = 2$ or 4), and n the length of the sequences.

The shape space covering radius for patches containing all common structures can thus be computed for the above given definition of *common*. The covering radius for natural sequences (**AUGC**) of chain length $n = 100$, for example, is $r_{cov} = 15$. This means that the maximum number of sequences which have to be searched in order to find a particular (common) structure is approximately 4×10^{24}. The analogous computation for **GC** only sequences of chain length $n = 30$ yields $r_{cov} = 6$.

The shape space covering conjecture was verified by means of a concrete example, **GC** only sequences of chain length $n = 30$. All approximately 10^9 sequences were folded and analyzed with respect to shape space covering (Grüner (1994)). The results derived from this reference sample ($[\mathrm{GC}]_{30}$) are shown in figure 7. The value derived from simple statistics given above (r_{cov}) is slightly smaller than the numerical data suggest. The difference can be explained by an inspection of the fine-structure of neutral networks which has its origin in the biochemical aspects of base-pairing logic.

A heuristic molecular explanation of shape space covering should make the newly found fundamental concept of (molecular) evolution more plausible. Sequences forming a given (RNA secondary) structure are found to be randomly distributed in sequence space. In other words it must be possible to mutate the polynucleotides in very few or in many positions of the sequence without altering structure. This is precisely so for typical RNA secondary structures. On the one hand we may change a single base in a loop or mutate a single base pair, for example from **GC** to **GU**, retain with substantial probability the structure, and thus find a neutral sequence (a sequence forming the same secondary structure) at Hamming distance $d_h = 1$ from the reference. On the other hand we can

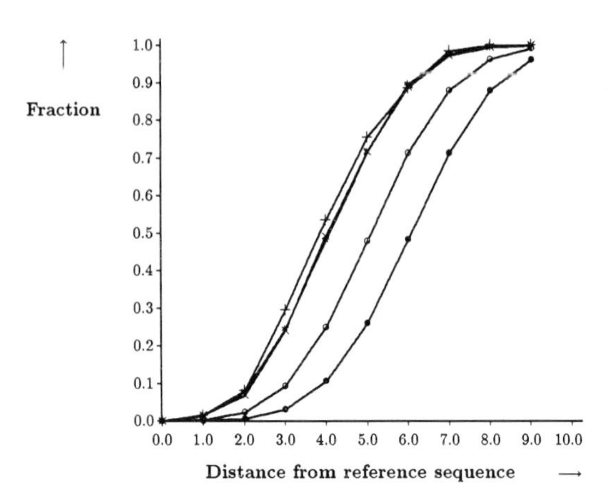

Fig. 7. Verification of shape space covering in the sequence space of **GC** only sequences of chain length $n = 30$. Environments of a large sample of arbitrarily chosen reference sequences are considered. The curves represent the average fraction of structures contained in spheres around these references. The five curves refer to the most common structure ($\times$), the top five ($*$), the top ten ($+$), the top 2625 most common structures ($\circ$) and all 22 718 common structures ($\bullet$).

also vary all nucleotides in an extended stacking region, fully retaining the base pairing principle, and again we are likely to find a sequence forming the same structure but now at large Hamming distance from the reference. The ultimate reason for shape space covering is thus to be seen in the existence of structural elements or modules that have a certain degree of autonomy in the folding process and sufficient resilience against variations in the sequence. It appears to be worth while to investigate in detail the sets of these neutral sequences in order to learn more about their regularities in sequence space.

5 Neutral Networks

The set of all sequences folding into the same secondary structure s is considered as a *neutral network* $\Gamma[s]$. In order to visualize a neutr al network all neighboring sequences are connected. Two classes of neighbors are defined which refer to single point mutations in single stranded parts of the structure and to base pair exchanges in the double helical stacks (see previous sections). The resulting graphs representing neutral networks were analyzed by random graph theory. Network properties were studied as a function of the average fraction of neutral neighbors in sequence space (i.e. the fraction of nearest neighbors folding into identical structures). These fractions are generally different for unpaired nucleotides ($\bar{\lambda}_u$) and base pairs ($\bar{\lambda}_p$). There is a sharp percolation threshold for $\bar{\lambda}$ that determines the structure of the graph representing the network (Reidys (1995), Reidys et al. (1995)):

$$\lambda_{cr} = 1 - \alpha^{-1/(1-\alpha)} .$$

Herein α is the number of alternatives at the corresponding position in the sequence. As shown in figure 5 we have $\alpha_u = 4$ and $\alpha_p = 6$ in natural RNA molecules. For $\bar{\lambda} < \lambda_{cr}$ the network is below the percolation threshold and consists of many components. Above threshold, $\bar{\lambda} > \lambda_{cr}$, we are dealing with a single component spanning the entire sequence space. In figure 8 we show

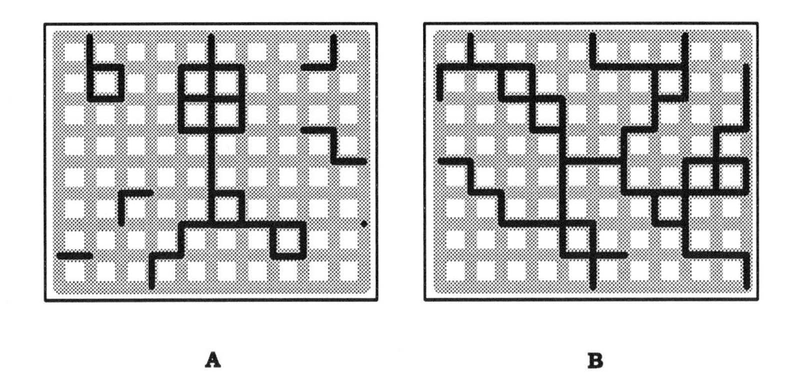

Fig. 8. Neutral networks and the percolation threshold. Distributions of RNA sequences folding into the same secondary structures are modeled by the theory of random graphs. Graphs representing neutral networks connect sequences which are nearest neighbors in sequence space and which fold into the same (secondary) structure. Two different situations are sketched schematically on a two-dimensional square lattice. (**A**) presents an example of a disintegrated network below the percolation threshold that is commonly split into a largest (giant) component and many small islands in sequence space. The sketch on the rhs. (**B**) shows a network above threshold which consists of a single component.

sketches of two characteristic examples of neutral networks one above and one below the percolation threshold.

A central result of the random graph approach deals with the structure of neutral networks below the percolation threshold. The generic decomposition of the network into components yields a *giant* component that is much larger than all other components. This largest component spans a fraction of sequence space which comes close to the maximum possible area. Comparison of the predictions from random graph theory with the reference sample derived from (numerical) folding of all **GC** sequences of chain length $n = 30$ ($[\mathrm{GC}]_{30}$) shows several characteristic differences. There are neutral networks whose sequences of components agree perfectly with those predicted by random graph theory but there are also other networks with two or four largest components of (almost) equal size. These deviations can be interpreted readily in terms of structural regularities and base pairing logic. For example, neutral networks with two or four (equally sized) largest components were found in structures with λ_u and λ_p above the threshold values when they have one or two structural elements that allow to form additional base pairs. These elements are, for example, stacking regions with tw o dangling ends, hairpin loops with five or more members or sufficiently large bulges, internal loops or multiloops. Structures with these elements are very likely to form an additional base pair whenever complementary bases are present in the appropriate positions. This happens less likely (and the structure in question is more likely to be formed as minimum free energy structure from the corresponding compatible sequence) when there is an excess of **G** or **C**, $\delta\#\mathbf{G}$ or $\delta\#\mathbf{C}$ respectively, in the base composition of the sequence (the $\mathbf{G}/\mathbf{C}$ ratio then being significantly different from one). One structural element that allows to form an additional base pair thus introduces a bias towards more **G** or more **C** in the average composition of the sequences just as it is found with the two components in a two-component neutral network. Two independent structural elements are superimposed in the relative $\mathbf{G}/\mathbf{C}$ content and thus we have

$$\delta\#\mathbf{G}\,\&\,\delta\#\mathbf{G}\ ,\ \delta\#\mathbf{G}\,\&\,\delta\#\mathbf{C}\ ,\ \delta\#\mathbf{C}\,\&\,\delta\#\mathbf{G}\ \text{and}\ \delta\#\mathbf{C}\,\&\,\delta\#\mathbf{C}\ .$$

The effects of the first and the last combination show a net **G** or **C** bias whereas compensation of differences brings the other two cases back into the middle of sequence space. Exactly this distribution of the four components is found by analyzing the data of the reference sample ($[\mathbf{GC}]_{30}$).

Mathematically the percolation phenomenon is manifested by two properties: density and connectivity of the random subgraphs. It turns out that $\lambda_{cr} = 1 - \alpha^{-1/(1-\alpha)}$ is a common threshold va lue for both properties. Here density means that for an arbitrary compatible sequence there is an adjacent sequence that is contained in the neutral network. We have already stated that the intersection between each two sets of compatible sequences is nonempty. This fact and the density of neutral networks above percolation threshold guarantee that each two neutral networks come close in sequence space. This has straightforward implications on the evolutionary optimization and we shall discuss this in detail in the following section.

6 Optimization on Landscapes

Sequence to structure mappings represent the first part of the unfolding of genotypes in order to yield the kinetic parameters of population dynamics. The second part being the evaluation of phenotypes. In molecular evolutionary biology this is tantamount to a mapping from sequences via structures into functions:

$$\textit{sequence} \quad\Longrightarrow\quad \textit{structure} \Longrightarrow \quad \textit{function}$$

The whole mapping thus assigns to every sequence a function which can be evaluated to yield the kinetic parameters, for example the fitness values. We are thus dealing with a mapping from sequence space into the real numbers which is commonly denoted as a *(fitness) landscape* (Schuster and Stadler (1994)). Evolutionary dynamics in simple systems following the Darwinian principle can be understood as an optimization problem or an adaptive walk on a fitness landscape (Kauffman and Levin (1987)).

A model system based on RNA sequence to secondary structure mappings was conceived about ten years ago and numerical computations were performed in order to study the basic mechanisms of optimization on realistic landscapes of molecular evolution (Fontana and Schuster (1987),Fontana et al. (1989)). Replication and mutation in populations of several thousand RNA molecules were simulated on the computer by means of an algorithm for stochastic kinetics of coupled chemical reactions (Gillespie (1977)). Kinetic rate constants and fitness values derived from them were obtained by evolution of RNA secondary structures according to a set of (predefined) rules. These studies have shown that the error threshold phenomenon of replication-mutation dynamics (Eigen et al. (1988),Eigen et al. (1989)) is observed also on realistic landscapes. At the same time these simulations indicated profound importance of neutral evolution in population dynamics.

Based on the recent knowledge on the internal regularities of sequence to structure mappings as expressed in the properties of neutral networks the computer simulations were performed in order to learn more about the mechanism of neutral evolution (Huynen et al. (1994)). Fitness values were assigned to structures and the preferred structure (for example a tRNA) got the highest fitness value. At sufficiently small error rates of replication the populations migrate essentially along the neutral network corresponding to the preferred structure. Simulations with different mutation rates confirmed the existence of a *phenotypic error threshold*: if the error rates exceed some threshold value the best structure is lost and the population drifts randomly in sequence space. These studies confirmed also that populations migrate on neutral networks by means of a diffusion like mechanism (Kimura (1983)). Populations split into subpopulations exhibiting a kind of autonomous behavior. This finding is related to similar observations made in simulations of evolution on flat landscapes of spin-glass type (Derrida and Peliti (1991)).

The overlap of two structures in sequence space (i.e. the set of sequences that are compatible to both structures) plays a decisive role in transitions between the neutral networks of two structures (Weber et al. (1995)). For different pairs of structures their overlap may essentially differ in size and topology but it is always nonempty. The overlap graph is the induced subgraph $(\mathcal{C}[s]\cup\mathcal{C}[s'])[\mathbf{C}[s]\cap\mathbf{C}[s']]$. A sketch of the overlap between two structures s and s' is shown in figure 9.

Pairs of structures with equal fitness are used to explore the mechanism of neutral evolution on an RNA folding landscape. Computer simulations were performed on a *double shape landscape* which is characterized by two privileged shapes (s and s') of higher and equal fitness (all other shapes having lower and equal fitness). Transitions between the neutral networks corresponding to the two shapes s and s' were found to occur via transient sequences of the overlap (E_{00}) or via sequences which differ from the overlap by a single base pair (E_{10}

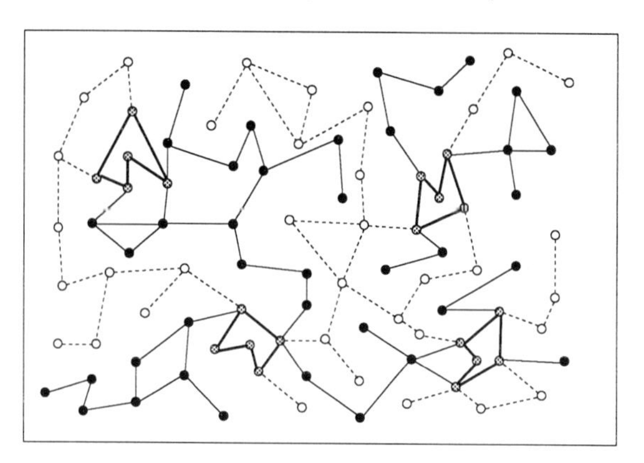

Fig. 9. Embedding of the overlap (grey points) in $\mathcal{C}[s] \cup \mathcal{C}[s']$. The overlap graph decomposes into islands that are connected by paths in $\mathcal{C}[s]$ (white points) and $\mathcal{C}[s']$ (black points) respectively.

or E_{01} denoting sequences that have one incompatible base pair with structure s or s', respectively). Another interesting feature of neutral evolution concerns the occurrence of fixation of shapes. A population is fixated on a neutral network at time t when sequences on the second network are absent, a population is transient if sequences from both networks are present. Figure 10 shows that the occurrence of larger amounts of sequences from the overlap or close to it (E_{00}, E_{10} or E_{01}) is strongly correlated with transitions between the two networks. If no such sequences can be found one of the two shapes is fixated. Alternating fixations of the population on either network and the frequency of transitions do not only depend on structure and size of the overlap, they depend also on structures and sizes of both networks as well as on the populations size. Larger populations imply higher search capacities and thus result in higher densities of transient states. An understanding of how populations migrate along pathways through sequence space thus requires detailed knowledge on the structure of neutral networks and their overlaps.

7 Evolutionary Biotechnology

Suggestions to apply of selection and evolutionary optimization in biotechnology in order to design new biomolecules were made already in the eighties (Eigen and Gardiner (1984), Kauffman (1986)). More recently, molecular evolutionary biology gave indeed birth to an applied discipline entitled *evolutionary biotechnology* or *applied molecular evolution* as it is commonly called overseas. In the core of these new research areas is the dynamics of evolutionary processes that we have reviewed here. The central idea is to make use of the Darwinian evolutionary principle of variation and selection in order to produce or *breed* biomolecules

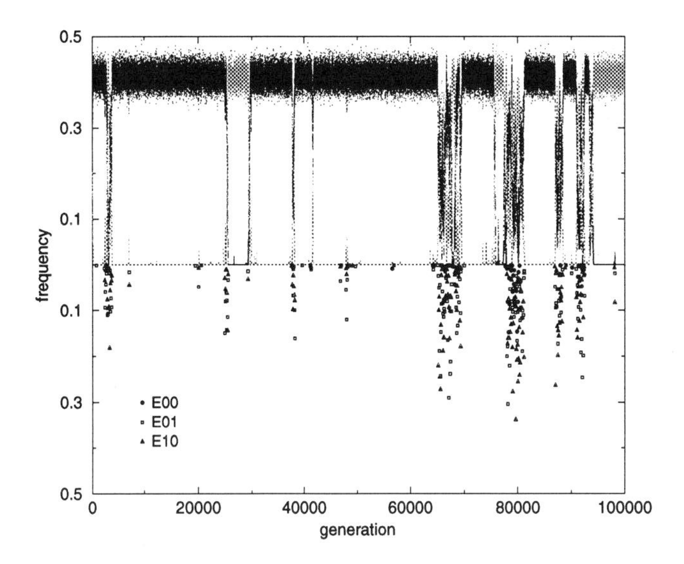

Fig. 10. A recording of a computer simulation demonstrating the mechanism of neutral evolution. The graph in the upper part presents the frequencies of sequences on two neutral networks whose shapes s = ...((((((((((....)))))))))) and s = ...((((((((((...)))))))))) (symbolic notation as shown in figure 5) have equal fitness (network $\Gamma[s]$ in black and network $\Gamma[s']$ in gray, respectively). The lower part shows the frequencies of sequences on the overlap of both shapes (E_{00}) as well as frequencies of the sequences having on incompatible base pair with s (E_{10}) or with s' (E_{01}), respectively.

with predefined properties in the test tube. The current state of the art allows to optimize primarily molecules that can be replicated. Evolutionary methods are thus restricted to polynucleotides, RNA or DNA (For recent reviews see for example (Schuster (1995), Joyce (1992), Kauffman (1994), Ellington (1994))).

Variation is introduced into polynucleotide sequences through replication with natural or increased error rates. Higher mutation rates are, for example, achieved by raising temperature or by using intercalating dyes which interfere with base recognition. An alternative approach starts from an ensemble of molecules obtained from random synthesis. Commonly, conventional automata are used to produce random DNA which is then transcribed into RNA. Two different selection strategies are usually pursued (figure 11):

- the *batch-technique* that couples the desired property with the selection process and works without isolation of individual molecules, and
- the technique of *molecular screening* which isolates the biological informa-

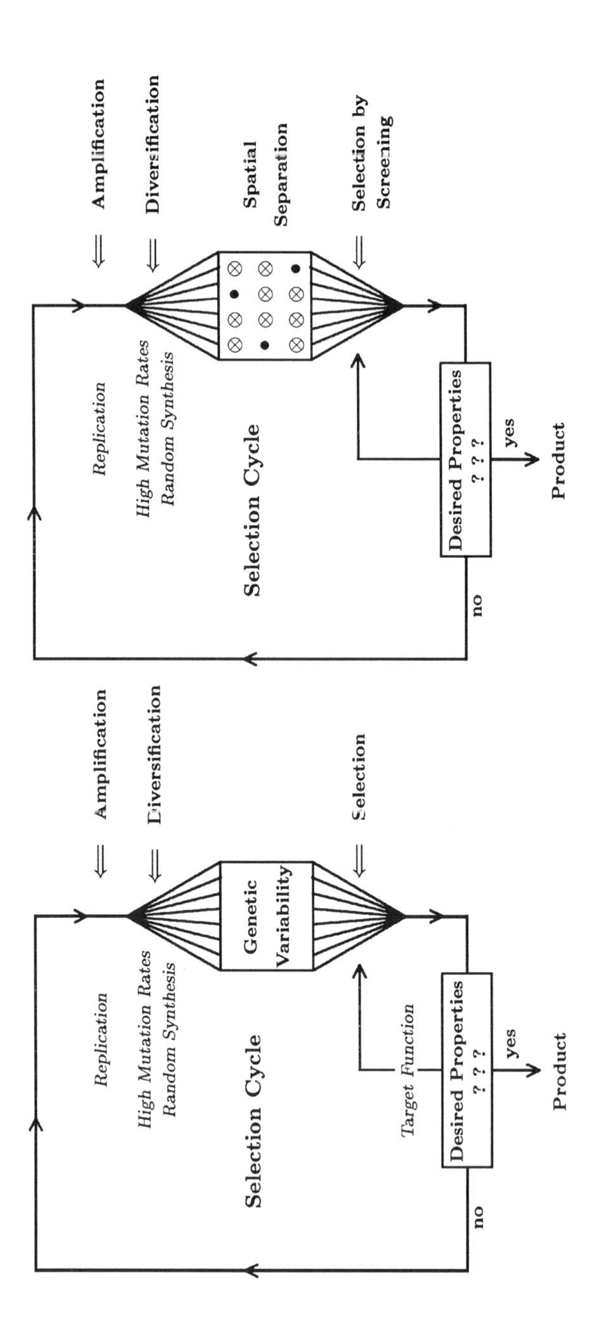

Fig. 11. Selection techniques in evolutionary biotechnology. Coupling of function and selection (l.h.s.) is contrasted with spatial separation and screening of molecules (r.h.s.).

tion carriers through spatial separation and performs testing on individual molecules or their clones (progenies of single molecules).

The two strategies require different biochemical and physical skills. The first approach requires chemical intuition and ingenious concepts in order to modify natural selection (that counts only numbers of fertile descendants) in such a way that it produces the molecules with the desired properties. The best studied and most successful technique produces RNA molecules with specific binding properties (Ellington (1994)). Target molecules are bound to the solid phase of a chromatographic column, a solution containing a variety of RNA molecules is passed through the column, and only those RNA molecules which bind to the column are used in the forthcoming selection cycles. The selection pressure is controlled by elution. The conditions for binding to the column are gradually made more stringent by changing the solvent. Some twenty selection cycles are usually sufficient to breed RNA molecules with highly specific binding properties.

An alternative strategy (figure 11) isolates individual molecules through spatial separation. Selection follows by screening at the molecular level. In order to perform evolution experiments efficiently and in sufficiently short times one has to apply massively parallel processing under automatic control. Application of microstructures based on silicon wafer technology (figure 12) provides miniaturized reaction vessels in numbers from some ten thousands to about one million on plates with up to 20 cm diameter (Schober et al. (1993)). Fast handling, processing, and screening of samples require special equipment like a micro pipette of liquid jetter type, a high-precision positioning device and a CCD-camera as recorder. Current technology allows manipulations of up to one thousand samples per minute. Under favorable circumstances the CCD-camera is able to detect a few hundred fluorescent molecules in each reaction vessel. The data given here show that a high-tech approach to evolutionary biotechnology is indeed in a position to work simultaneously with high numbers of very small samples.

Problems that become solvable by means of the microtechnology described here are manyfold and span a wide range from elaborate serial transfer experiments and molecular screening after spatial separation to the simultaneous software controlled parallel synthesis of high numbers of oligonucleotides or oligopeptides. Wafer technology appears to be particularly important for an evolutionary technique of protein design in the future. The principle of Darwinian evolution becomes applicable if proteins are evolved together with their messenger-RNAs. The key problem is indeed to provide tight coupling between the RNA and the protein derived from it by translation. A batch-system in which all RNA and protein molecules are contained in one solution would clearly not work. One possibility is to isolate individual RNA molecules prior to translation in compartments, for example in the reaction chambers on a wafer. After translation the proteins in the chambers may be subjected to screening for the desired properties. The use of a massively parallel technique is important since up to millions of variants have to be compartmentalized, translated and tested in order to perform successful searches. The messenger-RNA molecules in the reaction chambers that contain the best suited proteins are isolated and then

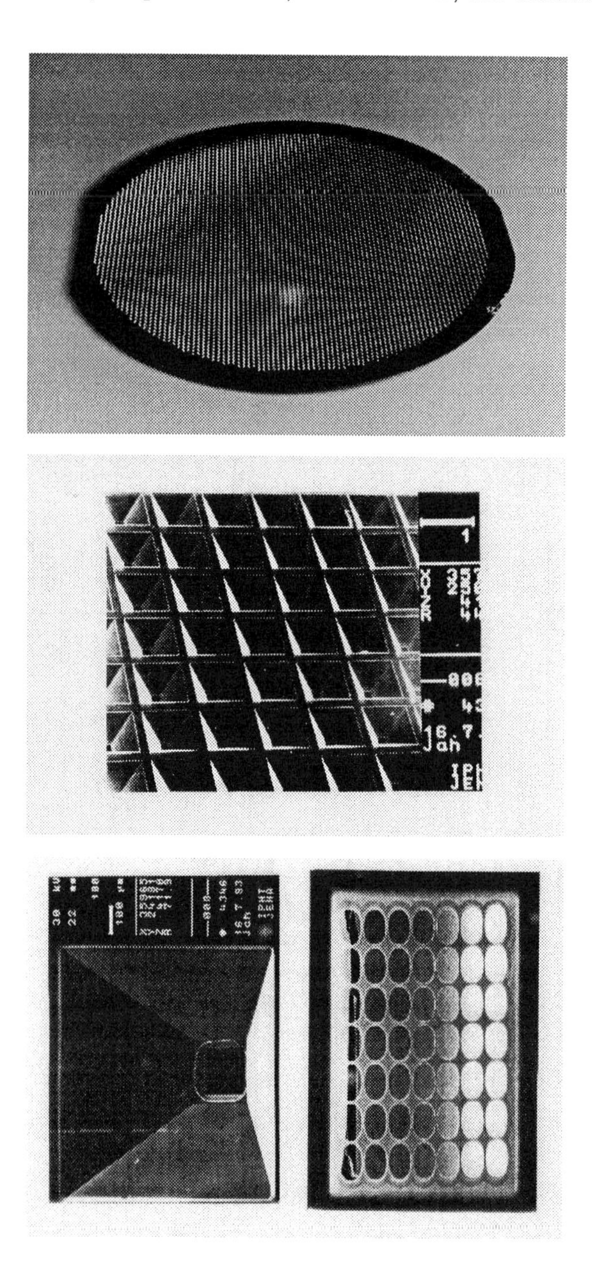

Fig. 12. Silicon wafer for massively parallel processing of tiny samples. The upper part shows a wafer of about 15 cm diameter with approximately ten thousand chambers. The lower parts present magnifications, in particular a single reaction chamber with a small sieve at the bottom that is used for filtering operations.

used for further selection cycles (as shown in figure 11). Variation is produced again through replication and mutation of RNA molecules. After an optimal gene has been produced it can be cloned and translated. Conventional genetic engineering techniques are available for protein production in larger quantities.

8 Concluding Remarks

Evolutionary dynamics visualized as an adaptive walk of a population on a complex multipeaked landscape optimizing its fitness thereby is equivalent to the conventional Darwinian scenario. This concept is based on several implicit assumptions, two of them are crucial for the validity of the model: (1) constant environment and (2) independent reproduction. The condition of constant environment may be relaxed to the requirement that environmental changes have to occur on a slower time scale than the adaptation of populations. In ecosystems part of the environmental change is caused by coevolving species. These changes are induced by the same adaptive mechanism and there is no *a priori* reason that they should occur more slowly than adaptation in the reference population. Landscape models of coevolution could easily fail therefore and should be considered with special care. Independent reproduction means that the rate constant of replication does not depend on the concentration of the reproducing entities. Host-parasite systems and symbiotic interactions are examples of such dependencies that lead to a breakdown of the landscape scenario. Current molecular evolution experiments are almost always dealing with constant environments and the replication rate constants are commonly independent of the concentrations of RNA molecules. In nature, however, none of these conditions is usually fulfilled.

What has the research on *in vitro* evolution of molecules contributed to the theory of evolution?

- Firstly, it has shown that replication and evolutionary adaptation are not exclusive privileges of cellular life. Polynucleotides were found to replicate with suitable enzymes in cell-free assays. Small oligonucleotides can be replicated even without protein assistance. Other replicators which are not based on nucleotide chemistry were found as well. Template chemistry starts to become a fascinating field in its own right. Still there are many unsolved problems in models for early and later evolution, in particular no satisfactory models can be given yet for the mechanisms leading to the origin of real novelties in biology often addressed as the great jumps in evolution. A recent paper rather states the problem than presents solutions (Szathmáry and Maynard-Smith (1995)) (For a ten years old analysis of the same problem in relation to questions concerning the origin of life see (Eigen and Schuster (1985))).
- Secondly, the old tautology debate on biological fitness has come to an end. Fitness in molecular systems can be measured independently of the survival in an evolution experiment. Evolutionary processes may be described and analyzed in the language of physics.

- Thirdly, the probability argument against a Darwinian mechanisms of evolution is invalidated by the experimental proof of target oriented adaptation found for evolution experiments with RNA molecules. Evolutionary search is substantially facilitated by the properties of sequence space and shape space, in particular by the fact that the evolutionarily relevant part of shape space is covered by a tiny fraction of sequence space.

Molecular evolution, in essence, has established the basic kinetic mechanisms of genetics. In the case of RNA replication and mutation the reaction mechanisms were resolved to about the same level of details as with other polymerization reactions in physical chemistry. Several aspects of a comprehensive theory of evolution, however, are still missing. For example, the integration of cellular metabolism and genetic control into a comprehensive theory of molecular genetics has not yet been achieved. Morphogenesis and development of multicellular organisms need to be incorporated into evolutionary theory.

Nevertheless the extension of molecular biology into biophysical and even organic chemistry as indicated here is predestined to find applications in biotechnology. Principles and mechanisms are taken from the living world, but new environments and controlled experimental conditions allow to proceed towards hitherto unexplored scientific territories. In a more speculative view of future developments in molecular evolutionary biology even the material carriers of biological information will be modified and new classes of polymers sharing only the capability of evolution through replication, mutation and selection with their natural relatives will be discovered, designed for and applied to a variety of different tasks.

Acknowledgments

Financial support by the Austrian *Fonds zur Förderung der wissenschaftlichen Forschung* (Projects S 5305-PHY, P 8526-MOB, and P 9942-PHY) and by the Commission of the European Communities (Contracts PSS*0396 and PSS*0884) is gratefully acknowledged.

References

Biebricher, C. K., Eigen, M. (1988): *Kinetics of RNA replication by Qβ replicase,* in E. Domingo, J. J. Holland, P. Ahlquist (eds.), RNA Genetics, Vol.I: RNA directed virus replication CRC Press: Boca Raton (Fl.), 1–21.

Biebricher, C. K., Luce, R. (1993): *Sequence analysis of RNA species synthesized by Qβ replicase without template,* Biochemistry **32**, 4848–4854.

Bonhoeffer, S., McCaskill, J. S., Stadler, P. F., Schuster, P. (1993): *RNA multi-structure landscapes,* Eur. Biophys. J. **22**, 13–24.

Dayhoff, M. O., Park, C. M. (1969): *Cytochrome-c: Building a phylogenetic tree,* in M. O. Dayhoff (ed.), Atlas of protein sequence and structure, National Biomedical Research Foundation: Silver Springs (Md.), Vol.4, 7–16.

Derrida, B., Peliti, L. (1991): *Evolution in a flat fitness landscape*, Bull. Math. Biol. **53**, 355–382.

Eigen, M. (1971): *Selforganization of matter and the evolution of biological macromolecules,* Naturwissenschaften **58**, 465–523.

Eigen, M. , Gardiner, W. (1984): *Evolutionary molecular engineering based on RNA replication*, Pure Appl. Chem. **56**, 967–978.

Eigen, M., McCaskill, J., Schuster, P. (1988): *The molecular quasispecies – An abridged account*, J. Phys. Chem. **92**, 6881–689.

Eigen, M., McCaskill, J., Schuster, P. (1989): *The molecular quasispecies*, Adv. Chem. Phys. **75**, 149–263.

Eigen, M., Schuster, P. (1985): *Stages of emerging life - Five principles of early organization*, J. Mol. Evol. **19**, 47–61.

Ellington A. D. (1994): *Aptamers achieve the desired recognition*, Current Biology **4**, 427–429.

Feinberg, M. (1977): *Mathematical aspects of mass action kinetics,* in L. Lapidus, N. R. Amundson (eds.), Chemical Reactor Theory. A Review, Prentice-Hall, Inc. Englewood Cliffs (NJ), 1–78.

Fontana, W., Konings, D. A. M., Stadler, P. F., Schuster, P. (1993): *Statistics of RNA secondary structures,* Biopolymers **33**,1389–1404.

Fontana, W., Schnabl, W., Schuster, P. (1989): *Physical aspects of evolutionary optimization and adaptation*, Phys. Rev. A **40**, 3301–3321.

Fontana, W., Schuster, P. (1987): *A computer model of evolutionary optimization*, Biophys. Chem. **26**, 123–147.

Fontana, W., Stadler, P. F., Bornberg-Bauer, E. G., Griesmacher, T., Hofacker, I. L., Tacker, M., Tarazona, P., Weinberger, E. D., Schuster, P. (1993): *RNA folding and combinatory landscapes,* Phys. Rev. E **47**, 2083–2099.

Tacker, M., Fontana, W., Stadler, P. F., Schuster, P. (1994): *Statistics of RNA melting kinetics,* Eur. Biophys. J. **23**, 29–38.

Gillespie, D. T. (1977): *Exact stochastic simulation of coupled chemical reactions*, J. Phys. Chem. **81**, 2340–2361.

Grüner, W. (1994): *Evolutionary optimization on RNA folding landscapes,* Doctoral Thesis, Universität Wien.

Hamming, R. W. (1986): *Coding and information theory*, 2nd Ed., Prentice Hall, Englewood Cliffs (NJ).

Hofacker, I. L., Fontana, W., Stadler, P. F., Bonhoeffer, L. S., Tacker, M., Schuster, P. (1994): *Fast folding and comparison of RNA secondary structures,* Mh. Chem. **125**, 167–188.

Huynen, M., Fontana, W., Stadler, P. F. (1994): Proc. Natl. Acad. Sci. USA, in press.

Joyce, G. F. (1992): *Directed molecular evolution*, Sci. Am. **267**(6), 48–55.

Kauffman, S. A. (1986): *Autocatalytic sets of proteins*, J. Theor. Biol. **119**, 1–24.

Kauffman, S. A. (1994): *Applied molecular evolution*, J. Theor. Biol. **157**, 1–7.

Kauffman, S. A., Levin, S. (1987): *Towards a general theory of adaptive walks on rugged landscapes,* J. Theor. Biol. **128**, 11–45.

Kimura, M. (1983): *The neutral theory of molecular evolution*, Cambridge University Press. Cambridge (U.K.).

Pley, H. W., Flaherty, K. M., McKay, D. B. (1994): *Three-dimensional structure of a hammerhead ribozyme,* Nature **372**, 68–74.

Reidys, C. (1995): *Neutral networks of RNA secondary structures,* Doctoral Thesis, Friedrich-Schiller-Universität Jena.

Reidys, C., Stadler, P. F., Schuster, P. (1995): *Generic properties of combinatory maps. Neutral networks of RNA secondary structures*, submitted to Bull. Math. Biol.

Schober, A., Günther, R., Schwienhorst, A., Lindemann, B. (1993): *Accurate high speed handling of very small biological samples*, BioTechniques **15**, 324–329.

Schuster, P., Fontana, W., Stadler, P. F., Hofacker, I. L. (1994): *From sequences to shapes and back: a case study in RNA secondary structures,* Proc. Roy. Soc. Lond. B **255**, 279–284.

Schuster, P., Stadler, P. F. (1994): *Landscapes: complex optimization problems and biopolymer structures,* Computers Chem. **18**, 295–324.

Schuster, P. (1995): *How to search for RNA structures. Theoretical concepts in evolutionary biotechnology,* J. Biotechnology **41**, 239–258.

Spiegelman, S. (1971): *An approach to the experimental analysis of precellular evolution,* Quart. Rev. Biophys **4**, 213–253.

Szathmáry, E., Maynard-Smith, J. (1995): *The major evolutionary transitions*, Nature **374**, 227–232.

Weber, J., Reidys, C., Forst, C., Schuster, P. (1995): Unpublished results. Jena (Germany).

Wright, S. (1932): *The roles of mutation, inbreeding, crossbreeding and selection in evolution,* in D. F. Jones (ed.), Proceedings of the Sixth International Congress of Genetics, Vol.1, 356–366. Ithaca(NY).

Smart Bacterial Colonies

Eshel Ben-Jacob[1], Inon Cohen[1] and Andras Czirók[1,2]

[1] School of Physics and Astronomy, Raymond & Beverly Sackler Faculty of Exact Sciences, Tel-Aviv University, Tel-Aviv 69978, ISRAEL
[2] Department of Atomic Physics, Eötvös University, Budapest, Puskin u 5-7, 1088 HUNGARY

1 Introduction

Many natural phenomena, in living and non-living systems alike, display the spontaneous emergence of patterns; growth of snowflakes, aggregation of soot particles, solidification of metals, formation of corals, growth of bacterial colonies and cell differentiation during embryonic development are just a few. The exciting developments of the past decade in the understanding of diffusive patterning in non-living systems [1, 2, 3, 4] hold a promise for a unified theoretical framework (for non-living systems), that might also pave the road towards a new understanding of processes in living systems.

Motivated by the above, we set out to study cooperative microbial behavior under stress. Bacterial colonies exhibit a far richer behavior than patterning of non-living systems, reflecting the additional levels of complexity involved. The building blocks of the colonies are themselves living systems, each with its own autonomous self-interest and internal degrees of freedom. At the same time, efficient adaptation of the colony to the imposed growth conditions requires adaptive self-organization – which can only be achieved via cooperative behavior of the individual bacteria.

The patterns exhibited by the colonies suggest that they should be viewed as adaptive cybernetic systems or multi-cellular organisms, possessing impressive capabilities for coping with hostile environmental conditions and survive them.

To achieve this, the bacteria have developed various communication channels [5, 6, 7]: from direct (by contact) bacterium-bacterium physical and chemical interaction, through indirect physical and chemical interactions via marks left on the agar surface and chemical (chemotactic) signalling, to genetic communication via exchange of genetic material [8].

The bacteria are "smart" as they develop, via designed genetic changes, new patterns which are better adapted to the growth conditions. It implies that the organization of the colony during complex patterning can directly affect the adaptive mutagenesis of the individual bacterium, and vice-versa [9, 10]. We present the new picture of genome cybernetics based on new conceptual elements – the cybernators [9, 10]. These elements can cause genetic changes in the genome of an individual bacterium but are regulated by the state of the colony as a whole. Thus they provide the colony with means to affect genetic changes in the genome of the individual bacteria, changes which in turn allow the colony more efficient self-organization in its attempt to cope with the environmental conditions.

2 Microevolution in a Petri-dish

Traditionally, bacterial colonies are grown in the lab on a thick substrate with a high nutrient level and intermediate agar concentration. Under such "friendly" conditions, the colonies typically develop simple compact patterns. However, in nature bacteria must regularly cope with adverse environmental conditions, one of which is low levels of nutrients. Drawing on our understanding of diffusive patterning in non-living systems, we can say that branching fractal patterns are expected when bacteria are grown on thin, poor substrate. The bacteria reproduction rate that determines the growth rate of the colony, is limited by the local level of nutrients concentration available for the bacteria. The latter is limited, for this poor substrate, by the diffusion of nutrients towards the colony. Hence, the growth of colonies appears to be similar to diffusion limited growth in non-living systems, such as solidification from a supersaturated solution, growth in a Hele-Shaw cell, electro-chemical deposition, etc. [1, 4]. From the study of diffusive patterning in non-living systems, we understand that the diffusion field drives the system towards decorated (on many length scales) irregular fractal shapes. Moreover, we now understand the competition between the action of the diffusion field and that of the microscopic effects (surface tension and surface kinetics) in the determination of the evolved pattern [1, 2, 3, 4, 11, 12, 13]. Generally speaking, the morphology changes from fractal-like [14, 15], when the diffusive instability dominates the growth, to compact patterns when the diffusive instability is weaker than the action of the microscopic effects.

In agreement with the above, Matsushita et al. [16, 17] have demonstrated fractal bacterial patterns during growth on low-nutrient substrates.

2.1 Bursts of Branching Growth

One might naively expect that whenever bacteria are grown on poor substrate, fractal patterns should be observed. However, this is not the case. For example, Ben-Jacob et al. have shown that the non-motile (non-motile on agar but motile in liquid) strain *Bacillus subtilis* 168 does not exhibit the expected branching patterns (Fig. 1a). When such bacteria are grown on poor substrate, they are subject to a selective pressure on the colony towards a branching colony. This, together with the fact that motility mutants are known, suggests that poor substrate provide selection for "branching strain". Indeed, as Ben-Jacob et al. have shown [18, 9], during growth of *Bacillus subtilis* 168 on low-nutrient agar, occasionally bursts are observed of a spectacular new mode of growth e xhibiting branching patterns (Fig. 1b). This new mode of growth was found to be inheritable, and the geometrical (morphological) properties are transferable by a single bacterium. We name this tip-splitting mode $\mathcal{T}$ morphotype. The term morphotype (replacing 'mode') was proposed by D. Gutnick (personal communication) to describe an inheritable morphological character of the colony that can be transferred by a single bacterium.

2.2 Cybernators and Cooperative Genetic Changes

Under the above conditions, the $\mathcal{T}$ morphotype colonies propagate much faster than colonies of the original *Bacillus subtilis* 168 (referred to as $\mathcal{B}$ morphotype). Hence the bacteria are "smart" enough to develop, via genetic changes, a new strategy for better adaptation to the environment. "Smart" because, as we assume, the burst of the new morphotype is not a result of an ordinary random mutation that was post-selected by the environment. Rather, it results from a novel designed genetic change which is even more complex than the adaptive mutations (see Ref. [19] for review about adaptive mutations). The bacteria of the new morphotype are not necessarily better adapted, as individuals, to the environment. But their new characteristics enable the colony to develop a pattern which is better adapted to the growth conditions. Thus a control mechanism from the macro-level of the colony to the micro-level of the individual bacteria must exist. In addition, since one bacterium might not be advantageous, a finite nucleation of the new morphotype must be formed for the morphotype to burst out. In Ref. [18] Ben-Jacob et al. concluded that: "Along the above assumptions, the colony organization (being the environment) can directly affect the genetic metamorphosis of the individuals. Hence, we expect to observe synchronized, autocatalytic and cooperative genetic variations of the colony, either spontaneous or in response to imposed growth conditions."

This framework is extended in Refs. [9, 10]. It is proposed to view the autonomous elements in the genome as cybernetic units (cybernators) in order to describe their functional role. An element may be, for example, a specific single macromolecule, a combination of molecules or even a collective excitation of the genome performing the specific function. Generally, it should be viewed as a conceptual unit regardless of its actual nature. The cybernators are cybernetic elements whose function is regulated by colony parameters such as growth kinetics, bacteria density, density variations, level of stress, etc. The crucial point is that, since the cybernators activity is regulated by the state of the colony, it can produce in the genome of the individual bacterium changes which are beneficial to the colony as a whole. Through cybernators, the colony organization can directly affect the genetic metamorphosis of the individuals in various forms, ranging from synchronized or autocatalytic genetic changes to cooperative ones. In this new picture, the genome is more than an information storage unit; it has the capability to function as a data processing and problem solving unit possessing its own direct channels for external information, including information about the state of the colony. In addition, it is capable of restructuring itself according to the outcome of the data processing. For these reasons, Ben-Jacob et al. proposed to view the genome as an adaptive cybernetic unit [18, 9].

Above we presented an hypothesis describing a cybernetic capacity of the bacteria, a capacity that serves to regulate three levels of interactions: the cybernator, the bacterium and the colony. The 'interest' of the cybernator serves

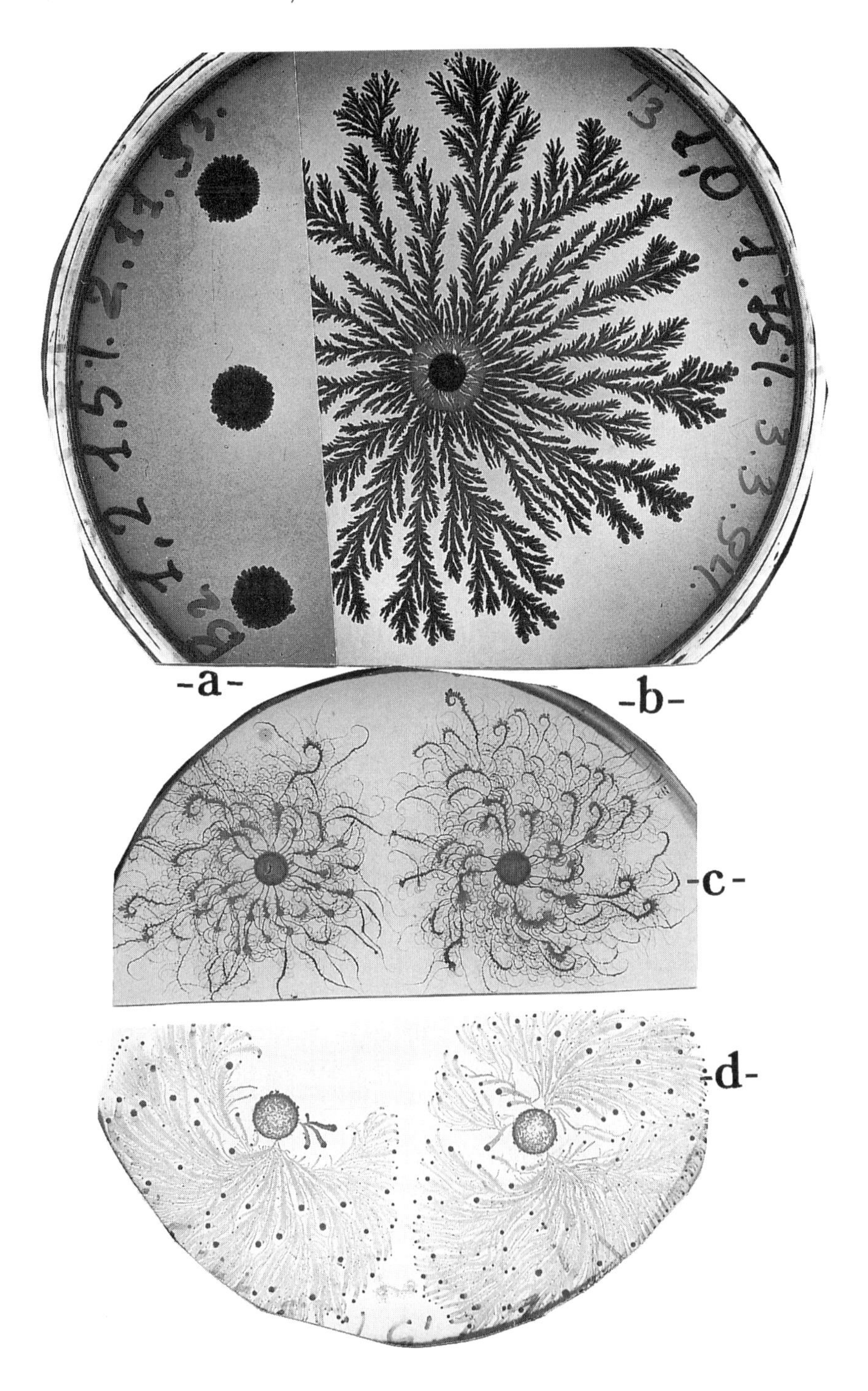

Fig. 1. Examples of the four morphotypes (each figure shows a section of a petri-dish). (a), (b), (c) and (d) are for $\mathcal{B}$, $\mathcal{T}$, $\mathcal{C}$ and $\mathcal{V}$ morphotypes, respectively.

the 'purpose' of the colony by genetically readjusting the genome of the single bacterium. The cybernator provides a singular feedback mechanism as the colony uses it to induce changes in the single bacterium, thus leading to a consistent adaptive self-organization of the colony.

2.3 Burst of the Chiral Morphotype

During growth of $\mathcal{T}$ morphotype on a soft substrate (about 1% agar concentration), bursts of new patterns that overgrow the original branching one are observed. The new patterns consist of thinner branches, all with the same handedness of strong twisting (Fig. 1c). Hence we refer to these new patterns as chiral ones. Again, the new morphological character is inheritable, transferable by a single bacterium, and is stable at a range of growth conditions, so we view it as a distinct morphotype – $\mathcal{C}$ (chiral) morphotype.

The $\mathcal{T} \to \mathcal{C}$ transitions on soft agar are in agreement with the working hypothesis that the transitions are to the faster morphotype – the one whose colony propagates faster [18, 9]. Indeed, the reverse $\mathcal{C} \to \mathcal{T}$ transitions are observed on hard agar where $\mathcal{T}$ morphotype is the faster one. Also in agreement with the hypothesis, *Bacillus subtilis* 168 colonies grown on soft agar show burst of $\mathcal{C}$ morphotype.

2.4 Bursts of the Vortex Morphotype

A fourth morphotype is shown in Fig. 1d. That morphotype was given the name $\mathcal{V}$ (vortex) morphotype, as its colonies consist of branches, each has a leading vortex. The latter is a collection of many bacteria, all move around a common center. The $\mathcal{V}$ morphotype was isolated from *Bacillus subtilis* in a manner similar to that of $\mathcal{T}$ morphotype, but at different growth conditions – most noticeably at temperature 30°C instead of 37°C.

Following the working hypothesis, transition experiments at the following three sets of growth conditions were performed: I. 37°C and soft agar (where $\mathcal{C}$ morphotype is fastest). II. 37°C and intermediate agar (where $\mathcal{T}$ morphotype is fastest). III. 30°C and hard agar (where $\mathcal{V}$ morphotype is fastest). The experiments were started with various strains of *Bacillus subtilis*. In all cases $\mathcal{B} \to \mathcal{C}$, $\mathcal{B} \to \mathcal{T}$ and $\mathcal{B} \to \mathcal{V}$ transitions were observed in conditions I, II and III respectively, all in agreement with the working hypothesis of the "fastest growing morphology" selection principle.

3 Observed Complex Patterns

Each of the three morphotypes exhibits all sorts of patterns as the growth conditions, i.e., the peptone level and agar concentration, are varied. Below we present only small fraction of the observed patterns. The kaleidoscope of shapes may be grouped into a number of "essential" patterns, each is characteristic for a range of growth conditions and is indicative of a specific biological feature being dominant. Thus observed patterns can be organized within a morphology diagram and there is a velocity-pattern correlation.

3.1 Patterns of the $\mathcal{T}$ Morphotype

At very high peptone levels (above about 10 g/l) and for typical values of agar concentration (1.5% to 2%) the patterns are compact. At somewhat lower but still high peptone levels (about 5–10 g/l) the patterns, which are very reminiscent of Hele-Shaw patterns, may be characterized as dense fingers (Fig. 2a). For intermediate peptone levels, branching patterns with lower fractal dimension (reminiscent of electro-chemical deposition) are observed (Fig. 2b). Surprisingly, at even lower peptone levels, the colonies again adopt a more organized structures of fine radial branches (Fig. 2c). The growth velocity measurements show distinct regimes of response, each corresponds to one of the essential morphology we have characterized above (for details see Refs. [9, 10]).

A closer look at an individual branch at high peptone levels reveals a phenomenon of density variations. These density variations are accumulations of bacteria in layers, hence referred to as 3-dimensional structures. The aggregates can form spots and ridges which are either scattered randomly, ordered in rows or organized in a leaf-veins-like structure (Fig. 2d).

Under the microscope, bacteria are seen performing a swimming in a wetting fluid. The latter is excreted by the bacteria and/or drawn by the bacteria from the agar. The bacterial movement is confined to this fluid. Wherever the bacteria are active, the envelope (i.e. the boundary of the fluid) propagates slowly, as a result of the bacterial movement.

The observations reveal also that the bacteria are more active at the leading tips of the branches, while further down the cells slow down and eventually sporulate.

3.2 Patterns of the $\mathcal{C}$ Morphotype

The patterns of $\mathcal{C}$ morphotype exhibit a behavior similar to that of the $\mathcal{T}$ morphotype. In general, they are compact at high peptone levels and become ramified (fractal) at low peptone levels. At very high peptone levels and high agar concentration, $\mathcal{C}$ morphotype conceals its chiral nature and exhibits branching growth similar to that of $\mathcal{T}$ morphotype .

The optical microscope observations indicate that again the bacteria move within a wetting fluid. The bacteria are long relative to $\mathcal{T}$ morphotype bacteria, and the movement appears correlated in orientation. The bacteria move forward and backward in a movement that seems to be between swimming and gliding. Each branch tip maintains its shape, and at the same time the tips keep twisting with specific handedness while propagating. Electron microscope observations do not reveal chiral structure on the bacterial membrane [20].

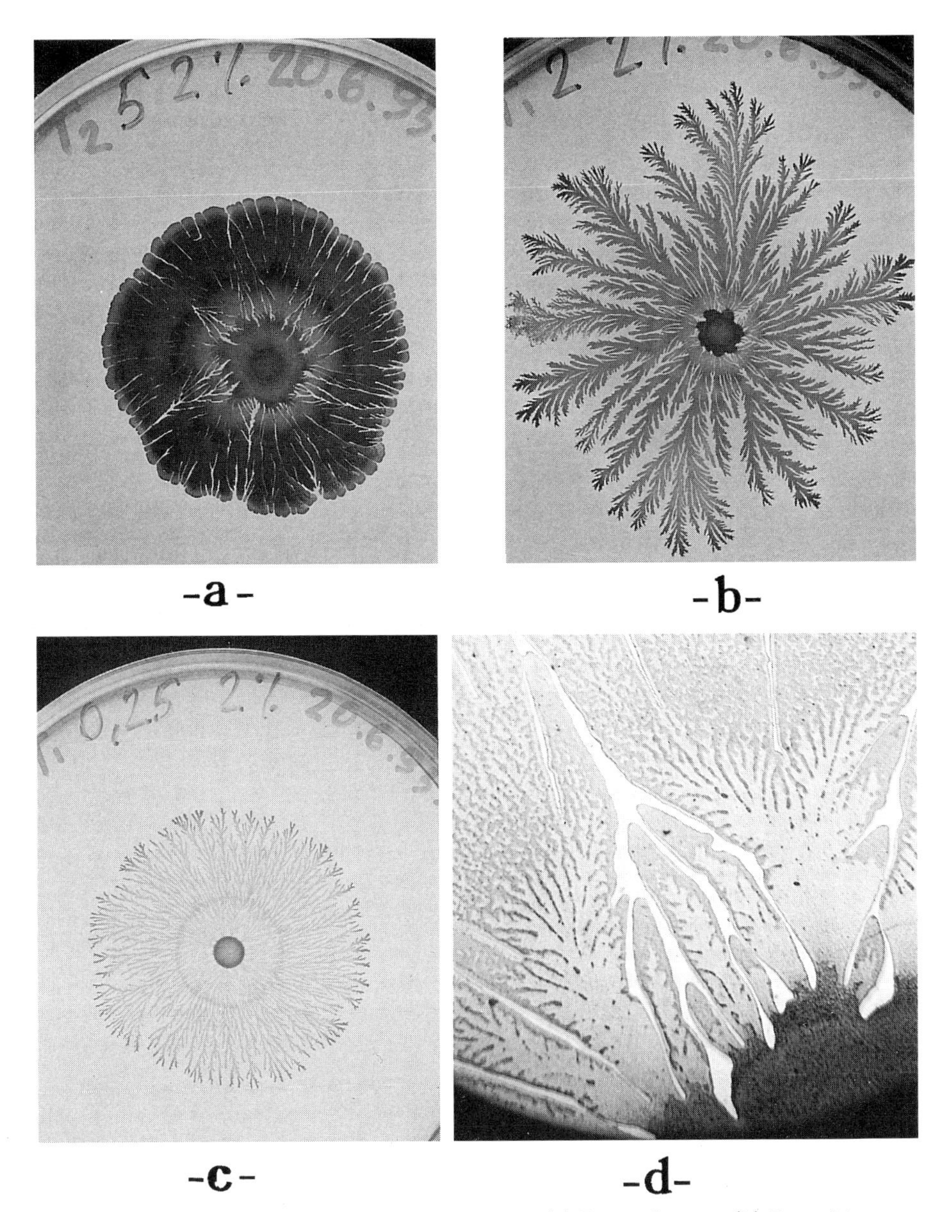

Fig. 2. Patterns exhibited by the $\mathcal{T}$ morphotype. (a) Dense fingers. (b) Branching pattern. (c) Fine radial branching. (d) Closer look at 3D structure.

3.3 Closer Look at the $\mathcal{V}$ Morphotype

Typically the $\mathcal{V}$ morphotype exhibits branching patterns. Each branch is produced by a leading droplet (a vortex) and has side branches, each with its own leading droplet. Each droplet consists of many bacteria that spin around a common center (hence named vortex) at typical velocity of $10\,\mu$m per second, being very sensitive to the growth conditions and the location in the colony. Depending on growth conditions, the number of bacteria in a vortex may vary from hundreds to millions. In many cases, the branches have well pronounced global twist (weak chirality) always with the same handedness. For some growth conditions (soft agar), the vortices do not form and the growth is tip-splitting-like [9].

During some stages of the growth we observed gilding of droplets which are very reminiscent of the "worm" motion of slime mold or schools of multicellular organisms. Typically, the collective movement is observed in the trail behind the leading vortex.

Microscope observations of the vortex morphotype reveal that the bacteria movement differs from the swimming of the $\mathcal{T}$ morphotype. The former is best described as a "bacterium-fluid" - the bacteria have a collective motion very much like that of a viscous fluid. The movement of a group of few bacteria, or even a single one, suggest a gilding mechanism. We neither observe tumbling as in T morphotype, nor movement forward and backward as in $\mathcal{C}$ morphotype. The motion is always forward, the bacteria are self propelled (self-driven particles) with a strong velocity-velocity interaction [21]: the bacteria tend to align their direction and move with the same velocity.

4 Modeling the Growth

4.1 The Generic Modeling Approach

How should we approach modeling of the complex bacterial patterning? With present computational power it is natural to use computer models as a main tool in the study of complex systems. However, one must be careful not to be trapped in the "reminiscence syndrome", described by J. D. Cowan (page 74 of [22]) as the tendency to devise a set of rules which will mimic some aspect of the observed phenomena and then, to quote J. D. Cowan "They say: 'Look, isn't this reminiscent of a biological or physical phenomenon!' They jump in right away as if it's a decent model for the phenomenon, and usually of course it's just got some accidental features that make it look like something." Yet, the reminiscence modeling approach has some indirect value. True, doing so does not reveal (directly) the biological functions and behavior. However, it does reflect understanding of geometrical and temporal features of the patterns, which indirectly might help in revealing the underlying biological principles. Another extreme is the "realistic modeling" approach, where one constructs an algorithm that includes in details all the known biological facts about the system. Such an approach sets a trajectory of ever including more and more details (vs.

generalized features). The model keeps evolving to include so many details that it loses all predictive power.

Here we try to promote another approach – the "generic modeling" approach [23, 24, 25]. We seek to elicit, from the experimental observations and the biological knowledge, the generic features and basic principles needed to explain the biological behavior and to include these features in the model. We will demonstrate that such modeling, with close comparison to experimental observations, can be used as a research tool to reveal new understanding of the biological systems.

4.2 The Communicating Walkers Model

The communicating walkers model is inspired by the diffusion-transition scheme used to study solidification from supersaturated solutions [26, 27, 28]. The former is also an hybridization of the "continuous" and "atomistic" approaches used in the study of non-living systems. The diffusion of the chemicals is handled by solving a continuous diffusion equation (including sources and sinks) on a triagonal lattice, while bacteria are represented by walkers allowing a more detailed description. In a typical experiment there are 10^9–10^{10} bacteria in a petri-dish at the end of the growth. Hence it is impractical to incorporate into the model each and every bacterium; instead, each of the walkers represents about 10^4-10^5 bacteria so that we work with 10^4-10^6 walkers in one run.

Although the food source in our experiments is peptone (and not a single carbon source), we represent the diffusion of nutrients by solving the diffusion equation for a single agent whose concentration is denoted by $C(\mathbf{r},t)$:

$$\frac{\partial C}{\partial t} = D_C \nabla^2 C - \sigma C_{\text{consumed}} \ , \tag{1}$$

where σ is the walkers density and C_{consumed} is the consumption of food by a walker, to be specified below. The equation is solved on the triagonal lattice using Dirichlet boundary conditions. We start the simulation with an uniform distribution C_0. The diffusion constant D_C is typically (depending on agar dryness) $10^{-4} - 10^{-6}\,\text{cm}^2/\text{sec}$.

We represent the metabolic state of the i-th walker by an 'internal energy' E_i. The rate of change of the internal energy is given by

$$\frac{dE_i}{dt} = \kappa C_{\text{consumed}} - \frac{E_m}{\tau_R} \tag{2}$$

Where κ is a conversion factor from food to internal energy ($\kappa \cong 5\cdot 10^3$ cal/g). E_m is the total energy loss over the minimal reproduction time τ_R. This energy loss is for all metabolic processes, excluding energy loss for cell division. C_{consumed} is

$$C_{\text{consumed}} \equiv \min\left(\Omega_C, \Omega'_C\right) \ , \tag{3}$$

i.e., the minimum between the maximal rate of food consumption Ω_C, and Ω'_C the rate of food consumption as limited by the locally available food. Assuming maximal reproduction time of 20 minutes and 3×10^{-12} g of food consumed by a single bacterium over this time, $\Omega_C \cong 2 \times 10^{-10}$ g/sec. At low level of food, the consumption rate is Ω'_C, which is proportional to the local food concentration $C((r),t)$ and inversely proportional to the bacterial density σ, for large σ. In the simulations we use about 10^5–10^6 lattice cells. As we show below, a typical simulated colony has a density of about 10 walkers per lattice cell. However, 1 g/l peptone in the experiments translates to about 10^{-7}–10^{-8} g per lattice cell which is sufficient to produce about one walker. Hence for such peptone level the growth is limited by nutrients diffusion towards the colony – diffusion limited growth. Moreover, walkers far behind the colony front do not have sufficient food and cannot "waste" energy in unessential processes (such as movement).

To incorporate the swimming of the bacteria into the model, at each time step each of the active walkers (motile and metabolizing) moves from its location $\mathbf{r}_i$ a step of size d at a random angle Θ to a new location $\mathbf{r}'_i$. If $\mathbf{r}'_i$ is outside the envelope of the wetting fluid, the walker does not move, but a counter on the segment of the envelope which would have been crossed by the movement $\mathbf{r}_i \to \mathbf{r}'_i$ is increased by one. When the segment counter reaches a specified number of hits, N_c, the envelope propagates one lattice step and an additional lattice cell is added to the colony. Note that N_c is related to the agar concentration, as more "collisions" are needed to push the envelope on a harder substrate.

Results of numerical simulations of the model are shown in Fig. 3a. One time step in the simulations corresponds to about 1 sec. A typical run is up to about $10^4 - 10^5$ time steps which translates to about two days of bacterial growth. Such simulations take on a RISK machine about 2.5 hours, which is shorter by a factor of 20 relative to "real life". As in the case of real bacterial colonies, the patterns are compact at high peptone levels and become fractal with decreasing food level. The branching pattern and the constant, in time, growth velocity are a manifestation of the diffusion field instability. However, while numerical results do capture some important features of the experimentally observed patterns, some other critical features, such as the ability of the bacteria to develop organized patterns at very low peptone levels (instead of more ramified structures as would be suggested by the diffusion instability), the details of the growth velocity as a function of peptone level, and the 3-dimensional structures at high peptone levels, are not accounted for by the model at this stage.

Chemotaxis. We propose that the above features are attained by varying the relative strength of three kinds of chemotactic mechanisms, as we describe below.

Generally, chemotaxis means changes in the movement of the bacteria in response to a gradient of certain chemical fields [29, 30, 7]. The movement is biased along the gradient either in the gradient direction or in the opposite direction. Usually chemotactic response means a response to an externally produced field like in the case of food chemotaxis. However, the chemotactic response can be also to a field produced directly or indirectly by the bacteria. We will refer to this case as chemotaxis signalling.

A bacterium senses the local concentration of a chemical via membrane receptors binding the chemical's molecules. The bacterium actually measure changes in the fraction of occupied receptors and not directly changes in the chemical's concentration. Thus, the term in the model representing the bacterial response to the chemical gradient should be multiplied by a prefactor, which is proportional to $(constant + chemical\ concentration)^{-2}$. This dependence on the chemical concentration is know as the "receptor law".

Repulsive signaling chemotaxis. We believe that the fine radial branching patterns at very low peptone levels result from the action of a long-range chemorepulsive signalling. From the study of non-living systems we know that in the same manner an external diffusion field leads to the diffusion instability, an internal diffusion field will stabilize the growth. Motivated by this we have proposed [23] that the stressed bacteria inside the colony emit a chemorepellent which acts as a stabilizing internal diffusion field. It is biologically reasonable that when lacking nutrients, before going through sporulation, bacteria first emit (either purposely or as a byproduct) a material causing other bacteria to move away, so that each bacterium will have more food. In Fig. 3b we demonstrate the dramatic effect of the repulsive chemotactic signalling.

Food chemotaxis. Chemotaxis towards high concentration of nutrients – food chemotaxis – is a well studied phenomenon in bacteria. We propose that this mechanism is the dominant chemotactic response for intermediate peptone levels. It should increase the growth velocity as the bacteria have an additional outward driving force. At the same time it should render the pattern fractal as it amplifies the diffusion instability. We demonstrate the validity of these predictions by inclusion of food chemotaxis in the communicating walkers model (Fig. 3c).

Attractive signaling chemotaxis. The characteristic length of both types of chemotaxis fields used above – nutrient chemotaxis and repulsive chemotaxis signaling – is longer then the typical branch width, even for the widest branches in dense fingering patterns. The length scale of the 3-dimensional structures observed at high peptone levels, on the other hand, indicates the existence of a dynamical process with a characteristic length shorter than the branch width, but much longer than the bacterial size. The accumulation of bacteria in the aggregates brings to mind the existence of short range self-attracting mechanism. Hence we propose that at high peptone levels the effects of the chemorepulsive signaling and the food chemotaxis weaken, and a new mechanism of attractive chemotaxis becomes the dominant one. In Fig. 3e we show an example of numerical simulations when chemoattractant is included.

Universality in bacterial patterns. We demonstrated the role of chemotaxis interplay in complex patterning of $\mathcal{T}$ morphotype. Is it a special ability of the $\mathcal{T}$ morphotype or do other bacteria employ similar strategies during their adaptive self-organization? We propose that it is an universal tactic and other

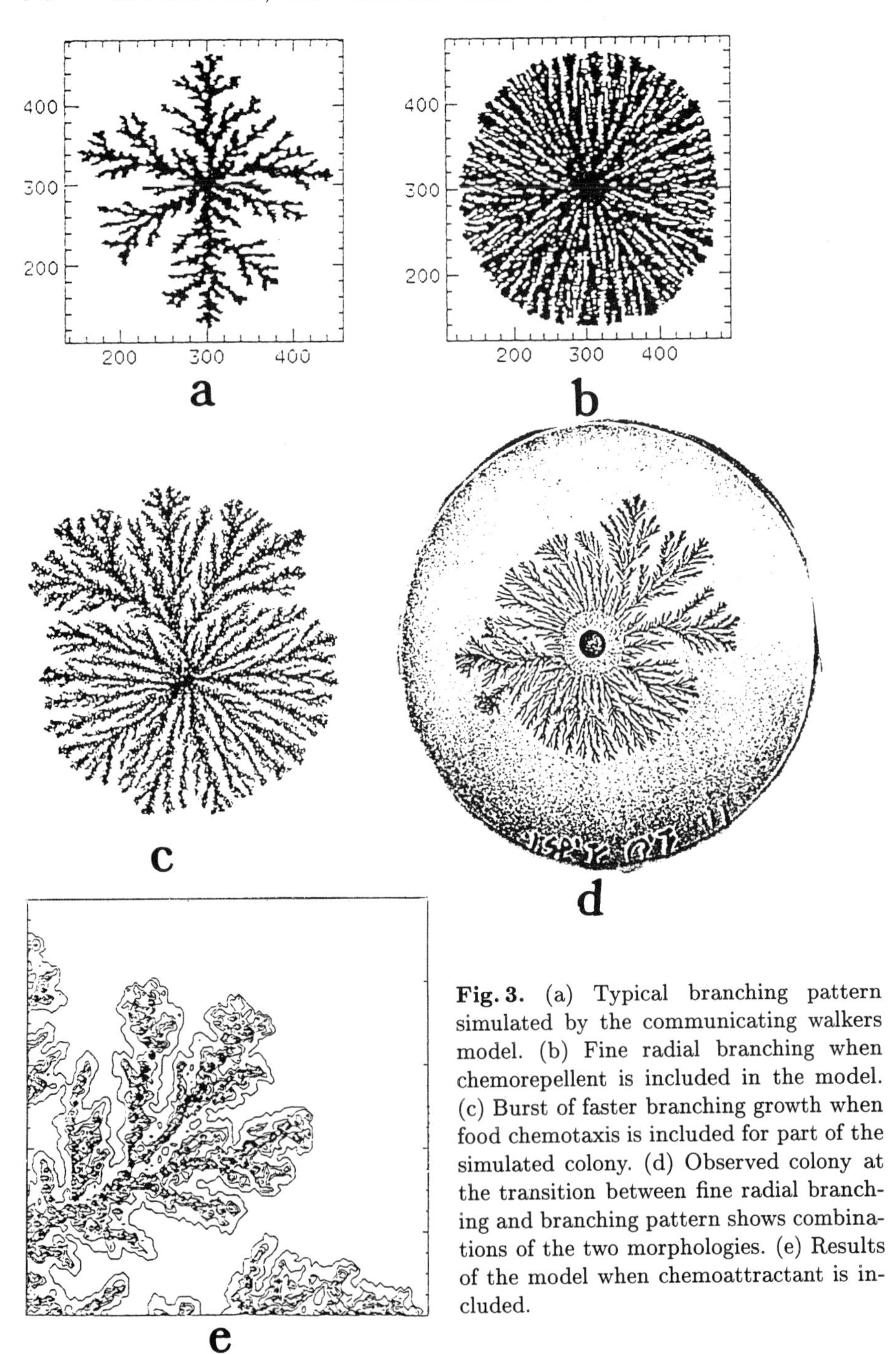

Fig. 3. (a) Typical branching pattern simulated by the communicating walkers model. (b) Fine radial branching when chemorepellent is included in the model. (c) Burst of faster branching growth when food chemotaxis is included for part of the simulated colony. (d) Observed colony at the transition between fine radial branching and branching pattern shows combinations of the two morphologies. (e) Results of the model when chemoattractant is included.

bacteria, such as *Esherichia coli*, *Salmonella typhimurium* and *Proteus mirabilis*, cope with hostile conditions in a similar manner, using an interplay of long-range and short-range chemosignalling as they form a profusion of complex patterns (see Refs. [31, 32, 33, 34] and references therein). In Sec. 4.4 we show the crucial role of such a strategy in the formation of vortex colonies.

4.3 From Flagella Handedness to the Macroscopic Chirality

We propose that the handedness of the flagella acts as a singular perturbation. It is amplified by the growth conditions and leads to the observed macroscopic chirality [35]. The situation is similar to the observed six-fold symmetry of snowflakes resulting from the molecular anisotropy of water.

It is known that flagella have specific chirality [36, 37, 38]. Ordinarily, as the bundles of flagella unfold, the bacteria tumble and end up at a new random angle (relative to the original one).The situation is different for quasi 2-dimensional motion (motion in the thin layer of fluid). We assume that in this case, of rotation on the agar surface, the turning has a well defined handedness of rotation.

To cause the observed chirality, the rotation must also be, on average, smaller than 90° relative to a specific direction and with a small stochastic part around the average value. We assume that a bacterium-bacterium co-alignment (orientational interaction) limits both the average rotation and its stochastic part. We further assume that the rotation is relative to the local mean orientation of the surrounding bacteria, $\Phi(\mathbf{r_i})$.

To test the above, we included the additional assumed features in the communicating walkers model [20]. To represent the bacterial orientation, we assign an angle θ_i to each walker. Every time step, each of the active walkers ($E_i > 0$) performs rotation to a new orientation θ'_i, which is derived from the walkers previous orientation θ_i by

$$\theta'_i = P(\theta_i, \Phi(\mathbf{r_i})) + Ch + \xi \tag{4}$$

Ch and ξ represent the new features of rotation due to tumbling. Ch is a fixed rotation and ξ is the stochastic part of the rotation, chosen uniformly from the interval $[-\eta, \eta]$. P is a projection function that represents the orientational interaction which acts on each walker to orient θ_i along the direction $\Phi(\mathbf{r_i})$. Once oriented, the walker advances a step d either in the direction θ'_i (forward) or in the direction $\theta'_i + \pi$ (backwards). As for the $\mathcal{T}$ morphotype, the movement is confined within an envelope which is defined on a triangular lattice. Results of the numerical simulations of the model are shown in Fig. 4.

4.4 Cooperative Formation of Vortices

The $\mathcal{V}$ morphotype bacteria do not swim, rather they glide with a strong velocity-velocity interaction. Hence we refer to the walkers in the $\mathcal{V}$ morphotype model as

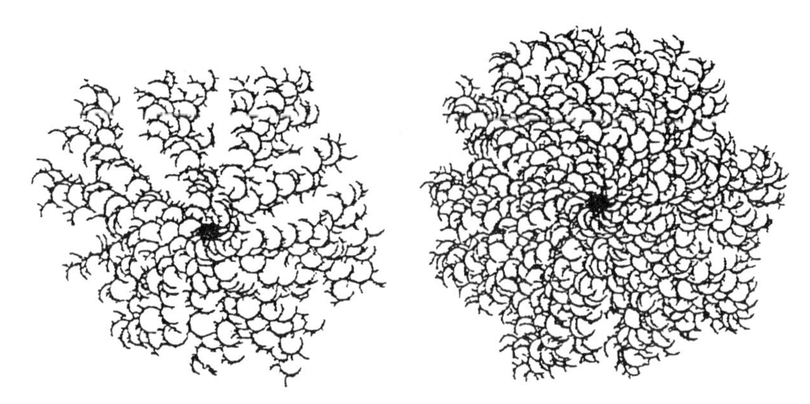

Fig. 4. Examples of simulations of chiral growth.

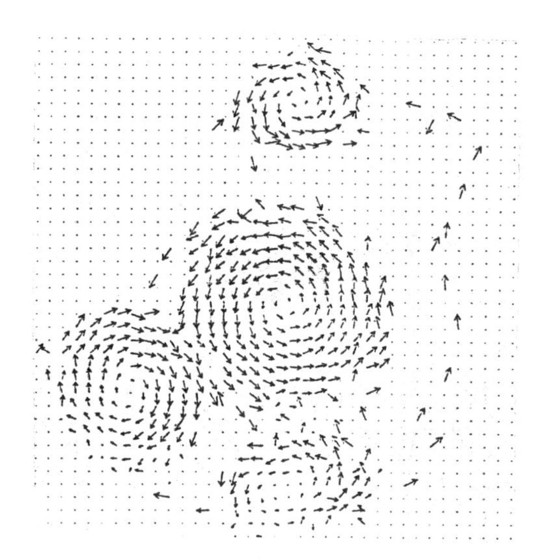

Fig. 5. Numerical simulations of vortices. The arrows indicate the direction and magnitude of the gliders' velocity.

self-propelled gliders, each now characterized by its velocity $\mathbf{v_i}$ whose equation of motion is

$$\frac{d\mathbf{v_i}}{dt} = \mu\left(\langle\mathbf{v}\rangle_{i,\epsilon} - \mathbf{v_i}\right) - \nu\mathbf{v_i} + F\frac{\mathbf{v_i}}{v_i} - \frac{\alpha_A}{v_i}\mathbf{v_i} \times (\mathbf{v_i} \times \nabla A) - \alpha_\sigma \nabla\sigma + \boldsymbol{\xi} \tag{5}$$

The terms on the right hand side of the equation are: (1) Interaction (drag force) with the averaged velocity over distance ϵ. (2) Friction with the substrate. (3) Constant propulsion force in the direction $\mathbf{v_i}$. (4) Hard-core interaction. (5)

Random noise. The coefficients depends on the level of the wetting fluid, the dynamics of which is described in Ref. [39].

Inclusion of these features in the model enables to capture the observed collective migration of the bacteria [21]. Yet, additional features have to be included to account for the vortex formation. Geometrical considerations suggest that a radial inward force can lead to vortex formation. Motivated by the demonstrated role of chemoattractant in bacterial colonies and our belief in universal strategies, we have proposed that a new kind of rotor chemoattractant can provide the required effect: the individual bacterium will weakly vary its velocity according to the local concentration of the chemotactic material. This will impose a torque or local vorticity on the collective motion. When a glider moves perpendicular to the chemical gradient, it is subjected to a force acting to twist its motion towards the high concentrations of the chemoattractant. Thus, in the case of positive "chemotaxis" an additional term $-\frac{1}{v}\mathbf{v} \times (\mathbf{v} \times \nabla C_A)$ is added to equation 5. To indicate the special nature of this chemotactic response we call it rotor chemotaxis.

In Fig. 5 we show that such rotor chemotaxis can indeed lead to the formation of stationary vortices (fixed in size and location). By changing the model parameters we can account for moving "worms" and other observed phenomena, all on a length-scale smaller than an individual branch. During the growth of the colony, these elements organize to form the observed global pattern. We have shown that food diffusion and consumption, together with long-range chemorepellent can provide the means for the colony self-organization [39].

5 Conclusions

If proven to be real transitions, and not a result of contamination, the observed morphotype bursts are the most fascinating and important of all the phenomena we have described in this chapter. The changes involved in the transitions are so dramatic that it is extremely hard to accept any explanation different from contamination. After all, more than once during the evolution of modern biology did contaminations play "dirty tricks" leading to false conclusions.

There are old observations of *Bacillus mycoides* colonies which exhibit patterns similar to those of the chiral morphotype [40]. In addition, more than half a century ago, observations of vortices formed by *Bacillus circulans* were reported (a movie of *Bacillus circulans* was produced [41]). Are the $\mathcal{C}$ morphotype a contamination of *Bacillus mycoides* and the $\mathcal{V}$ morphotype a contamination of *Bacillus circulans*? For reasons detailed in Ref. [10], we belive they are not contaminations, although we still do not have a conclusive proof for this. The consequences of an alternative explanations would be most meaningful. J. Lederberg said he was convinced these were contaminations, but if proven otherwise he would have to reexamine his views with respect to bacterial mutations [42].

Ben-Jacob et al. [18, 9, 10] proposed that cooperative processes involving exchange of genetic information are needed for the morphotype adaptation or adaptive morphogenesis. The above is supported by very recent experiments demonstrating that genetic communication is indeed required for adaptive mutations [43, 44].

Accepting that adaptive morphogenesis is a cooperative genetic process, it is natural to accept that it involves major changes in the bacterial membrane, both to render it more permeable to genetic materials and to ease the motility. This directly exposes a bacterium to a higher risk of various hazards, including exposition to antibiotics and bacteriophages. The motility also indirectly increases the vulnerability of the bacteria as it enables growth of thin branches with much lower densities of bacteria. Therefore, "smart bacteria" will couple the adaptive morphogenesis with resistance to antibiotics. Indeed, the new morphotypes do show such resistance.

Imposed by stress, the genome probably performs a "search" for old abilities at its disposal that under normal conditions are turned off as they are not very efficient. For example, it is known that *Bacillus subtilis* has a non-active β-Galactosidase gene. If turned on during the search, it can explain the $lacZ^+$ character of the new morphotypes. Similarly, it is easy to imagine how various auxotrophic characters are turned on and off during the search.

The above discussion is not an attempt to explain the morphotypes transitions, nor to prove their reality. We simply wanted to show that all fits nicely together if we are willing to accept cooperative genetic processes or co-morphogenesis.

It is definitely premature to present the implications of our new and far-from-established theoretical framework of cooperative evolution (for more details see Ref. [10]). We would just like to emphasize that the new picture of cooperative evolution is entirely different from Darwinism. In cooperative evolution there is no more "every one competing against anyone else", which is a reasonable strategy when you rely on pure luck. Now a cooperative effort is done to "shape up" the luck. The new picture has some features common with Lamarkism in the sense that current conditions (together with information about the past) affect mutations, but it should not be mistaken as a return to Lamarkism.

To conclude, a genetic engineering and biotechnology revolution is generally recognized. We hope we were successful in convincing the reader that a new revolution, yet to be recognized, is emerging and we enter a new era, one which involves a shift from the pure reductionistic point of view to a rational holistic one.

Acknowledgments

This chapter describes the results of a research venture with many collaborators. O. Shochet has been a "key player" in the collaboration until he finished his PhD thesis in the summer of 1995. Modeling of complex patterning of $\mathcal{T}$, $\mathcal{C}$ and $\mathcal{V}$ morphotypes was done in collaboration with T. Vicsek. The morphotype transitions experiments are performed with D. Gutnik, R. Rudner and E. Freidkin. The picture of genome cybernetics has been developed with A. Tenenbaum. We are most thankful to I. Brains for her technical assistance. We are grateful to D. Gutnik, R. Rudner, J. Shapiro and E. Ron for many discussions and their help in guiding us in the microbiology world. The research endeavor described here was supported in part by a grant from the Israel-USA binational foundation BSF 92-00051 and a grant from the German Israel foundation GIF 090102.

References

1. E. Ben-Jacob and P. Garik. The formation of patterns in non-equilibrium growth. *Nature*, 343:523–530, 1990.
2. J.S. Langer. Dendrites, viscous fingering, and the theory of pattern formation. *Science*, 243:1150–1154, 1989.
3. D. A. Kessler, J. Koplik, and H. Levine. Pattern selection in fingered growth phenomena. *Adv. Phys.*, 37:255, 1988.
4. E. Ben-Jacob. From snowflake formation to the growth of bacterial colonies. part i: Diffusive patterning in non-living systems. *Contemp. Phys.*, 34:247–273, 1993.
5. J. Adler. A method for measuring chemotaxis and use of the method to determine optimum conditions for chemotaxis by *Esherichia coli* . *J. Gen. Microbiol.*, 74(1):77–91, 1973.
6. P. Devreotes. *Dictyostelium discoideum*: a model system for cell-cell interactions in development. *Science*, 245:1054–1058, 1989.
7. J. M. Lackiie, editor. *Biology of the chemotatic response*. Cambridge Univ. Press, 1986.
8. J. A. Shapiro. Adaptive mutation: Who's really in the garden. *Science*, 268:373–374, 1995.
9. E. Ben-Jacob, A. Tenenbaum, O. Shochet, and O. Avidan. Holotransformations of bacterial colonies and genome cybernetics. *Physica A*, 202:1–47, 1994.
10. E. Ben-Jacob, I. Cohen, and A. Czirók. Smart bacterial colonies: From complex patterns to cooperative evolution. *Fractals*, 1995. (in press).
11. H. Müller-Krumbhaar and w. Kurz. In P. Haasen, editor, *Phase Transformation in Materials*. VCH-Verlag, Weinheim, 1991.
12. E. A. Brener and V. I. Mel'nikov. Pattern selection in two-dimensional dendritic growth. *Adv. Phys.*, 40(1):53–97, 1991.
13. E. Ben-Jacob, O. Shochet, and R. Kupferman. über die vielfalt und dynamik diffusionsbedingter musterbildung. In A. Deutsch, editor, *Muster des Ledendigen: Faszination inher Entstehung und Simulation*. Verlag Vieweg, 1994.
14. T. Vicsek. *Fractal Growth Phenomena*. World Scientific, New York, 1989.
15. J. Feder. *Fractals*. Plenum, New York, 1988.
16. M. Matsushita and H. Fujikawa. Diffusion-limited growth in bacterial colony formation. *Physica A*, 168:498–506, 1990.
17. H. Fujikawa and M. Matsushita. Fractal growth of *Bacillus subtilis* on agar plates. *J. Phys. Soc. Jap.*, 58:3875–3878, 1989.
18. E. Ben-Jacob, H. Shmueli, O. Shochet, and A. Tenenbaum. Adaptive self-organization during growth of bacterial colonies. *Physica A*, 187:378–424, 1992.
19. P. L. Foster. Adaptive mutaion: The uses of adversity. *Annu. Rev. Microbiol.*, 47:467–504, 1993.
20. E. Ben-Jacob, O. Shochet, A. Tenenbaum, I. Cohen, A. Czirók, and T. Vicsek. Communication, regulation and control during complex patterning of bacterial colonies. *Fractals*, 2(1):15–44, 1994.

21. T. Vicsek, A. Czirók, E. Ben-Jacob, I. Cohen, O. Shochet, and A. Tenenbaum. Novel type of phase transition in a system of self-driven particles. *Phys. Rev. Lett.*, 75:1226–1229, 1995.
22. J. Horgan. From complexity to perplexity. *Sci. Am.*, June 95:74–79, 1995.
23. E. Ben-Jacob, O. Shochet, A. Tenenbaum, I. Cohen, A. Czirók, and T. Vicsek. Generic modelling of cooperative growth patterns in bacterial colonies. *Nature*, 368:46–49, 1994.
24. D. A. Kessler and H. Levine. Pattern formation in *dictyostelium* via the dynamics of cooperative biological entities. *Phys. Rev. E*, 48:4801–4804, 1993.
25. M. Y. Azbel. Survival-extinction transition in bacteria growth. *Europhys. Lett.*, 22(4):311–316, 1993.
26. O. Shochet, K. Kassner, E. Ben-Jacob, S.G. Lipson, and H. Müller-Krumbhaar. Morphology transition during non-equilibrium growth: I. study of equilibrium shapes and properties. *Physica A*, 181:136–155, 1992.
27. O. Shochet, K. Kassner, E. Ben-Jacob, S.G. Lipson, and H. Müller-Krumbhaar. Morphology transition during non-equilibrium growth: II. morphology diagram and characterization of the transition. *Physica A*, 187:87–111, 1992.
28. O. Shochet. *Study of late-stage growth and morphology selection during diffusive patterning.* PhD thesis, Tel-Aviv University, 1995.
29. J. Adler. Chemoreceptors in bacteria. *Science*, 166:1588–1597, 1969.
30. H. C. Berg and E. M. Purcell. Physics of chemoreception. *Biophysical Jornal*, 20:193–219, 1977.
31. J. A. Shapiro and D. Trubatch. Sequential events in bacterial colony morphogenesis. *Physica D*, 49:214–223, 1991.
32. E. O. Budrene and H. C. Berg. Complex patterns formed by motile cells of *escherichia coli. Nature*, 349:630–633, 1991.
33. Y. Blat and M. Eisenbach. Tar-dependent and -independent pattern formation by *Salmonella typhimurium . J. Bac.*, 177(7):1683–1691, 1995.
34. J. A. Shapiro. The significances of bacterial colony patterns. *BioEssays*, 17(7):597–607, 1995.
35. E. Ben-Jacob, I. Cohen, O. Shochet, A. Czirók, and T. Vicsek. Cooperative formation of chiral patterns during growth of bacterial colonies. *Phys. Rev. Lett.*, 75:2899–2902, 1995.
36. M. Eisenbach. Functions of the flagellar modes of rotation in bacterial motility and chemotaxis. *Molec. Microbio.*, 4(2):161–167, 1990.
37. J. B. Stock, A. M. Stock, and M. Mottonen. Signal transduction in bacteria. *Nature*, 344:395–400, 1990.
38. C. H. Shaw. Swimming against the tide: chemotaxis in *agrobacterium. BioEssays*, 13(1):25–29, 1991.
39. A. Czirók, E. Ben-Jacob, I. Cohen, and T. Vicsek. Cooperative formation of vortices. (to be published).
40. T. H. Henrici. *The Biology of Bacteria: The Bacillaceae.* D. C. Heath & company, 3rd edition, 1948.
41. G. Wolf, editor. *Encyclopaedia Cinematographica.* Institut für Wissenschaftlichen Film, Göttingen, 1968.
42. J. Lederberg, personal communication.
43. J. P. Rasicella, P. U. Park, and M. S. Fox. Adaptive mutation in *Esherichia coli* : a role for conjugation. *Science*, 268:418–420, 1995.
44. T. Galitski and J. R. Roth. Evidence that f plasmid transfer replication underlies apparent adaptive mutation. *Science*, 268:421–423, 1995.

Complementarity of Physics, Biology and Geometry in the Dynamics of Swimming Micro-Organisms

John O. Kessler[1] and Nick A. Hill[2]

[1] Physics Department, University of Arizona, Tucson, AZ 85721, U.S.A.
[2] Department of Applied Mathematical Studies, University of Leeds, Leeds LS2 9JT, U.K.

1 Introduction

Concentrated populations of motile micro-organisms that are initially uniformly distributed in space rarely remain so. Individual organisms may interact with one another directly through consumption and emission of metabolites. The chapter by E. Ben-Jacob describes such situations when the organisms are located on solid or gel substrates. The interactions then cause aggregation or dispersal, modifications in behaviour, and even elicit expression of particular genetic traits. Macroscopically viewed, the interactions among organisms result in sharply-patterned, spatio-temporal variations of organism population density.

A rather different sort of patterning often occurs in populations of micro-organisms that swim. These bioconvection patterns (Wager 1911; Pfennig 1962; Pedley and Kessler 1992a,b) consist of spatial and temporal variations in the concentration of organisms, causally coupled with convection of the water in which they live. In these situations, direct interaction among the organisms, e.g. algal cells or bacteria, are not significant, at least to first order. These remarkable patterns (Figs. 1 and 2) result from (a) the behaviour of individual cells within their dynamic environment, (b) physical forces and conservation principles that break symmetry, supply nonlinearity and may directly orient the organisms' distribution of swimming velocities, and (c) geometric constraints that determine modal structure as well as directionality. In spite of their partly mechanistic origin, bioconvection patterns can serve to improve viability and otherwise benefit the microbial population that generates them (Kessler 1989; Kessler and Hill 1995).

2 Hydrodynamics

Algal and bacterial cells, of the kind that participate in bioconvection, are from 1% to 15% more dense than water. A small element of a fluid of density ρ_w gm cm^{-3} that contains n cells cm^{-3}, each of density ρ_c gm cm^{-3} and volume v cm^3 has a mean density

$$\bar{\rho} = \rho_w + (\rho_c - \rho_w)nv \; . \tag{1}$$

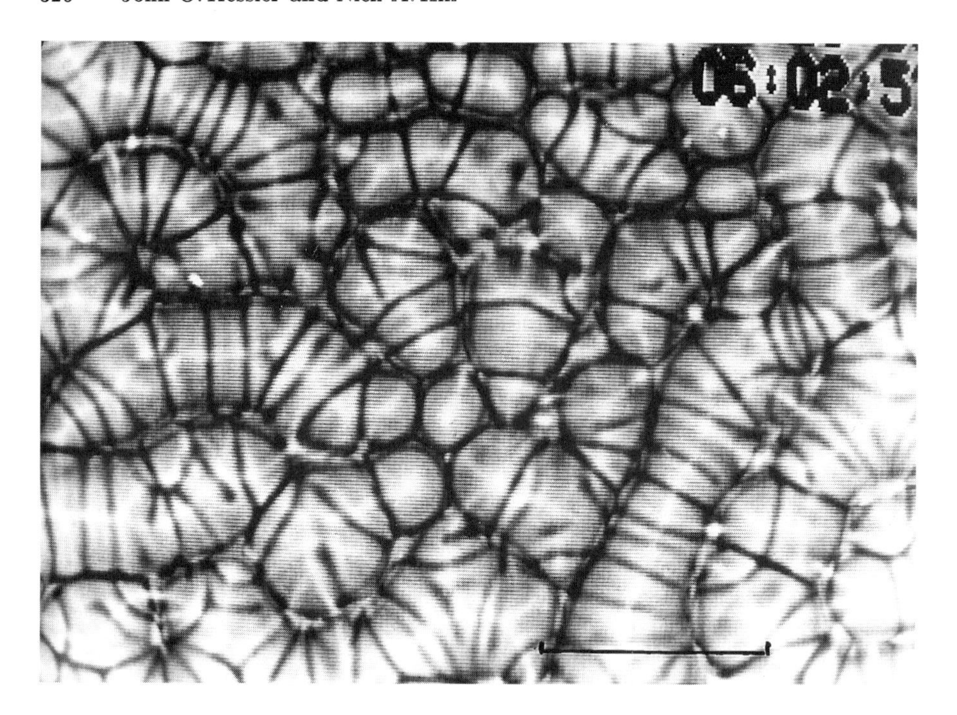

Fig. 1. Bioconvection pattern of *B. subtilis* seen in plan view, under dark field illumination. Locations where the cell concentration is high are light; black indicates low cell concentration. The black bar represents 1.0 cm. The mean concentration of cells is $\sim 10^9\,\mathrm{cm}^{-3}$; the depth of fluid is $\sim 3\,\mathrm{mm}$. The dark and bright cross bands as well as the convection rolls themselves continually move and change shape. The apparent three-dimensionality is an optical illusion.

When n is a function of location, $\overline{\rho}$ also varies. The local gravitational force, per volume, on an element of fluid is then

$$\mathbf{k}(\mathbf{x},t) = (\rho_\mathrm{c} - \rho_\mathrm{w})v[n(\mathbf{x},t) - \bar{n}]\mathbf{g}\ , \tag{2}$$

where $\bar{n}$ is a mean concentration and $\mathbf{g}$ is the acceleration due to gravity. Bioconvection is driven by the spatial distribution of forces $\mathbf{k}(\mathbf{x},t)$. It will become apparent that in the fully-developed nonlinear regime the flow of the fluid helps to determine $n(\mathbf{x},t)$.

What is the origin of nonuniformity in $n(\mathbf{x},t)$? Consider a layer of fluid that initially contains swimming organisms which are uniformly distributed. If the distribution of swimming directions is isotropic (ignoring special wall effects) the concentration of cells, n, will remain uniform except for statistical fluctuations. Suppose now that all the organisms develop an upward bias in their motion. When the population tends to swim upward, it accumulates at the top boundary of the fluid, the air-water interface. If this unstable density stratification is sufficiently great, then it will result in the onset of Rayleigh-Bénard-like convection (Acheson 1990). The reason for upswimming will be discussed later.

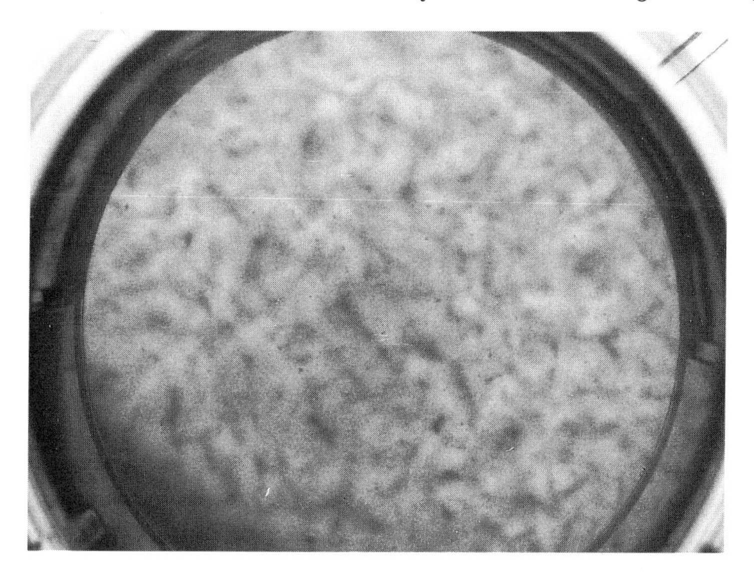

Fig. 2. Bioconvection pattern of *Pleurochrysis carterae*. The diameter of the petri dish is 3.6 cm, the depth of fluid is $\sim 4\,\mathrm{mm}$; it contains $10^5 - 10^6$ cells cm^{-3}. Diffuse illumination from below.

Another mechanism that can drive convection depends on lateral accumulation. Assume once more that there is an initially uniform concentration of cells. Now let the swimmers be biased to swim horizontally. The details will now depend on particular circumstances, but sideways swimming in a confining geometry again results in a spatial modulation of n, and hence a nonuniform distribution of mass density. Again the associated forces, $\mathbf{k}$, generate convection, this time without a threshold, i.e. with no parameter equivalent to a critical Rayleigh number.

It seems appropriate here to emphasize that an entirely physical dynamic phenomenon, the gravitationally-driven motion of fluid, is due to small deviations from symmetry in the behaviour of a population of microbes.

In reality, a mixture of upward and sideways swimming occurs. If the inhabitants of the fluid were to swim entirely without any upward bias, viscous damping would eventually stop the convective motion. It is upward swimming of the micro-organisms that generates the energy required for maintaining bioconvection. The upward swimming generates the gravitational potential energy that balances viscous dissipation. The primary source of that energy is biochemical, the metabolites consumed by bacteria and the photons consumed by the algae.

In the continuum approximation, the hydrodynamics of bioconvection is summarized by a set of partial differential equations. Conservation of momentum is described by the Navier-Stokes equation, driven by the density offset due to organisms. In the Boussinesq approximation,

$$\bar{\rho}\frac{\mathrm{D}\mathbf{u}}{\mathrm{D}t} = \mu\nabla^2\mathbf{u} - \nabla p + (\rho_\mathrm{c} - \rho_\mathrm{w})nv\mathbf{g}\ , \tag{3}$$

where μ is the viscosity of water and $\mathbf{u}(\mathbf{x},t)$ is the fluid velocity. Conservation of fluid volume requires

$$\nabla \cdot \mathbf{u} \approx \mathbf{0} \,. \tag{4}$$

This expression tends toward equality as the volume fraction of organisms, nv, tends to 0.

Assuming that the birth and death rates of organisms are negligible, the conservation of organisms is given by

$$\frac{\partial n}{\partial t} = -\nabla \cdot \left[n\mathbf{u} + \int_{\mathrm{V}} nf(\mathbf{V},Q)\mathbf{V}\,d\mathbf{V} \right] \equiv -\nabla \cdot \mathbf{j} \,. \tag{5}$$

The integral expression is the flux of organisms associated with their locomotion. The normalized probability density f specifies the correlated distributions of swimming speeds and directions. $\mathbf{V}$ is the swimming velocity vector, and the quantity Q symbolizes the directional and stochastic influences on $f(\mathbf{V},Q)$.

Previous theoretical treatments of bioconvection have generally assumed that the organism flux can be expressed as

$$\mathbf{j} = n\mathbf{u} + n\mathbf{V}_{\mathrm{c}} - D\nabla n \,, \tag{6}$$

where $\mathbf{V}_{\mathrm{c}}$ is the mean swimming velocity, and the diffusion term models the stochastic aspects of locomotion. Use of the transport integral in (5) is a method for including actual or modelled experimental data in the theory. In the past, bioconvection of swimming algae and bacteria has usually been modelled using (3) – (5), but with the flux expansion given in (6). Pedley and Kessler (1992a) provide a summary.

The requirement that viscous damping be offset by the rate of potential energy generation created by swimming can be summarized by

$$\frac{1}{T}\int_0^T dt \int_{\mathrm{volume}} d\mathbf{x} \left[-\mathbf{j}\cdot\mathbf{g}v(\rho_{\mathrm{c}} - \rho_{\mathrm{w}}) - \frac{\mu}{2}\left(\frac{\partial u_i}{\partial x_j} + \frac{\partial u_j}{\partial x_i}\right)^2 \right] = 0 \,. \tag{7}$$

The first term is the distributed rate of energy production due to upswimming, where $\mathbf{j}$ is given by (6) or by the integral in (5). The second term is the viscous dissipation (Acheson 1990, p. 216). This energy accounting requires that the integrals be over the volume occupied by the convecting system and over a "long" time interval, T, since the energy balance need be neither local nor instantaneous. An assessment of the energy balance given by (7) requires knowledge of $\mathbf{u}(\mathbf{x},t)$ and $n(\mathbf{x},t)$ over the entire volume. However, if one compares $\mu(\Delta u/\lambda)^2$ to an estimate of the upswimming term, one obtains approximate equality for shallow patterns as in Fig. 1. In such cases, the typical observed velocity of particles entrained in a convection roll implies that $\Delta u \approx 10^{-2}\,\mathrm{cm\,s^{-1}}$. The pattern wavelength is $\lambda \approx 10^{-1}\,\mathrm{cm}$. The magnitude of each term is approximately $10^{-4}\,\mathrm{erg\,cm^{-3}\,s^{-1}}$ or $10^{-13}\,\mathrm{erg\,cell^{-1}\,s^{-1}}$. The power dissipated by each cell's swimming is approximately $6\pi\mu \times (10^{-4}\,\mathrm{cm}) \times (10^{-3}\,\mathrm{cm\,s^{-1}})^2$ or $2\times 10^{-11}\,\mathrm{erg\,cell^{-1}\,s^{-1}}$; thus the creation of gravitational potential energy by the swimmers is a small fraction of their locomotion energy budget.

3 The Distribution of Swimming Velocities

Directionality of swimming is a central theme in the discussion of bioconvection. Understanding directional locomotion also provides insights into the biology of organisms, including various aspects of sensing stimuli and responding to them, search strategies, swimming modes and their control mechanisms. The orienting interactions between organisms and the fields of force in which they live are an important ingredient in the theory of bioconvection and in the modelling of organism fluxes in microbial ecology. The competition between two or more orienting influences (Haeder 1987; Kessler et al. 1992) is not even generally recognized as a problem, although it is ubiquitous in nature. The consequences of directional swimming have been investigated using cell accumulation as a measure (Kessler 1985a; Boon and Herpigny 1984). Explicit measurements of $f(\mathbf{V}, Q)$ are now under way.

In a symmetric environment the distribution of swimming directions is random. The distribution of swimming speeds can relate to population factors, e.g. age-mix, or to some intrinsic stochastic property of the motile cells. When symmetry is broken in a manner that can be detected by the cells, they alter the statistics of their behaviour. For example, algal cells might, on average, swim more quickly toward a light source than away from it. By suitable phasing of the flagella, their swimming apparatus, they may also change the distribution of swimming directions. In the absence of physical forces or torques acting on the organisms, any modification of the distribution of swimming velocities must be generated via their internal biophysical-chemical workings. Some examples of such chains of sensing and responding (Armitage 1992) are swimming up or down a gradient of chemical concentration (chemotaxis) and towards or away from a light source (phototaxis, photokinesis).

The orientation of swimming velocities of algal cells is correlated with the direction of gravity ("gravitaxis" or "geotaxis") , as shown in Fig. 3, and amply demonstrated elsewhere (Kessler 1985a,b, 1986a,b). "Sensing and responding" is not required in that case, since a cell can continue swimming normally while the orientation of its body is determined by the gravitational torque arising from the asymmetric distribution of mass within the cell's body. Furthermore, the shear stress applied by the fluid in which the organism swims may introduce another torque. The shear and gravity torques may balance when an algal cell is oriented at an angle θ, say, with the vertical direction so that when the cells swim, their average direction is θ. This oriented locomotion, called gyrotaxis, can be demonstrated by showing that cells swim toward and accumulate near the axis of a downward Poiseuille flow (Kessler 1985a,b, 1986a,b). Purely passive alignment by gravity and shear can account for gravitaxis and gyrotaxis. However, it is still possible that there is also some active orienting response. Cells of *B. subtilis* are apparently not gravitactic but shear must have an orienting effect.

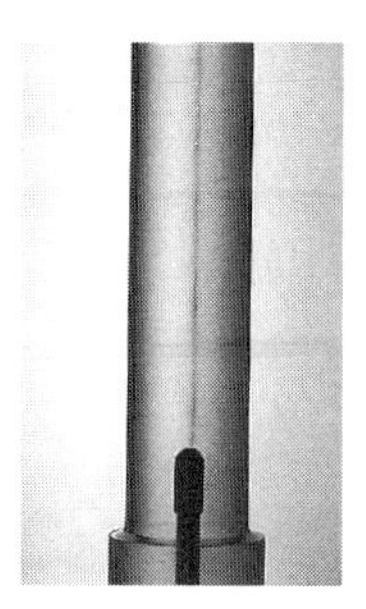

Fig. 3. Gyrotactic focusing. A culture containing the biflagellated, motile algae *Chlamydomonas nivalis* flows slowly downward. The cells then swim toward the axis of the flow. They appear here as a dark line. The nozzle at the bottom collects the focussed cells. The diameter of the tube is 1.5 cm and the flow rate is ~ 0.1 cm s^{-1}. The swimming direction is determined jointly by upward gravitaxis and the viscous torque due to vorticity (Kessler 1985a).

4 Consumption

The force of gravity, rotational viscous drag, directional illumination, and gradients of chemical concentration that might be imposed from external sources, all induce asymmetry in $f(\mathbf{V}, Q)$. All these orienting influences, except gravity, are modified and even generated by the presence of the population of living swimming organisms. The shaping of the dynamics and symmetry of the population's environment is a principal source of the nonlinearity that governs the geometry of bioconvection.

The consumption of photons by algae, through absorption and scattering will be discussed separately. Patterns that originate because of shading of some members of a population by others depend on the constitutive equations or rules of phototaxis.

The bacteria *B. subtilis* consume oxygen. Their swimming speed decreases, and eventually vanishes, as the oxygen concentration decreases. If a layer of water whose upper interface is open to air contains a high concentration of these cells, consumption of oxygen throughout the volume, and supply from the air above, generate an upward gradient of oxygen concentration. The equation describing conservation of oxygen is

$$\frac{\partial c}{\partial t} = -\nabla \cdot [c\mathbf{u} - D_0 \nabla c] - n\gamma \, , \tag{8}$$

where c is the concentration of oxygen, D_0 is its diffusivity, and the consumption rate γ per bacterial cell, may depend on c when the concentration is low. The diffusivity D_0 may be slightly affected by the presence of the cells.

The upwardly directed oxygen concentration gradient is a consequence of consumption and supply. Gravity determines the symmetry of the problem, through the location of the oxygen-supplying interface. Here gravity interacts quite indirectly with the swimming cells, in causing the anisotropy of $f(\mathbf{V}, Q)$.

The complementarity of physics, geometry and biology is especially well illustrated here by the development of the conditions that set the stage for bacterial bioconvection. Physics, i.e. through the gravitational force, requires that water lies below air. The diffusion and conservation of oxygen are also purely physical effects. Geometry plays its part via the solution of (8) and its associated boundary conditions. For example, the aspect ratio, width/depth, for a given volume of bacterial culture determines whether most of the cells will swim toward oxygen, or whether it is only the top layer that swim, the rest becoming immobile for lack of oxygen before the oxygen gradient, ∇c, reaches them. Biology, i.e. the metabolism of cells and their ability to sense and respond to changes in $c(\mathbf{x},t)$ completes the trilogy.

A mathematical version of these events, using (5), (6) and (8), together with various assumptions that simplify the biology, has been presented by Hillesdon et al. (1995). Quantitative data are required if one wishes to understand the biology of the protagonists and the dynamical system that they operate. Although ansätze like (6) are adequate for producing reasonable mathematical descriptions of bioconvection (Childress et al. 1975; Kessler 1985b, 1986a,b; Pedley et al. 1988; Hill et al. 1989; Pedley and Kessler 1990, 1992a; Hillesdon and Pedley 1995), a more accurate theory requires the details provided by $f(\mathbf{V}, Q)$. The phenomenology of individual cells sensing and responding is also of intrinsic interest. We have therefore begun measuring $f(\mathbf{V}, Q)$ under various circumstances. Some representative results are presented here.

5 Distributions of Swimming Directions and Speeds

Statistics on large sets of trajectories of swimming algae and bacteria have been gathered using a video camera attached to a microscope. The images are analyzed by a system supplied by the Motion Analysis Corporation, modified in accord with various specialised requirements. For measurements on gravitaxis and rotational orbits the microscope's optical axis was horizontal. Trajectories of bacteria were measured in flat closed microslides, lumen 0.1 mm, supplied by Vitro Dynamics.

We present a few examples of the data obtained. They are chosen to illustrate the power of the method, and to give some insight into the organisms' behaviour. Results such as these provide information on the biology of directed locomotion, including stochastic aspects. They are also the fundamental ingredient for realistic modelling of bioconvective dynamics in the laboratory and in nature.

5.1 Bacteria: *B. subtilis*

The data were obtained with cultures of these common soil bacteria containing $\sim 10^9$ cells cm^{-3}. The experimental arrangement did not permit gravity-driven flows. An air interface served as a source of oxygen. When the cells are far from the interface, consumption uniformly depletes the oxygen concentration. The cells then swim more and more slowly, and then stop altogether. The criterion for uniformity is that the location of measurement be many diffusion lengths away from the interface. The consumption rate, γ, is (Berg 1983) approximately 10^6 molecules cell^{-1} s^{-1}, the initial oxygen concentration $C_0 \sim 10^{-17}$ molecules cm^{-3}, and the diffusivity of oxygen in water is $D = 2 \times 10^{-5}\,\mathrm{cm^2\,s^{-1}}$. Then the depletion time $\tau \sim C_0/\gamma n \sim 10^2$ s as observed, and the diffusion length $L \sim (D\tau)^{1/2} \sim 10^{-2}$ cm.

The measurements in Figs. 4 and 5 were made 2 mm away from the air source. Figure 4 shows the decline of swimming speed. Figure 5 shows the distribution of speeds before significant depletion of oxygen has occurred. We are attempting to fit the data in such histograms with curves of the form $V^a \exp(-bV^c)$. The smooth line is one such fit, but more data will have to be analyzed before definite values of the constants can be presented. Finding a generally applicable functional form for the speed distribution would be useful for calculations. It is likely to yield information on the operation of cells, much like the Maxwell distribution informs about the ideal gas.

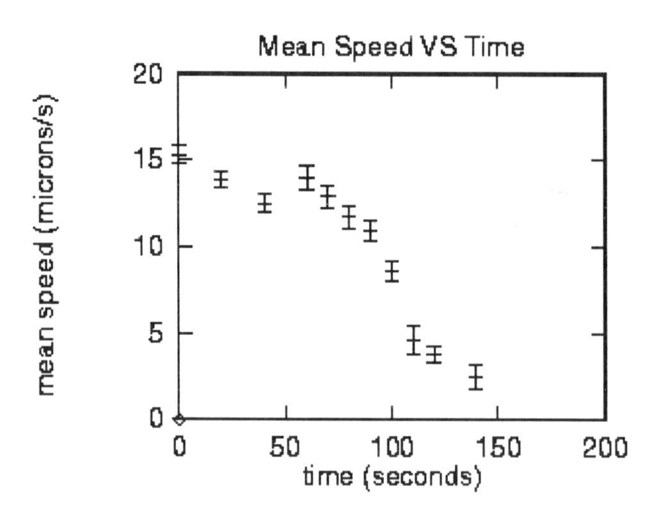

Fig. 4. The collapse of mean swimming speed of *B. subtilis* due to decreasing concentration of oxygen. The oxygen source is 3 mm away from the location of these measurements, too far to make up for consumption during the time available. The error bars are the standard error of the mean. The speeds were determined by dividing the distance between head and tail of a track by the elapsed time. Because this method averages over the noise and the tortuosity of tracks it yields lower mean speeds than it would obtain using a frame by frame method.

When the location of measurements is closer than L to the oxygen source, then the oxygen diffusion front reaches the cells being measured whilst they still

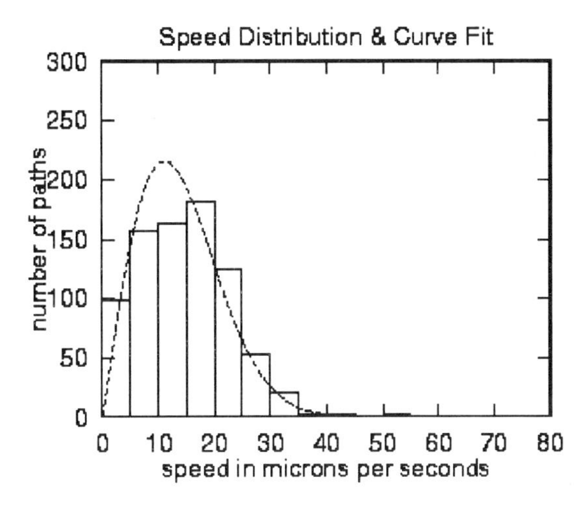

Fig. 5. The distribution of swimming speeds of well-aerated cells of *B. subtilis*. The speeds were determined as in Fig. 4. The smooth curve is a fit to this histogram, using the function $V^{1.1}\exp[-V^2/a]$, where $a = 284\mu\text{m}^2\,\text{s}^{-2}$.

swim normally. We then find that the swimming direction is skewed up-gradient. The cells also swim more quickly upgradient than down (Kessler et al. 1995). The angular distribution is also shown in that paper.

5.2 Algae: *Pleurochrysis carterae*

P. carterae are coccolithophorids, a major component of the phytoplankton in tropical oceans. These cells produce intricate calcium carbonate structures and are responsible for worldwide chalk deposits (Lee 1989). Concentrated populations form bioconvection patterns.

Measurements of the angular distribution function of the swimming trajectories of a ten day old culture of *P. carterae* are shown in Fig. 6. This distribution clearly shows both gravitaxis, i.e. upswimming, and the angular spread around the upward direction. For these cells one component of the angular spread around the mean direction is stochastic; another part of it can be accounted for by the helicity of the trajectories. The speed distribution is shown also.

Quantitative analysis of algal dynamics requires quantitative data on the joint orienting action of gravity and vorticity. To attack this problem, we have constructed an apparatus where algae are suspended in a fluid that rotates steadily and rigidly (except for regions in the immediate neighborhood of suspended objects). The vorticity everywhere within such a fluid is twice the rotation rate. Then, if the axis of rotation is perpendicular to **g**, it is possible to observe the joint effect, gyrotaxis, in the form of distribution functions. Figure 7 shows the distribution of angles relative to **g** in the rotating system. It should be noted that in that system the gravity vector rotates with the angular velocity

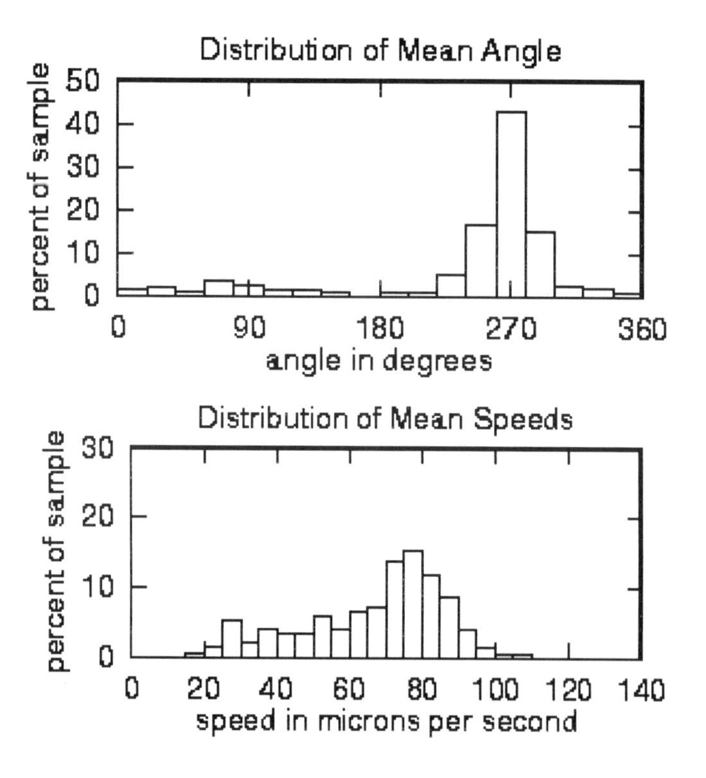

Fig. 6. Gravitaxis and swimming speed of *Pleurochrysis carterae.* These measurements were taken in a stationary cuvette, thickness 1mm between glass coverslips. The upper histogram plots number of cells swimming within angular sectors. The angle 270° is vertically upward. The lower histogram is the distribution of swimming speeds, measured on a frame by frame basis. Projection effects may account for some of the low speed part of the spectrum.

which is $60^\circ\,\mathrm{s}^{-1}$ in this case. Evidently the cells tend still to swim along the $\mathbf{g}$ axis, but not quite in phase.

Further measurements as a function of rotation rate, and for different algal species, will provide a quantitative description of gyrotaxis and a comparison with theory (Kessler 1985a,b; Pedley and Kessler 1987). Measurements of the joint action of light, gravity, and vorticity will become possible. One may also expect to gain some insights concerning the behaviour of phytoplankton in the waves of their natural environment.

6 Phototaxis and Shading

Many free-swimming micro-organisms such as algae (which need to photosynthesize) and their predators, detect and swim towards or away from sources of light. These behaviours are known as positive and negative phototaxis, respectively.

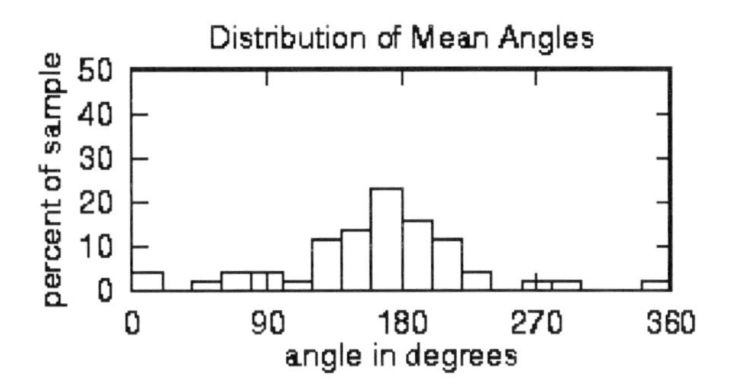

Fig. 7. The distribution of swimming directions of *P. carterae* relative to gravity, measured in a reference frame that rotates at the same rate ω as the fluid. Note that $\mathbf{g}$ rotates with rate $-\omega$. At each successive pairs of video frames, equivalent to successive "instants of time", the cells' orientation along their trajectories is compared with the direction of $\mathbf{g}$. The histogram represents the ensemble of directions, over each trajectory, and for all trajectories in the set. In this plot 270° represents the direction of $\mathbf{g}$. The data imply that the direction of the cells' velocity lags $\mathbf{g}$ by about 90°, at this particular rotation rate $\sim 1\ \mathrm{rad\,s^{-1}}$.

These active responses often dominate passive orienting mechanisms, e.g. gravitaxis due to the cells being bottom-heavy (Kessler et al. 1992). The cells are able to detect the light with an organelle and, because almost all micro-organisms rotate as they swim, this photoreceptor scans the environment continuously and so provides the cell with the data from which it is able to compute the direction of the light source (Foster and Smyth 1980; Hill and Vincent 1993).

Typically species of algae, such as *Chlamydomonas nivalis, C. reinhardtii* and *Euglena gracilis*, are both positively phototactic when the light is weak and negatively phototactic when the light is strong, because the UV B radiation in bright sunlight can kill the cells by bleaching. Thus there is an optimum or critical light intensity, $I = I_c$, at which the mean cell swimming direction for a population of cells,

$$\mathbf{V}_c = V_s \overline{p}(I)\mathbf{k} \, , \tag{9}$$

is zero, i.e. $\overline{p}(I_c) = 0$, and cells will tend to aggregate at places where $I = I_c$. Here $\mathbf{k}$ is the unit vector pointing towards the light source and V_s is the mean cell swimming speed. For light intensities close to I_c, we can linearise $\overline{p}(I)$ about I_c so that the cell conservation equation (5), using the simplified form of the flux (6), becomes

$$\frac{\partial n}{\partial t} = -\nabla \cdot [n\mathbf{u} + n\Lambda(I_c - I)V_s\mathbf{k} - D\nabla n] \, , \tag{10}$$

where Λ is a linearisation constant.

The patterns formed by *C. nivalis* and other algae are strongly influenced by light and, when a suspension of cells is brightly illuminated from the side, the patterns form 'windrows', like sinking curtains, which are aligned nearly parallel to the light. This suggests that individual cells are aggregating in regions where they are shaded by the cells between them and the light source. This 'self-shading' by the suspension of cells as it 'consumes' the photons is another example of the local changes in their environment that the population can effect.

Since the suspensions are dilute, with mean volume fractions of the order of 0.1% – 1%, the Lambert-Beer Law (Herdan 1960) can be used to model the weak scattering and absorption of light by the algal cells as follows. The average light intensity, $I(\mathbf{x})$, received by the photoreceptor of a cell at a position $\mathbf{x}$ in the fluid is

$$I(\mathbf{x}) = I_s \exp\left(-\alpha \int_{\mathbf{r}} n(\mathbf{x})\, d\mathbf{r}\right) , \tag{11}$$

where I_s is the intensity of the light source, $\alpha \ll 1$ is the absorption coefficient, and $\mathbf{r}$ is the vector from the cell to the source. This expression (11) is used in $\overline{p}(I)$ in (9) and (10) for $\mathbf{V}_c$ to complete the system of equations. Further simplifications to make the equations more amenable to analysis can be achieved by linearising (11) and realising that in experiments I_s is close to I_c.

As an example of pattern formation, Hill and Vincent (1995) have considered an experiment in which an initially uniform suspension of purely phototactic cells, in a horizontal layer which is at rest, is illuminated from above by a distant source. When I_s is a little greater than I_c, the cells will swim into a concentrated horizontal layer centred on the line $\mathbf{x} = \mathbf{x}_c$, where the light is sufficiently attenuated by self-shading that $I(\mathbf{x}_c) = I_c$. This dense layer is susceptible to a Rayleigh-Bénard-like instability and proves to be an interesting example of penetrative convection, in which the bulk fluid motion in the unstable bottom layer below $\mathbf{x} = \mathbf{x}_c$ penetrates into the stable fluid layer above. Vincent (1995) has extended this model to incorporate the effects of gyrotaxis and analysed a similar stability problem, and work is currently under way to develop a full numerical computer program for bioconvection incorporating the solution of the Fokker-Planck equation for the probability distribution function of the cells' swimming velocity.

7 Concentrated Swarms

Other parts of this chapter are concerned with collective behaviour, at a relatively low volume fraction, of cells that do not interact directly with one another. Here we discuss the situation arising when, due to oriented locomotion, cells accumulate near a boundary. Very large concentrations often occur in that case. Gravitactic algal cells that swim upward in a wet sand pile, or in a wad of cotton wool immersed in the cell culture, accumulate at the air interface (Kessler 1986b). The cell bodies are then close packed. They may interact with one another. Their flagella vigorously stir the water, presumably enhancing the interfacial exchange of gas between air and water.

Algae accumulate to high concentrations in these situations because porous media introduce strong viscous damping. Convection modes are very slow or do not occur at all. 'Slow' in this context means that cells swim upward through the porous medium more quickly than they are advected downward by whatever stream does arise.

Bacterial accumulations are observable in the vicinity of the air water interface in shallow flat capillaries, arranged so as to eliminate buoyancy driven convection. The cells swim as far as geometry allows into the meniscus that separates air and water. They are nearly close packed there. Great surges and streams of cells occur; they imply local cell body contacts and flagellar interactions between colliding bacteria. It is very likely that vortical interactions between the flagellar bundles of adjacent cells and cell groups cause intermittent streaming, as seen in Fig. 8. Larger scale fluid dynamic interactions among groups of cells also occur at some distance away from the meniscus. Here, surges or waves of dimensions $\geq 100\times$ cell sizes (Fig. 9) arise and decay in times of the order of 1–10 seconds. These collective phenomena again imply hydrodynamic interactions among cell groups. Such mesoscale motions augment diffusive mixing by advection, folding and stirring, thus filling the gap between the scales associated with bioconvective and diffusive transport.

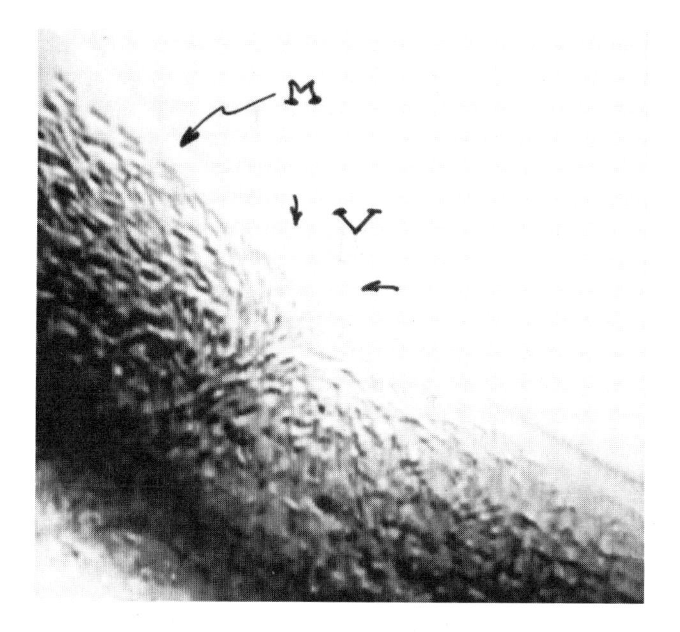

Fig. 8. Concentrated *B. subtilis* near the air-water interface, demonstrating collective motions under nearly close-packed conditions. These cells have swum up the oxygen gradient created by consumption. They accumulate near the meniscus. The air/water/glass triple boundary is indicated by M. The letter V and associated arrows show a rapid stream comprising a large number of cells. The individual cells, seen near M, are $\sim$ 4–5 μm long.

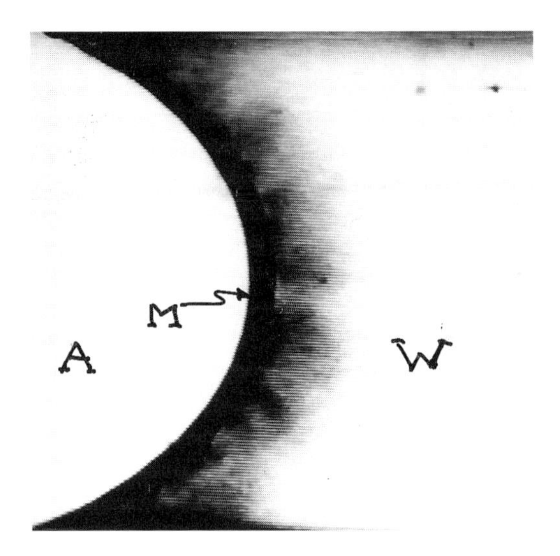

Fig. 9. Concentrated *B. subtilis* cells near the air water interface. The height of this image is 1 mm. The dark blotches near the interface are accumulations of cells surging back and forth. The forms of these wavy accumulations change continually. They are a manifestation of cell interactions and collective fluid dynamics in the absence of gravity effects. Chemical signalling between cells other than O_2 consumption may occur in these situations. See also Fig. 8. The meniscus is M, water is indicated by W, and air by A.

8 Conclusion

The behaviour of individual micro-organisms is a function of their biological ancestry and of the physics and biochemistry of their environment. The only invariant of that environment is the force of gravity which acts to break symmetry and to catalyze particular modes of individual and collectively driven motion. The other properties of the organisms' habitat are dynamic; they depend on energy input by the organisms, on their consumption of molecules or photons, and on the physical laws and geometric constraints that govern the dynamics of fluid systems. The immediate environment of an individual cell affects that cell's behaviour. The sum of behaviours, collected and incorporated into the physical and geometrical properties of the entire system creates the individuals' environment. This complex dynamical system demonstrates biophysical complementarity. It is a model for primitive, biological pattern formation among simple swimming cells. It generates great enhancement of transport by adding the advective distribution of molecules to the diffusive. It generates continua of local properties, such as light intensity and molecular concentrations that may be of use in the development of new or optimal associated biological responses.

By providing a few examples of experimentally measured probability densities for speeds and directions of locomotion of various species in various environments, we have shown that at least the ingredients of the problem are manageable and a likely subject for future investigations. We have provided a mathematical framework for continuing the investigation of patterns shown in the figures, once the individual behaviours are better characterised. We have also indicated that the collective behaviour of bacteria at high concentrations implies an entirely new approach to hydrodynamic modelling of concentrated multicellular bacterial assemblies.

An analogy with conventional physics may be appropriate here. The properties of isolated atoms are of great interest. The development of quantum mehanics was based on insights due to atomic physics. The analogue here is the biologically complex behaviour of individual swimming cells in a well-characterised environment. Many atoms placed in close proximity can form a solid or liquid. The properties of the condensed phase depend on the properties of the individual atoms but transcend them through generation of collective properties that modify the atoms' local environment and produce new system-wide dynamics. The transition from collections of individual organisms to dynamical systems seems strongly analogous.

Acknowledgements

This work was supported by the generosity of Ralph and Alice Sheets, provided through the University of Arizona Foundation, and in part by NASA grant NAG 9442. We should like to thank Charles Roberts, Robert Strittmatter, Diahn Swartz and David Wiseley who contributed greatly to all phases of this work. Discussions with Neil Mendelson have been very instructive. Finally we wish to thank Mitzi de Martino for help with this manuscript.

References

Acheson, D.J. (1990): *Elementary Fluid Dynamics* (Clarendon Press, Oxford)

Armitage, J.P. (1992): Behavioral responses in Bacteria. Ann. Physiol. **54**, 683–714

Berg, H.C. (1983): *Random Walks in Biology* (Cambridge University Press, Cambridge)

Boon, J.P., Herpigny, B. (1984): Formation and evolution of spatial structures. Springer Lecture Notes in Biomathematics **55**, 13–29

Childress, S., Levandowsky, M., Spiegel, E.A. (1975): Pattern formation in a suspension of swimming micro-organisms. J. Fluid Mech. **69**, 595–613

Foster, K.W., Smyth, R.D. (1980): Light antennas in phototactic algae. Microbiological Reviews **40**, 572–630

Häder, D.-P. (1987): Polarotaxis, gravitaxis and vertical phototaxis in the green flagellate *Euglena gracilis*. Arch. Microbiology **147**, 179–183

Herdan, G. (1960): *Small particle statistics, 2nd edition* (Butterworth and Co., London)

Hill, N.A., Pedley, T.J., Kessler, J.O. (1989): The growth of bioconvection patterns in a suspension of gyrotactic micro-organisms in a layer of finite depth. J. Fluid Mech. **208**, 509–43

Hill, N.A., Vincent, R.V. (1993): A simple model and strategies for orientation in phototactic microorganisms. J. Theor. Biol. **163**, 223–235

Hillesdon, A.J., Pedley, T.J., (1995): Pattern formation in suspensions of oxytactic bacteria: linear theory. *Submitted to* J. Fluid Mech.

Hillesdon, A.J., Pedley, T.J., Kessler, J.O. (1995): The development of concentration gradients in a suspension of chemotactic bacteria. Bull. Math. Biol. **57**, 299–344

Kessler, J.O. (1985a): Hydrodynamic focusing of motile algal cells. Nature **313**, 218–220

Kessler, J.O. (1985b): Cooperative and concentrative phenomena in swimming micro-organisms. Contemp. Phys. **26**, 147–166

Kessler, J.O. (1986a): The external dynamics of swimming micro-organisms. In: Round, F.E., Chapman, D.J. (eds.) *Progress in Phycological Research* **4** (Biopress, Bristol) 257–307

Kessler, J.O. (1986b): Individual and collective fluid dynamics of swimming cells. J. Fluid Mech. **173**, 191–205

Kessler, J.O. (1989): Path and Pattern — the mutual dynamics of swimming cells and their environment. Comments Theor. Biol. **1**, 85–108

Kessler, J.O. (1992): Theory and experimental results on gravitational effects on monocellular algae. Adv. Space Res. **12**, 33–42

Kessler, J.O., Hill, N.A., Häder, D.-P. (1992): Orientation of swimming flagellates by simultaneously acting external factors. J. Phycol. **28**, 816–822

Kessler, J.O., Hill, N.A. (1995): Microbial Consumption Patterns. In: P.E. Cladis, P. Palffy-Muhoray (eds.) *Spatio-Temporal Patterns in Nonequilibrium Complex Systems* (Addison-Wesley, Reading MA) 635–647

Kessler, J.O., Hoelzer, M.A., Pedley, T.J., Hill, N.A. (1994): Functional patterns of swimming bacteria. In: Maddock, L., Bone, Q., Rayner, J.M.V. (eds.) *Mechanics and Physiology of Animal Swimming* (Cambridge University Press) 3–12

Kessler, J.O., Strittmatter, R.P., Swartz, D.L., Wiseley, D.A., Wojciechowski, M.F. (1995): Paths and Patterns: the biology and physics of swimming bacterial populations. In: *Biological Fluid Dynamics*, (Society for Experimental Biology Symposium **49**, Cambridge University Press) (in press)

Lee, R.E. (1989): *Phycology* (Cambridge University Press) Chapter 11

Pedley, T.J., Kessler, J.O. (1987): The orientation of spheroidal microorganisms swimming in a flow field. Proc. R. Soc. Lond. **B 231**, 47–70

Pedley, T.J., Kessler, J.O. (1990): A new continuum model for suspensions of gyrotactic micro-organisms. J. Fluid Mech. **212**, 155–182

Pedley, T.J., Kessler, J.O. (1992a): Hydrodynamic phenomena in suspensions of swimming micro-organisms. Ann. Rev. Fluid Mech. **24**, 313–358

Pedley, T.J., Kessler, J.O. (1992b): Bioconvection. Sci. Progress **76**, 105–123

Pedley, T.J., Hill, N.A., Kessler, J.O. (1988): The growth of bioconvection patterns in a uniform suspension of gyrotactic micro-organisms. J. Fluid Mech. **195**, 223–237

Pfennig, N. (1962): Beobachtungen über das Schwärmen von *Chromatium okenii.* Arch. Mikrobiol. **42**, 90–95

Vincent, R.V. (1995): *Mathematical Modelling of Phototaxis in Motile Microorganisms* (Ph.D. thesis, University of Leeds, Leeds, England)

Vincent, R.V., Hill, N.A. (1995): Bioconvection in a suspension of phototactic algae. *Submitted to* J. Fluid Mech.

Wager, H. (1911): On the effect of gravity upon the movements and aggregation of *Euglena viridis*, Ehrb., and other microorganisms. Phil. Trans. R. Soc. Lond. **B 201**, 333–390

Mass Extinctions vs. Uniformitarianism in Biological Evolution

Per Bak and Maya Paczuski

Department of Physics, Brookhaven National Laboratory, Upton NY 11973 USA

1 Introduction

The theory of uniformitarianism, or gradualism, was formulated in the last century by the geophysicist Charles Lyell (1830) in his tome, *Principles of Geology*. According to this theory, all change is caused by processes that we currently observe which have worked at the same rate at all times. For instance, Lyell proposed that landscapes are formed by gradual processes, rather than catastrophes like Noah's Flood, and the features that we see today were made by slow persistent processes with time as the "great enabler" that eventually makes large changes. Uniformitarianism is a "linear" theory where the amount of change is proportional to the amount of time that has elapsed.

At first sight, Lyell's uniformitarian view is reasonable. The laws of physics are generally expressed in terms of smooth, continuous equations. Since these laws should describe all observable phenomena, it is natural to expect that the phenomena which we observe should also vary in a smooth and gradual manner. The opposing philosophy, catastrophism, claims that change occurs through sudden cataclysmic events. Since catastrophism smacks of creationism, with no connection to the natural sciences as we know them, it has been largely rejected by the scientific community.

Charles Darwin (1910) adapted Lyell's ideas of gradualism in an uncompromising way. According to his theory, evolution proceeds through random mutations followed by selection of the fitter variants. This slow process takes place at all times and all places at a steady rate. Darwin took it for granted that such a process would necessarily force evolution to follow a smooth, gradual path. Consequently, Darwin denied the existence of mass extinctions where a large fraction of species would abruptly disappear.

1.1 Avalanches and Punctuated Equilibrium

However, we know that many natural phenomena evolve intermittently (Bak and Paczuski 1995; Paczuski, Maslov, and Bak 1996). The dynamics may follow a step-like pattern with long, dormant plateaus where little change takes place interrupted by sudden bursts, or avalanches, of concentrated activity. The magnitudes of the bursts may extend over a wide range. Even though uniformitarianism, as opposed to catastrophism, has historically dominated both geology and paleontology, prototypical examples of intermittent behavior lie in these two domains.

Earthquakes: For instance, the crust of the earth accommodates large, devastating earthquakes in which hundreds of thousands of people are killed. Most of the time the crust of the earth appears to be stable. These periods of stasis are punctuated by earthquakes or avalanches taking place on a fault structure that stores information about the history of the system.

In fact, the size distribution of earthquakes follows a simple power law known as the Gutenberg-Richter law (1956). The power law demonstrates that earthquakes have no typical scale; otherwise the distribution would have a signature at that scale. The smooth variation from small to large events suggests that a common dynamical mechanism is responsible for all earthquakes, regardless of their size. Volcanic eruptions constitute another intermittent phenomenon in geophysics. Solar flares, pulsar glitches, and the formation of neutron stars are examples of intermittent behavior in astrophysics. All these phenomena are examples where avalanches of activity exhibit power law distributions similar to the Gutenberg Richter law. There is no way to accommodate the power law distribution for earthquake sizes within the framework of a linear theory such as uniformitarianism.

A Gutenberg-Richter Law for Extinctions: One might, therefore, suspect that Darwin's use of uniformitarianism in a theory of evolution may also need to be reexamined. In fact, about twenty years ago, Gould and Eldredge (1977) proposed that biological evolution takes place in terms of punctuations, where many species become extinct and new species emerge, interrupting periods of low activity or stasis. Figure 1 shows the record of extinction events as recorded by J. J. Sepkoski (1993). These extinction events in biology are analogous to earthquakes in geology. Note the spikes of extinction events spanning a range of magnitudes. The largest events are associated with the Cambrian explosion 500 million years ago, and the Permian extinction 200 million years ago. Raup (1986) has plotted similar data as a histogram (figure 2) where each column shows the number of 4 million periods with a given extinction intensity. The smooth variation from the smallest to the largest extinctions indicates a common mechanism. Actually, punctuated equilibrium usually refers to the intermittent dynamics of a single species, where morphological change is concentrated in short time periods interrupted by long periods of stasis.

1.2 External Shocks: "Bad Luck"

The extinctions of species appear to take place simultaneously across families; they "march to the same drummer". This could be explained if mass extinctions were caused by large, exogenous cataclysms, i. e. if extinctions were due to "bad luck" rather than "bad genes". For example, in the most prominent theory, Alvarez, Alvarez, and Michel (1980) suggest that the Cretaceous extinction event where the dinosaurs disappeared was caused by a meteor hitting the earth some 55 million years ago. Indeed, a large crater was observed near the Yucatan peninsula in Mexico. However, in order for an exogenous event such as a meteor

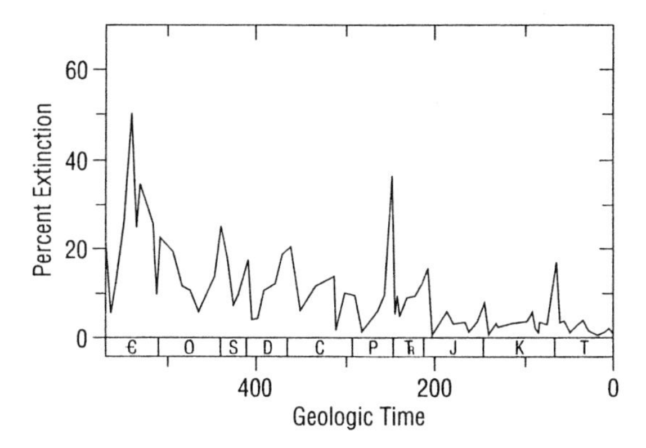

Fig. 1. Temporal pattern of extinctions recorded over the last 600 million years, as given by J. Sepkoski (1993). The ordinate shows an estimate of the percentage of species that went extinct within intervals of 5 million years

to wipe out an entire species, it must have a global effect over the entire area that the species occupies; otherwise the impact would be insufficient to cause extinction, except for species with small local populations. In addition, extinctions of species take place all the time without an external cause. Extinctions are taking place right now! These extinction events are obviously not caused by a meteor. Some are known to be intrinsic to evolution, being caused by humans.

In his book *Bad Genes or Bad Luck*, Raup (1982) distinguishes between bad luck, extinctions from external sources, and bad genes, extinctions due to intrinsically poor fitness. Whether or not external shocks play an important role in evolution, it is important to understand the dynamics of biological evolution in the absence of these shocks.

1.3 Evolution of Isolated vs. Many Interacting Species

In early theories of evolution, by Fisher (1932) and others, evolution of a single species in isolation was considered. Individuals within each species mutate, leading to a distribution of fitnesses, and the fitter variants were selected. This leads to a fitness which always increases smoothly *ad infinitum*. Many biologists appear content with this state of affairs, and rarely is the need for a more comprehensive theory expressed. For instance, Maynard Smith (1993), in his book *The Theory of Evolution* notices with great satisfaction that nothing important has changed in the 35 years intervening between his first and second editions.

However, Fisher's picture of a species evolving in isolation does not appear to us to be able to explain any of the intricacy, diversity, and complexity of real life. This is because evolution is a cooperative phenomenon. Species form ecologies where they interact with each other in a global ecology with predator-prey relationships and food webs. For example, humans depend on the oxygen emitted by trees and other plants. It is quite likely that the interaction of many

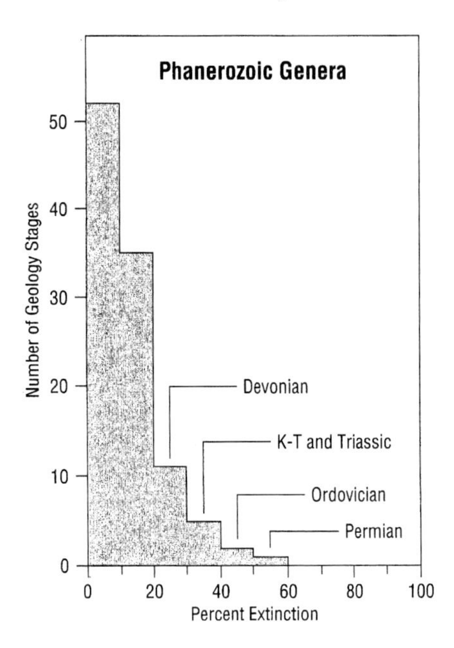

Fig. 2. Histogram of extinction events as shown by Raup (1986). The histogram is based on the recorded time of extinction of 2316 marine animal families

species in a global ecology plays a more important role in evolution than the specific behavior of a single or a few species in isolation.

Our approach is to consider biology as a large, dynamical system with very many degrees of freedom. Interactive systems like biology may exhibit emergent behavior which is not obvious from the study of typical local interactions and detailed behaviors of two or three species. Indeed, the population dynamics of a few interacting species has been described in terms of coupled differential equations, known as Lotka-Volterra, or replicator equations. These equations may lead to interesting, chaotic behavior, but of course not to mass extinction or punctuated equilibrium. Traditional evolutionary theory may be able to explain the behavior of a few generations involving a few hundred rats, but it can not explain evolution on the largest scale in which trillions of organisms interact throughout billions of years.

2 Self-Organized Criticality

A few years ago, Bak, Tang, and Wiesenfeld (1987, 1988) suggested that large dynamical systems may organize themselves into a highly poised "critical" state where intermittent avalanches of all sizes occur. The theory, self-organized criticality (SOC), is a nonlinear theory for how change takes place in large systems. It has become a general paradigm for intermittency and criticality in Nature. Evolution to the critical state is unavoidable, and occurs irrespective of the interactions on the smallest scales which we can readily observe. A visual example is a pile of sand on a flat table, onto which sand is gradually added. Starting from a flat configuration, the sand evolves into a steep pile which emits sandslides, or avalanches, of all sizes. This idea has been successfully applied to a variety of geophysical and astrophysical phenomena, in particular to earthquakes where it provides a natural explanation for the Gutenberg-Richter law. It is now broadly accepted that earthquakes are a self-organized critical phenomenon (Newman, Turcotte, and Gabrielov 1995). One may think of SOC as the underlying theory for catastrophism.

Can this nonlinear picture be applied to biological evolution? Even if we accept Darwin's mechanism for evolution, it is difficult to extract its consequences for macroevolution. In contrast to the basic laws of physics which are described by equations such as Newton's equations, or Maxwell's equations, there are no "Darwin's Equations" to solve, as one of our editors, Henrik Flyvbjerg, once pointed out. It may seem rather hopeless to try to mathematically describe macroevolution without the fundamental microscopic equations. On the other hand, we know from our experience with many body collective phenomena that statistical properties of the system at large scales may be largely independent of small scale details. This is called "universality." The interactions are more important than the details of the entities which make up the system. Universality belies the usual reductionist approach in the physical sciences where features at large scales are explained in terms of models at successively smaller scales with more and more details included. Universality is a way to throw out almost all of these details.

Thus, our studies are based on abstract mathematical models. The models cannot be expected to reproduce any specific detail that may actually be observed in nature, such as humans or elephants. The confrontation between theory and reality must take place on a statistical level. This is not unusual in the natural sciences. Quantum mechanics and thermodynamics are inherently statistical phenomena. Chaotic systems are unpredictable, so comparison with experiment or observations must also be on the statistical level. Indeed, the Gutenberg-Richter law is a statistical law which can be explained in terms of grossly over-simplified SOC models for earthquakes. One might hope to be able to do the same for biology.

We shall argue that life may be a self-organized critical phenomenon. Periods of stasis where evolutionary activity is low are interrupted by coevolutionary avalanches where species become extinct and new species emerge. Good genes during periods of stasis are no guarantee for survival. Extinctions may take place not only due to "bad luck" from external effects, like meteors, but also due to bad luck from freak evolutionary incidents endogenous to the ecology. Biological evolution operating in the critical state can explain a variety of empirical observations, including the lifetime distribution of species and the occurrence of mass extinctions.

3 Co-evolutionary Avalanches

Stuart Kauffman of the Santa Fe Institute was among the first to suggest that life might be a self-organized critical phenomena where evolution takes place in terms of co-evolutionary avalanches of all sizes. Together with Sonke Johnsen (Kauffman and Johnsen 1991) he studied complex models of very many species forming an interactive ecology, the NKC-models. In these models, each species evolves in a rough fitness landscape, with many local peaks, employing a picture invented by Sewall Wright 50 years ago Wright (1982) (figure 3) in his seminal work, *The Shifting Balance Theory.* Populations are modified by means of mutation and differential selection towards higher fitness. Random mutations allow individuals to cross barriers, formed by troughs of lower fitness and move to other maxima. Then they might initiate a population at or near the new maximum.

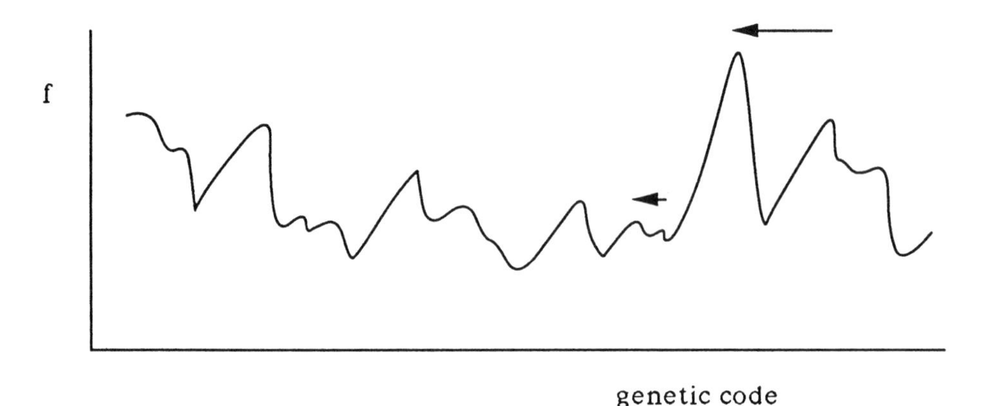

Fig. 3. Fitness landscape. Note that the species with low fitnesses have smaller barriers to overcome in order to improve their fitness than species with high fitnesses

Each species can be thought of as a group of individuals in the vicinity of some fitness peak, and may be represented by the genetic code of one of those individuals. In Kauffman's models, the genetic code is represented by a string of N bits or genes (0011011....11101000). Each configuration has a fitness associated with it, which can be calculated from an algorithm, the NK-algorithm. The contribution to the fitness from each gene or "trait" depends on the state of K other genes. The fitness depends on the coupling between genes. The NK models are generalized spin glass models, invented by physicists to describe metastability and frozen behavior of magnetic systems with random interactions.

The elementary single step is what could be called a "mutation of a species". Despite the fact that this notation may raise a red flag among biologists, it will be used throughout this chapter. In evolution, this step is made by random mutations of individuals followed by selection of the fitter variant, and subsequent transformation of the entire population to that variant. The landscape is traced out as the bits are varied. By randomly mutating one bit at the time, and accepting the mutation only if it leads to a higher fitness, the species will eventually reach a local peak from which it can not improve further from single mutations. Of course, by making many coordinated mutations the species can transform into something even more fit, but this is very unlikely. A species can not spontaneously develop wings and fly.

However, the fitness landscape of each species is not rigid; it is a rubber landscape modified when other species in the ecology change their properties. For instance, the prey of some predator may grow thicker skin (or become extinct), so that the fitness of the predator is reduced. Within the landscape picture, this corresponds to a deformation where the fitness peak the predator previously occupied has vanished. The predator might find itself down in a valley instead of up on a peak. Then it may start climbing the fitness landscape again, for instance by developing sharper teeth, until it reaches a new local peak. This may in turn affect the survivability of other species, and so on.

Kauffman and Johnsen represented the interdependence of species in terms of their model. Mutating one of the N genes in a species affects K genes within the species and also affects the fitnesses of C other species. This is called the NKC model. Now, starting from a random configuration, all the species start climbing their landscapes, and at the same time start modifying each other's landscapes. Their idea was that this would eventually set the system up in a poised state where evolution would happen intermittently in bursts. However, this failed to occur.

Either of two scenarios would emerge. i) If the number of interactions, C, is small, the ecology evolves to a frozen state where all species rest on a local peak of their respective landscapes, and no further evolution takes place. A random, external environment is introduced by randomly flipping genes. This initiates avalanches where a number of species evolve. However the avalanches are small, and the ecology soon comes to rest in a new frozen state. ii) If the the number C is large, the ecology goes to a highly active chaotic state, where the species perpetually evolve without ever reaching a peak. In this case, the coevolutionary avalanches never stop. Only if the parameter C is carefully tuned does the ecology evolve to a critical state.

Kauffman and Johnsen argued that the ecology as a whole is most fit at the critical point. "The critical state is a nice place to be," Kauffman claims. However, it can be proven that the NKC models do not self-organize to the critical point (Flyvbjerg and Lautrup 1992; Bak, Flyvbjerg, and Lautrup 1992). Divine intervention is needed. Apart from the question as to what type of dynamics may lead to a critical state, the idea of a poised state operating between a frozen and a disordered, chaotic state makes an appealing picture for the emergence of complex phenomena. A frozen state cannot evolve. A chaotic state cannot remember the past. This leaves the critical state as the only alternative.

4 A Simple Model for Evolution

Bak and Sneppen (Bak and Sneppen 1993; Sneppen et al. 1995) introduced a simple model to describe the main features of a global interactive ecology. In one version of the model, L species are situated on a one dimensional line, or circle. Each species interacts with its two neighbors, to the left and to the right. The system can be thought of as a long food chain. Instead of specifying the fitness in terms of a specific algorithm, the fitness of the species at position i is simply given as a random number, f_i, say between 0 and 1. The fitness is not specified explicitly in terms of an underlying genetic code, but is chosen as a random function of these variable. Probably not much is lost since we do not know the actual fitness landscapes anyway.

At each time step the least fit species is selected for mutation or extinction. This is done by finding the smallest random number in the system. By inspection of the fitness landscape in figure 3, it is clear that species located on low fitness peaks have a smaller distance to find better peaks than species with higher fitness. The barriers to find better peaks can be thought of as the number of coordinated mutations needed. So the time it takes to reach a higher peak increases exponentially with the size of the barriers and can become astronomically large if the genetic mutation rate is low. This justifies the selection of the least fit species as the next in line for mutation.

The mutation of a species is represented by replacing its random number with a new random number. One might argue that the new random number should be higher than the old one, but this does not change the behavior, so for mathematical simplicity we replace the old random number with a completely new random number between 0 and 1. One might think of this elementary event either as an extinction of a species occupying a certain ecological niche followed by the replacement with another species, or as a pseudo-extinction where a species mutates. As far as our mathematical modeling is concerned, this doesn't make any difference. The mutation of the species at site i results in a change in the physical properties of that species, so that it affects the fitnesses of its two neighboring species. For simplicity, this is modelled by choosing a new, randomly selected, fitness for the neighbors also. One might argue their fitness should only be affected slightly, say less than 1/10, or that their fitness should generally be worsened. Again, the details of the model do not affect the resulting outcome, so we choose a completely new random number.

To summarize: *At each time step in the simulation, the smallest fitness, and the fitness of the two neighbors are each replaced with new random fitnesses. This step is repeated again and again. That's all!*

What could be simpler than replacing some random numbers with some other random numbers? Despite the simplicity, the outcome of the process is nontrivial. One might suspect that the model would be so simple that it would easily yield to mathematical analysis, but the complexity of its behavior sharply contrasts with its simple definition. We shall see that modified versions of the model are more tractable.

In particular, a multi-trait evolution model (Boettcher and Paczuski 1996), which behaves similarly to the Bak-Sneppen model, can be completely solved. Instead of each site i having a single fitness f_i it has many fitnesses associated with its M different traits that evolve independently. The introduction of many internal traits is consistent with paleontological observations indicating that evolution within a species is "directed"; morphological change over time is concentrated in a few traits, while most others traits of the species are static (Kaufmann 1993). The multi-trait model includes the Bak-Sneppen model when $M = 1$ and is solvable when $M \to \infty$.

4.1 The Self-Organized Critical State

Figure 4 shows a snapshot of the fitnesses after billions of updates for a Bak-Sneppen ecology with 300 species. Most of the fitnesses are above a critical value, or gap, $f_c = 0.67002$ (Paczuski, Maslov, and Bak 1996). Note however a localized area in the species space where the fitnesses are below this gap. The next species to mutate is the one with the lowest fitness, #110. The fitness of this species and its two neighbors are updated. It is very likely that the next species to mutate is nearby. Subsequent evolution takes place within this localized burst. After a while, there will be no more species below the gap, and a new burst, or avalanche, will start somewhere else in the system.

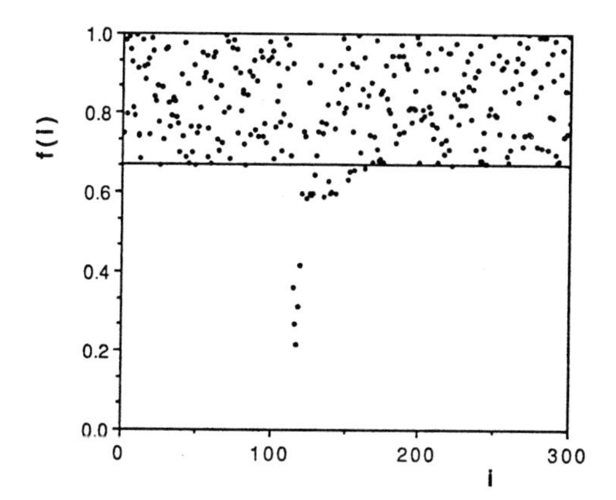

Fig. 4. Snapshot of fitnesses f for 300 species during an avalanche in the evolution model. Most of the f values are above the critical value. The cluster of active species with $f < f_c$ participate in the avalanche and undergo frequent changes (Paczuski et al. 1996)

During the avalanche, the species are mutating again and again, until eventually a self-suspended, stable network of species has been reestablished. The size, s, of the burst can be measured as the total number of mutations separating instances with no species in the gap. Figure 5 shows the distribution of burst sizes. There are avalanches of all sizes, with a power law distribution

$$P(s) \sim s^{-\tau} \text{ where } \tau \simeq 1.07 \quad . \tag{1}$$

The power law for large sizes shows that the ecology has self-organized to a critical state. The large avalanches represent mass extinction events, like the Cambrian explosion (Gould 1989). During the large avalanches, nature tries one combination after another until a relatively stable configuration is reached, and a period of stasis in this area of the global ecology begins.

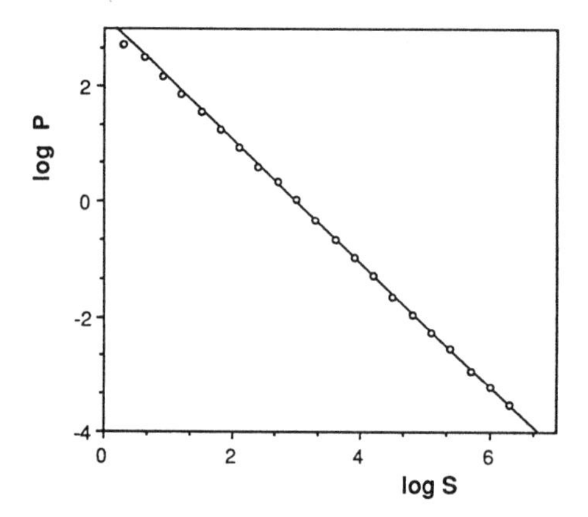

Fig. 5. Distribution of the size of avalanches in the critical state for the one dimensional evolution model. The straight line on the log-log plot indicates a power law with an exponent $\tau \simeq 1.07$ (Paczuski et al. 1996)

As a consequence of the interaction between species, even species that possess well-adapted abilities, with high barriers, can be undermined in their existence by weak species with which they interact. For instance, a species with fitness above the critical gap never mutates on its own. However, eventually it may be hit by "bad luck" because a mutation among its neighbors destroys its pleasant and stable life. A species can go extinct through no fault of its own and for no apparent external "reason" such as a meteor. Nature is not fair! A high fitness is only useful as long as the environment defined by the other species in the ecology remains intact.

Figure 6 shows the accumulated number of mutations of an arbitrary single species in the model. Note the relatively long periods of stasis, interrupted by bursts of activity. One might imagine that the amount of morphological change of a species is proportional to the the total number of mutations, so the curve shows punctuated equilibrium behavior. The large jumps represent periods where very many subsequent mutations take place at a rapid pace, because an ecological co-evolutionary avalanche happens to visit the species. Thus, the big jumps between "useful" or highly fit states are effectuated by cumulative small jumps through intermediate states which could exist only in the temporary environment of a burst. The curve is a Cantor set, or Devil's staircase, invented by the mathematician Georg Cantor in the last century. The length of the period of stasis is the "first return time" of the activity for a given species. That quantity also has a power law distribution (Maslov, Paczuski, and Bak 1995). In evolution, this time can be thought of as the lifetime of a species before it undergoes a (pseudo)extinction. In fact, the distribution of lifetimes of fossil genera (Sepkoski 1993) appears to follow a power law with a characteristic exponent $\simeq 2$ (Sneppen et al. 1995).

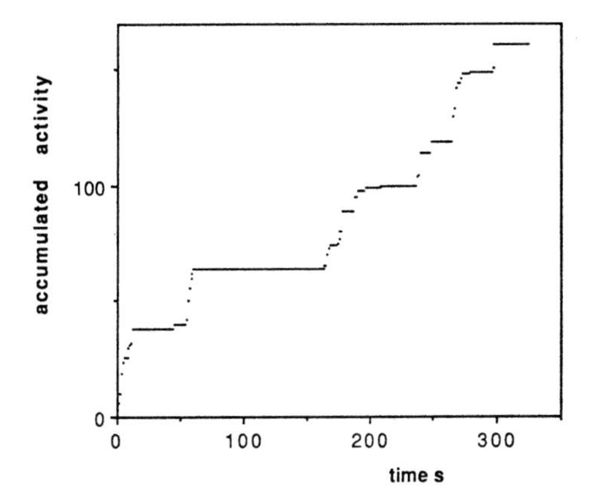

Fig. 6. Accumulated number of mutations for a single species, or a single ecological niche in the stationary state. The curve exhibit punctuated equilibrium behavior, with periods of stasis interrupted by intermittent bursts. The curve is a Cantor set, or a Devil's staircase

As a consequence of the power law distribution of burst sizes, most of the evolutionary activity occurs within the few large avalanches rather than the many smaller "ordinary" events. Self-organized criticality can thus be thought of as a theoretical justification for catastrophism.

4.2 Comparison with the Fossil Record

The time unit in the computer simulations is a single update step. Of course, this does not correspond to time in real biology. Based on the rugged fitness landscape picture, the time-scale for traversing a barrier of size b is exponential in the barrier height, which is roughly proportional to the fitness, so $t_i \sim \exp(f_i/T)$. Here T is an effective temperature which represents an average mutation rate in the genetic code. In real biology, there is no search committee locating the least fit species, but mutation takes place everywhere at the same time. In the limit where the effective temperature T approaches zero, the minimal fitness is always selected next for mutation. Punctuated equilibrium behavior can exist only where the mutation rate is slow; otherwise there would not be long periods of stasis. A system with a high mutation rate will not have sufficient memory to develop complex behavior, since any new development will be erased in a relatively short time span.

Sneppen et al. (1995) performed a simulation where at each time step a mutation takes place everywhere with probability $p = exp(-f_i/T)$. Figure 7 shows the resulting space-time plot of the activity, with $T = 0.01$. Note the temporal separation of the avalanches, which show up as connected black areas. The information in this diagram can be presented differently. In figure 8, the time scale has been coarse grained in a simulation over 8000 steps into 60 equal time intervals. The total amount of events in each time step is plotted as a function of time. Note the similarity with Sepkoski's plot, figure 1. During each time period, there are generally many avalanches. One can show that the resulting distribution for the total number of events in each period approaches a Pareto-Levy distribution, which has power law tails for large events. The information in Raup's histogram of Sepkoski's data is too sparse to test whether or not it represents a Pareto-Levy distribution.

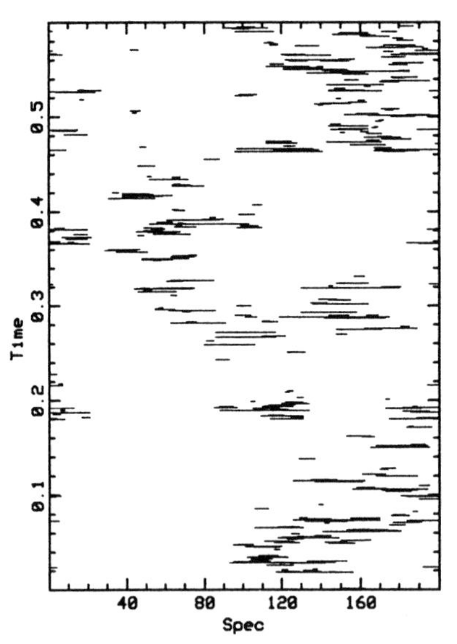

Fig. 7. Space time plot of the activity. The horizontal axis is the species axis. The time at which a species mutates is shown as a black dot. The avalanches appear as connected black areas. Calculation was done for a value of the mutation parameter $T = 0.01$ (Sneppen et al. 1995)

In the Bak-Sneppen model, the number of species is conserved. It does not allow for speciation where one species branches into two species, but might be justified as a consequence of competition for resources. Vandewalle and Ausloos (1995) have constructed a model where phylogenetic trees are formed by speciation, starting from a single species. This model also evolves to the critical state, with extinction events of all sizes.

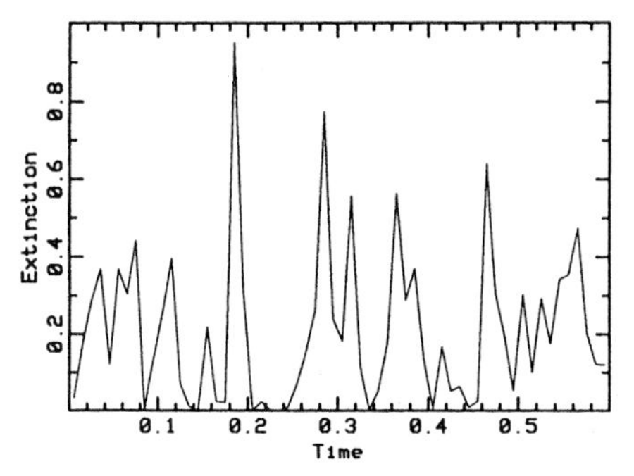

Fig. 8. Temporal pattern of evolution, with $T = 0.01$. Note the similarity with Sepkoski's plot, figure 1 (Sneppen et al. 1995)

Of course, it would be interesting if one could perform controlled experiments on evolution. The second best is to construct artificial evolution in the computer. Ray (1992) and Adami (1995) have done just that. They created a world of replicating and mutating program segments. Adami found that evolution in this artificial world exhibits punctuated equilibria, with a lifetime distribution of organisms following a power law, indicating that the system self-organizes into a critical state.

4.3 External Effects

To what extent is evolution described by such simple models affected by external events, such as temperature variations, volcanic eruptions, neutrino bursts, or meteors? Schmultzi and Schuster (1995) studied a model where extinction takes place when the fitness of a species falls below a random number drawn from an independent distribution. The random number represents the effect from external sources. Self-organized criticality and punctuated equilibria were also found. Newman and Roberts (1995) took a similar approach. The species were assigned not only barriers for spontaneous mutation, but independent random fitnesses. At each point in time, species with fitnesses less than a randomly varying external perturbation would go extinct, in addition to species mutating spontaneously. The external perturbations initiate avalanches. They found a power law distribution of co-evolutionary avalanches, with an exponent $\tau \simeq 2$.

5 Theory

Theoretical developments to mathematically describe the behavior of these abstract computer models have taken two routes. The first is a phenomenological approach that we have undertaken in collaboration with Sergei Maslov (Paczuski, Maslov, and Bak 1994, 1996). We have made a unified theory of avalanche dynamics which treats not only evolution models but also other complex dynamical systems with intermittency such as invasion percolation, interface depinning, and flux creep. Complex behavior such as the formation of fractal structures, $1/f$ noise, diffusion with anomalous Hurst exponents, Levy flights, and punctuated equilibria can all be related to the same underlying avalanche dynamics. This dynamics can be represented as a fractal in d spatial plus one temporal dimension. In particular, the slow approach to the critical attractor, i.e. the process of self-organization, is governed by a "gap" equation for the divergence of avalanche sizes as the gap in the fitness distribution approaches the critical value, starting from the flat gapless initial distribution. Figure 9 shows the minimum value of the fitness vs. the number of update steps. The envelope function of this curve is the gap. Avalanches of activity below the gap separate the instances where the gap grows. Clearly, the avalanches become bigger and bigger as time goes on.

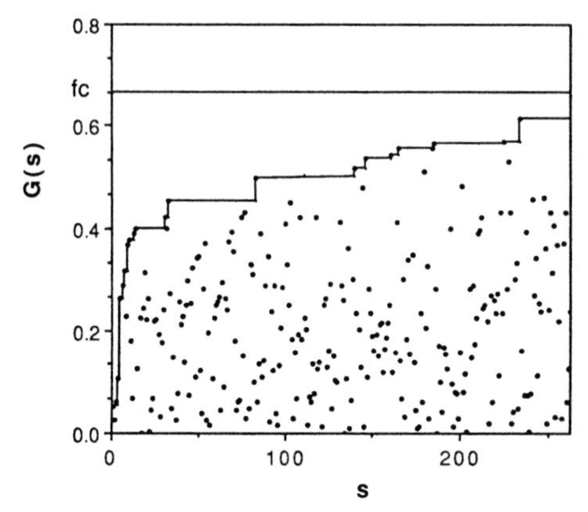

Fig. 9. The self-organization process for a small system. f_{min} vs time is shown (crosses). The full curve shows the gap, which is the envelope function of f_{min}. On average, the avalanche size grows as the critical point is approached, and eventually diverges, as the gap approaches the critical value 0.6700. (Paczuski et al. 1996)

We have developed a scaling theory that relates many of the critical exponents describing various physical properties to two basic exponents characterizing the fractal attractor. The phenomenological theory does not provide information about the values of those exponents.

The second approach has been aimed at obtaining exact results for specific models. For the multitrait model (Boettcher and Paczuski 1996), an explicit equation of motion for the avalanche dynamics can be derived from the microscopic rules. Exact solutions of this equation, in different limits, proves the existence of simple power laws with critical exponents that verify the general scaling relations mentioned above. Punctuated equilibrium is described by a Devil's staircase with a characteristic exponent $\tau_{\mathrm{FIRST}} = 2 - d/4$ where d is the spatial dimension. Actually, for the multi-trait evolution model, the distribution of avalanche sizes is known exactly when $M \to \infty$. It is

$$P(s) = \frac{\Gamma\left(s+\frac{1}{2}\right)}{\Gamma\left(\frac{1}{2}\right)\Gamma(s+2)} \sim s^{-3/2} \text{ for } s \gg 1 \quad . \tag{2}$$

This distribution of sizes is the same as for the random neighbor models in which each species interacts with 2 randomly chosen other species in the ecology rather than with near neighbors on a regular grid (Flyvbjerg et al. 1993; deBoer et al. 1995). The power law has a characteristic exponent $\tau = 3/2$ rather than $\tau = 1.07$ for the Bak-Sneppen chain.

In the multitrait model, avalanches propagate via a "Schrödinger" equation in imaginary time with a nonlocal potential in time. This nonlocal potential gives rise to a non-Gaussian (fat) tail for the subdiffusive spreading of activity. For the chain, the probability for the activity to spread beyond a species distance r in time s decays as $\sqrt{\frac{24}{\pi}} s^{-3/2} x \exp\left[-\frac{3}{4}x\right]$ for $x = (\frac{r^4}{s})^{1/3} \gg 1$ (Paczuski and Boettcher 1996). This anomalous relaxation comes from a hierarchy of time scales, or memory effect, that is generated by the avalanches. In addition, a number of other correlation functions characterizing the punctuated equilibrium dynamics have been determined exactly. For instance, the probability for a critical avalanche to affect a species at distance r is exactly $12/((r+3)(r+4))$ in one dimension.

6 Acknowledgments

This work is supported by the Division of Materials Science, U. S. Department of Energy under contract # DE-AC02-76CH00016. We are grateful for discussions and collaborations leading to the results summarized here with K. Sneppen, H. Flyvbjerg, and S. Maslov. We thank Mike Fuller for biologically relevant comments on our paper.

References

Adami, C. (1995): Self-organized criticality in living systems. Phys. Lett. A **203**, 29–32

Alvarez, L. W., Alvarez, F. A., Michel, H. W. (1980): Extraterrestrial cause for the Cretaceous–Tertiary extinction Science **208**, 1095–1108

Bak, P., Flyvbjerg, H., Lautrup, B. (1992): Coevolution in a rugged fitness landscape. Phys. Rev. A **46**, 6724-6730

Bak, P., Paczuski M. (1995): Complexity, contingency, and criticality. Proc. Natl Acad. Sci. USA **92** 6689-6696

Bak, P., Sneppen, K. (1993): Punctuated equilibrium and criticality in a simple model of evolution. Phys. Rev. Lett. **71**, 4083–4086

Bak, P., Tang, C., Wiesenfeld K. (1987): Self-organized criticality. An explanation of 1/f noise. Phys. Rev. Lett. **59**, 381–384

Bak, P., Tang, C., Wiesenfeld K. (1988): Self-organized Criticality. Phys. Rev. A **38**, 364–374

Boettcher S., Paczuski M. (1996): Exact results for spatiotemporal correlations in a self-organized critical model of punctuated equilibria. Phys. Rev. Lett. **76**,348-351

Darwin, C. (1910): *The Origin of Species* (Murray, London), 6th edition.

deBoer, J., Derrida, B., Flyvbjerg, H., Jackson, A. D., Wettig, T. (1994): Simple model of self-organized biological evolution. Phys. Rev. Lett. **73**, 906–909

Fisher, R. A. (1932): *The Genetical Theory of Natural Selection*

Flyvbjerg. H., Bak P., Jensen, M. H. Sneppen, K. (1995): A self-organized Critical Model for Evolution, in *Modelling the Dynamics of Biological Systems.* E. Mosekilde and O. G. Mouritsen (Eds.) (Springer Berlin, Heidelberg, New York)pp 269–288

Flyvbjerg, H., Lautrup, B. (1992): Evolution in a rugged fitness landscape. Phys. Rev. A **46**, 6714–6723

Flyvbjerg, H., Sneppen K., Bak P. (1993): Mean field theory for a simple model of evolution. Phys. Rev. Lett. **71**, 4087–4090

Gould, S. J. (1989): *Wonderful Life* (Norton, New York)

Gould, S. J., Eldredge N. (1977): Punctuated equilibria: The tempo and mode of evolution reconsidered. Paleobiology **3**, 114–151

Gutenberg B., Richter C. F. (1956): Ann. Geofis. **9**, 1

Kauffman, S. A. (1993): *The Origins of Order: Self-Organization and Selection in Evolution* (Oxford University Press, Oxford)

Kauffman, S. A., Johnsen, S. J. (1991): Coevolution to the edge of chaos: Coupled fitness landscapes, poised states, and coevolutionary avalanches. J. Theo. Biology **149**, 467–505

Maslov, S., Paczuski, M., Bak, P. (1994): Avalanches and $1/f$ noise in evolution and growth models. Phys. Rev. Lett. **73**, 2162–2165

Maynard Smith, J. (1993): *The Theory of Evolution* (Cambridge University Press, Cambridge)

Newman, M. E. J., Roberts, B. W. (1995): Mass extinctions: Evolution and the effects of external influences on unfit species. Proc. Roy. Soc. London B **260**, 31

Newman, W. L., Turcotte D. L., Gabrielov A. M. (1995): Log-periodic behavior of a hierarchical failure model. Phys. Rev. E **52**, 4827

Paczuski, M., Maslov, S., Bak, P. (1994): Field theory for a model of self-organized criticality. Europhys. Lett. **27**, 97–100; Erratum **28**, 295–296

Paczuski, M., Maslov, S., Bak, P. (1996): Avalanche dynamics for evolution, growth, and depinning models. Phys. Rev. E **53**, 414

Paczuski, M., Boettcher, S. (1996): in preparation

Raup, D.M. (1982): *Extinction. Bad Genes or Bad Luck?* (Oxford University Press, Oxford)

Raup, D. M. (1986): Biological extinction in history. Science **231**, 1528–1533

Ray, T. (1992): in *Artificial Life II* ed. C. G. Langton (Addison-Wesley, Reading Massachussetts)

Schmultzi, K., Schuster, H. G. (1995): Introducing a real time scale into the Bak-Sneppen model. Phys. Rev. E **52**, 5273

Sepkoski, J. J. Jr. (1993): Ten years in the library: New data confirm paleontological patterns Paleobiology **19**, 43–51

Sneppen. K. , Bak P., Flyvbjerg H., Jensen, M. H. (1995): Evolution as a self-organized critical phenomena. Proc. Natl. Acad. Sci. USA **92**, 5209–5213

Vandewalle, N., Ausloos, M. (1995): Self-organized criticality in phylogenetic-like tree growth. J. de Phys. (Paris) **5**, 1011–1025

Wright, S. (1982): Character change, speciation, and the higher taxa. Evolution **36**, 427–443

Subject Index

Subject Index